Farming System
and Sustainable Agriculture

Related Books

Title	Author
• A Grain Saved is a Grain Produced	*S. Mohan*
• Agricultural Drainage Engineering	*H.V. Parmar*
• Agricultural Marketing	*S.B. Verma*
• Agricultural Microbiology	*R.P. Pareek*
• Agriculture and Fertilizers	*O.C. Bockman*
• Agricultural Extension a Glimpse	*S. Jananai*
• Agroforestry for Climate Resilience and Rural Livelihood	*Inder Dev*
• Agroforestry for Sustained Productivity in Arid Regions	*J.P. Gupta*
• Combating Desertification Land Degradation and Climate Change: Management of Dry Lands	*T.S. Chouhan*
• Commercial Beekeeping	*Dharm Singh*
• Communication Skills and Personality Development	*J.R. Kadam*
• Compensatory Afforestation in Rajasthan: An Evaluation	*G. Singh*
• Crop Improvement and Mutation Breeding	*A.K. Sharma*
• Crop Management 2nd Ed	*U.S. Walia*
• Crop Production in Salt Affected Soils	*I.C. Gupta*
• Crop Production in Waterlogged Saline Soils	*S.K. Gupta*
• Drainage Engineering: Principles and Practices	*S.K. Gupta*
• Drought Proofing and Water Resource Management	*T.S. Chouhan*
• Dryland Resources and Technology (Vol. 8)	*L.L. Somani*
• Educating Farmers About Agriculture	*Raghunath Singh*
• Farm Mechanization for Production	*Dhirendra Khare*
• Food Crops: Evolution, Diversity and Advances	*D.A. Patil*
• Fundamentals of Human Ethics and Agriculture Extension	*D.K. Meena*
• Fundamentals of Organic Farming and Gardening: An Instructors Guide	*A. McGregor*
• Introduction to Crops of India 2nd Ed	*N.R. Das*
• Introduction to Forestry & Agroforestry	*K.T. Parthiban*
• Kharpatvar Prabandhan (Weed Management)	*J.P. Tiwari*
• Krishi Jagat: JET Evam Krishi Paryavekshak Pariksha Upoagi	*M.L. Jat*
• Spice Crops Production Technology	*P.C. Das*
• Sustainable Agriculture	*R.S. Meena*
• Sustainable Agriculture: From Common Principles to Common Practice	*F.J. Hani*
• Sustainable Development of Dryland Agriculture in India	*R.P. Singh*
• Sustainable Dryland Farming	*A.M. Dhopte*

Farming System and Sustainable Agriculture

Dr. S.S. Walia
Dr. U.S. Walia
Department of Agronomy
Punjab Agricultural University,
Ludhiana-141 004 (Pb), INDIA

SCIENTIFIC
PUBLISHERS

Published by:

SCIENTIFIC PUBLISHERS (INDIA)
5A, New Pali Road, P.O. Box 91
Jodhpur 342 001, India
E-mail: info@scientificpub.com
Website: www.scientificpub.com

Print 2026

ISBN: 978-93-88812-90-0
eISBN: 978-93-88812-92-4

Printed in India

PREFACE

Agriculture is the largest enterprise in India which is at present and will continue to be the lifeline of the Indian economy in the foreseeable future. However, due to urbanization, agricultural land is shrinking and human population is increasing day by day and so there is a need for vertical increase in agricultural produce to feed the increasing population. Due to changing climatic conditions, there is a need for reorientation of the presently practised agricultural technologies. At the same time there is a need to save/conserve the natural resources. Crop yield can be improved with the adoption of improved production and protection technologies for raising field crops. For better crop profitability, judicious use of inputs like viable seeds of improved varieties, timely sowing, optimum plant stand, required amount of fertilizers, irrigation water and pesticides (insecticides, fungicides and herbicides) should be done.

This book contains detailed and comprehensive information regarding components of farming system, historical background of framing system, allied enterprises of Integrated Farming System (IFS), models of IFS in different climatic zones, role of IFS in livelihood security, resource cycling and flow of energy, role of crop residues in different farming systems and farming system of drylands agriculture. It also discusses in detail the latest concept of agronomy like conservation agriculture, sustainable agriculture, scope of hydroponics in agriculture, scope of organic farming in agriculture, role of agroforestry in farming system. etc.

CONTENTS

1 FARMING SYSTEM AND ITS COMPONENTS

The Indian economy is predominantly rural and agriculture oriented. In agriculture, 85% of the holdings are less than two hectares and the declining trend in the average size of the farmer holdings, poses a serious problem. Majority of them are dry lands, which depend on erratic monsoon rains. The rest of the area is cultivated with supplemental irrigation. The farmers concentrate mainly on crop production, which is invariably subjected to a high degree of uncertainty income and employment.

In India the cultivable land is 143.8 million hectares and there is very little possibility of extending it further. Therefore, to meet the requirement of food grains for increasing population, the only option open is effective utilization of time and space in agriculture. The time concept relates to increasing the intensity of cropping under assured irrigated conditions, whereas space utilization pertains to building up of vertical dimension through multi-tier cropping and farming system approach. Thus by making use of these time and space concept either in irrigated or in rained areas, the productivity per unit area per unit time can be substantially enhanced. Therefore the only way to increase agricultural production in the small/marginal units of farming is to increase the productivity per unit time and area. This may be achieved by selecting early maturing varieties with equal yields or by improving agronomic techniques like judicious use of fertilizers, pesticides and other inputs.

System approach

Management practices are developed for individual crops and recommendation are made for individual crops. The residual effect of individual crops is not considered in crop based recommendation. In this resources are not utilized efficiently. To a farmer management practices for all crops are considered that are to be grown on a piece of land. Therefore system approach is applied to agriculture for efficient utilization of all resources, maintains sustainability in production and obtaining higher net returns.

Farming is a dynamic biological and open system with human or social involvement. Being primarily biological with a high degree of dependence on weather variables and changing socio-political environments. Farming is inherently more risky than any other system. Farming system can also mean different things to different people. Farming is the process of harnessing solar energy in the form of economic plant and animal products and system implies a set of inter related practices/processes organized into a functional entity.

A system consists of several components which depend on each other. A system is defined as a set of elements or components that are inter- related and interacting among themselves. System could be defined as an organized unitary whole composed of two or more inter dependent and interacting parts, components or subsystems delineated by identifiable boundary or its environmental super system.

Farm and farming system

- **Farm:** It is an organized economic unit in which crop and animal production is carried out with purpose of producing economic net returns.
- **Farming system:** It is a part of farm system which crop production, animal production and combination of both *i.e* mixed farming.
- **Farm area:** Productive cropped area/ fields pastures, plantations, fishponds, forest etc. + indirectly productive land- farm building, ditches, roads etc. + fallow land + unproductive land.

"Farming system" therefore designates a set of agricultural activities organized while preserving land productivity, environmental quality and maintaining desirable level of biological diversity and ecological stability. The emphasis is more on a system rather than on gross output.

Farming system specially refers to a group combination of enterprises in which the products and or the byproducts of one enterprise serve as the inputs for production of other enterprise.

Farming system takes into account the combined needs of the family, the economic factors like relative profitability of the technically feasible enterprises, availability of off farm resources, infrastructure and institutions such as irrigation, marketing facilities including storage and transportation and credit besides the agro biological consideration namely interdependence, if any among various technically feasible enterprises and the performance of individual farmers.

Farming is defined as the way in which the farm resources are allocated to the needs and priorities of the farmers in his local circumstances which include-

I. Agro climatic condition such as the quantity, distribution and reliability of rainfall, soil type, topography, temperature etc.
II. Economic and institutional circumstances like market opportunities, prices, institutional and infrastructure facilities and technology.

CLASSIFICATION OF FARMING SYSTEMS

1. **According to the size of the farm:**
   (a) Collective farming
   (b) Cultivation farming
      (i) Small-scale farming
      (ii) Large-scale farming.

2. **According to the proportion of land, labour and capital investment:**
   (a) Intensive cultivation
   (b) Extensive cultivation
3. **According to the value of products or income:**
   (a) Specialized farming
   (b) Diversified farming
   (c) Mixed farming
   (d) Ranching
   (e) Dry farming
4. **According to the water supply:**
   (a) Rain fed farming
   (b) Irrigated farming
5. **According to type and intensity or rotation:**
   (I) Type of rotation:
      (a) Lay system:
         (i) unregulated lay farming
         (ii) regulated lay system
      (b) Field system
      (c) Perennial crop system
   (II) Intensity of the rotation:
      (a) Shifting cultivation
      (b) Lay or fallow farming
      (c) Permanent cultivation
      (d) Multiple cropping
6. **Classification according to degree of commercialization:**
   (a) Commercialized farming
   (b) Partly commercialized farming
   (c) Subsistence farming
7. **Classification according to degree of nomadic:**
   (a) Total nomadic
   (b) Semi nomadic
   (c) Partial nomadic
   (d) Transhumant
   (e) Stationary animal husbandry
8. **Classification according to implements used for cultivation:**
   (a) Spade farming.

(b) Hoe farming.

(c) Mechanized or tractor farming.

FARMING SYSTEM COMPONENTS

In integrated farming system, it is always emphasized to combine cropping with other enterprises/activities, many enterprises are available and these includes cattle maintenance i.e sheep or goat rearing, poultry, piggery, rabbit rearing, bee keeping etc. Any one or more can be combined with the cropping.

Cattle maintenance:

1. **Buffaloes:** Important dairy breeds of buffalo are Murrah, Mehsana, Zefarabadi, Godavari.

   Feeding: Cattle feed generally contains fibrous coarse low nutrient straw material. Roughage is basic for cattle ration and includes legumes and non-legume hays, straw and silage of legume and grasses. Per day requirement @ 1 kg concentrate per 2 litre of milk and green fodder (20-30 kg), straw 5-7 kg and water- 32 litre per animal.

2. **Sheep rearing:** Sheep are well adapted to many areas. They are excellent gleaners and make use of much of waste feed. Consume roughage, converting a relative cheap food into a good cash product. Housing is not expensive.

   Feeding: Require 1-2 kg of leguminous hay per day. Protein supplied through concentrate as groundnut cake.

3. **Goat rearing:** In India, activity of goat rearing takes place under different environments. The activity is also associated with different systems such as crop or animal based, single animal or mixed herd small or large scale. Goat is mainly reared for meat, milk, hide and skin. Goat meet is preferred in India, A goat on hoof fetches a better price than a sheep on hoof.

   Feeding: Per head nutrients requirement of goat is low. Hence they are suitable for resources poor small farmers with marginal grazing lands they eat plants and leaves of tree, which any other animals do not touch. Goat needs fodder, concentrate of maize, groundnut cake etc. and drink clean and fresh water.

4. **Poultry:** Poultry is one of the fastest growing food industries in the world. Poultry meat accounts for about 27% of total meat consumed worldwide. Poultry industry in India is relatively a new agricultural industry. The average global consumption is 120 eggs per person/year and in India it is only 32-33 eggs per capita/year. To meet the nutritional requirement the per capita consumption is estimated at 180 eggs and 9 kg meat/year.

   Feeding: The feed conservation efficiency of the bird is superior to other animals. About 60-70% of the total expenditure on poultry farming is spent on the poultry feed. Hence, use of cheap and efficient ration will give maximum profit by using grains like maize, barley, oats, wheat, rice-broken etc. Apart from feed, a few minerals are also given.

5. **Duck rearing:** Ducks account for about 7% of the poultry population in India. They are popular in water logged States like West Bengal, Orissa, A.P, Tamil Nadu. They have production potential of about 130-140 eggs/bird/year. These can rear in marshy riverside wetland. Duck farming can be a better alternative.

   Feeding: Eating fallen grains in harvested paddy fields, small fishes and other aquatic materials, a variety of crop residues and insects in the farm.

6. **Turkey rearing:** Turkey is a robust bird and can be reared in humid tropics. It actively feed on a variety of crop residues and insects in the farm.

7. **Piggery:** Pigs are maintained for production of pork. Their reproduction rate is high and growth is very quick.

8. **Rabbit rearing:** In India it is of recent origin though hunting of wild rabbits for meat is not uncommon. Rabbit can be easily reared with relatively less concentrate feed with high production rate.

9. **Beekeeping:** Beekeeping is one of the most important agro-based industries which do not require any raw material like other industries. Nectar and pollen from flower are the raw material, which are available in plenty in nature. Beekeeping can be started with a single colony. *Honey collection*: Honey should be of good quality. Qualities such as aroma, color, consistency and floral sources are important. Honey is an excellent energy food with an average of about 3500 calories/kg. It is directly absorbed into the human blood stream requiring no digestion.

10. **Aquaculture:** Ponds serve as domestic requirement of water, supplementary irrigation to crop and pisciculture with the traditional management, farmers obtain hardly 300-400 of wild and culture fish/ ha/ year. However, polyfish culture with the stocking density of 7500 fingerlings and supplementary feeding will boost the total biomass production.

    Species: Cattle, rohu, common carp, silver carp, and grass carp (feed on aquatic plants).

    Management: Pond depth 1.5- 2.0 m, water should be slightly alkaline, pH (7.5-8.5). If the pH is less than 6.5, it can be adjusted with addition of lime, higher pH (>8.5) can be reduced with addition of gypsum. Application of fresh dung may also reduce high pH in the water.

    The fish are to be nourished with supplementary feeding with rice bran and oilseed cakes. This will enable faster growth and better yield. Each variety of crops stocked to 500 fingerlings with the total of 5000- 8000/ha. This gives 2000 to 5000 kg/ha of fish annually.

11. **Sericulture:** It is the keeping of silk moths and their larvae for the production of silk. Sericulture is defined as a practice of combining mulberry cultivation; silkworm rearing and silk reeling. Sericulture is a recognized practice in India. The total area under mulberry is 240 thousand ha in the country. It plays an important role in socio- economic development of rural poor in some areas. Climate condition is favorable for mulberry and rearing of silk

worms throughout the year. Karnataka is the major silk producing state in India (Temperature- 21 to 30°C), in Kashmir climate suitable is from May to October.

Moriculture: Cultivation of mulberry plants is called moriculture. The crop yield is well for 12 years. Yield of mulberry leaves is 30- 40 t/ha/year.

Rearing: Eggs are allowed to be laid over a cardboard. In bamboo tray rice husk is spread. Tender chopped mulberry leaves are added to the tray. The hatched out larvae are transferred to the leaves. It is important to change the leaves every 2-3 hours during the first 2-3 days. The cocoon is made of silk. The cocoons required for further rearing are kept separately and moths are allowed to emerge from them.

12. **Mushroom cultivation:** Mushroom is an edible fungi with great diversity in shape, size and colour. Essentially mushroom is a vegetable that is cultivated in protected farms in a highly sanitized atmosphere; mushroom contains 90% moisture. It is rich in quality protein, fairly good source of vitamin C and B complex. It is a rich source of mineral like Ca, P, K & Cu. They contain less of fat and carbohydrates and are considered good for diabetic and blood pressure patients.

    Types of mushroom: i) Oyster mushroom, ii) Paddy straw mushroom- *Volvarilla volvacea*, iii) White button mushroom- *Agaricus gisporus* (Var, A-11, Horst V3).

13. **Biogas Plant:** Biogas is a clean, unpolluted and cheap source of energy, which can be obtained by a simple mechanism and little investment. The gas is generated from cow dung during anaerobic decomposition. Biogas generation is a complex bio-chemical process, cellulolytic material are broken down into methane and CO_2 by different groups of micro-organisms. It can be used for cooking purpose, burning lamps, *etc*. Biogas near to kitchen and cattle shed to reduce cost of gas transfer and cow dung transport and also sufficient sunlight is important for maintaining optimum temperature.

    Biogas slurry: Slurry is obtained after the production of biogas. It is enriched manure; another positive aspect of this manure is that even after weeks of exposure to the atmosphere the slurry does not attract fleas and worms. Dry slurry contains about 1.8% N, 1.1% P and 1.5% K.

CROPPING SYSTEMS

Cropping systems, an important component of a farming system, represents a cropping pattern (*i.e.* the proportion of area under various crops at a point of time in a unit area) used on a farm and their interaction with farm resources, other farm enterprises and available technology, which determine their makeup.

It is defined as the order in which the crops are cultivated on a piece of land over a fixed period or cropping system is the way in which different crops are grown. In the cropping systems, sometimes a number of crops are grown together or they are grown separately at short intervals in the same field.

Classifications of cropping system

Depending on the resources and technology available, different types of cropping systems are adopted on farms, which are as below.

1. **Mono-cropping or single cropping:** Mono-cropping refers to growing only one crop on a particular piece of land year after year e.g., growing only *Rabi* crops in dry lands or on lands situated in river basins which often remain flooded during rainy season. This is due to climatologically and socio-economic conditions or due to specialization of a farmer in growing a particular crop. Groundnut or cotton or sorghum is grown year after year due to limitation of rainfall. Flue-cured tobacco is grown in Guntur (A.P.) due to specialization of a farmer in growing a particular crop. Rice crop is grown, as it is not possible to grow any other crops, in canal irrigated areas, and under waterlogged conditions.
2. **Monoculture:** Practice of repetitive growing only i.e one crop irrespective of its intensity, for example rice after rice in Kerala, West Bengal and Orissa.
3. **Sole cropping:** One crop variety grown alone in pure stand at normal density.
4. **Multiple cropping or polycropping:** It is a cropping system where two or three crops are grown annually on the same piece of land using high input without affecting basic fertility of the soil. Growing two or more crops on the same piece of land in one calendar year is known as multiple cropping. It is the intensification of cropping in time and space dimensions. It includes intercropping, mixed cropping and sequence cropping.

Multiple cropping is a philosophy of maximum crop production per unit area of land with minimum of soil deterioration.

- Rice-potato-green gram
- Rice-mustard-maize
- Rice-potato-sesame
- Jute-rice-potato

Cropping intensity is more than 200% when the farm as a whole is considered; the Multiple Cropping Index (MCI) is determined by the number of crops and total area planted divided by the total arable area. When the value is three or more, it is said to be most promising farm. This is also called as intensive cropping.

Polyculture: Cultivation of more than two types of crops grown together on a piece of land in a crop season.

Relay cropping: Growing the succeeding crop in the standing crop when it is about to mature or sowing of the next crop immediately after the harvest of the standing crops. Or it is a system of cropping where one crop hands over land to the crop in quick succession.

Overlapping cropping: Succeeding crop is sown before the harvesting of the preceding crops. Here lucerne and berseem are broadcasted in standing paddy crop just before they are ready for harvesting.

Advantages:

- Minimum tillage is needed for relay cropping and primary cost of cultivation is less.
- Weed infestation is less, as land is engaged with crops year round.
- Crop residues are added in the soil and thus more organic matter.
- Residual fertilizer of previous crops benefits succeeding crops.

CROPPING PATTERN

The selection of crops and their varieties is to be made depending on the soil and rainfall situation in the rained areas. The photo insensitive crops and varieties with shorter duration should be chosen to escape drought of different intensities. There are wide variations, location to location in water availability periods in dryland areas. Thus depending upon water availability following are the different crops and cropping patterns to suit different climatic situations.

Cropping pattern for rained areas:

<table>
<tr><td rowspan="2">Monoculture</td><td>Scarcity zone</td><td colspan="2">Pearl millet, Red gram, Green gram, Black gram, Horse gram and Groundnut
Rabi : Sorghum and Safflower</td></tr>
<tr><td>Assured rainfall</td><td colspan="2">Cotton, Sorghum, Red gram, Black gram, Green gram, Soybean and Sunflower</td></tr>
<tr><td rowspan="3">Double cropping</td><td>Scarcity zone</td><td>Kharif: Moong/Urad, Bajra and Sunflower</td><td>Rabi: Safflower, Sorghum and Gram</td></tr>
<tr><td>Assured rainfall zone</td><td>Paddy, Soybean, Moong/Urd and Sunflower</td><td>Safflower, Sorghum and Gram</td></tr>
<tr><td>Irrigated areas</td><td>Sorghum, Maize and Groundnut</td><td>Wheat, Sorghum, Gram and Sunflower</td></tr>
</table>

Stable intercropping systems for rainfed areas:

Scarcity zone: Bajra + Tur in 2:1 row proportion

Assured rain: Sorghum + Moong/Urd in 2: 1 row proportion.

Cotton + Moong/Urd in 1: 1 row proportion

Cotton + Tur in 8: 2 row proportion

Sorghum + Tur in 2: 1 row proportion.

Tur + Moong/Urd in 1: 3 row proportion.

Grassland or pasture management: Most of the marginal lands are not able to sustain arable crops particularly during the drought years. Such lands can be developed into dependable pastures by following soil and water conservation measures like contour trenches and contour furrows. Controlled grazing may also help in building the forage resource.

At times, native pastures are stocked with low productive and less palatable species. These pastures lack legume component, thus, making the pasture lands nutritionally deficient. Artificial renovation of such pastures is likely to provide forage of good quality as well as sufficient quantity. In rained areas, different legumes from the genera *Dolichos, Leucaena, Clitoria, Cassia* and *Stylosanthes* have been found to do well with or without grasses like *Cenchrus ciliaris*. But *Stylosanthes* has been found to be excellent in all situations with regard to persistence, nutritive value and palatability. Different grasses from the genera *Dichanthium, Cenchrus, Lasiurus, Chloris, Urochloa, Panicum*, and *Pennisetum etc*. have been observed doing well. *Cenchrus ciliaris* has been found to be good in most of the situations. The pastures are easily established if they are seeded at the beginning of rainy season. Seeds of *Cenchrus ciliaris* @1.0 kg. *Stylosanthes hamata* @4.0 kg and *Stylosanthes scabra* @1.0 kg per hectare may be used as seed moistures. The seed moisture may be broadcasted on a drizzling day. After that, light raking of the soil may improve germination chances considerably.

Research investigations have revealed that application of 20-25 Kg N increases dry matter yield of grass species considerably. Similarly, 30-40 kg of P_20_5 gives good response of legume component. For the establishment of pasture as well as for getting increased forage production the access of livestock to pastures should be controlled so that grazing pressure could be minimized.

Cropping pattern in different regions of India

- The most important element of farming in India is the production of grains and the dominant food-chain is grain-man. On this basis, the country may be divided broadly into five agricultural regions.
- The rice region extending from the eastern part to include a very large part of the north-eastern and the south-eastern India, with another strip along the western coast.
- The wheat region, occupying most of the northern, western and central India.
- The millet-sorghum region, comprising Rajasthan, Madhya Pradesh and the Deccan Plateau in the centre of the Indian Peninsula.
- The temperate Himalayan region of Kashmir, Himachal Pradesh and Uttar Pradesh and some adjoining areas. Here potatoes are as important as cereal crops (which are mainly maize and rice), and the tree-fruits form a large part of agricultural production.

Difference between cropping system and cropping pattern

S. No.	Cropping Pattern	Cropping System
1	Cropping pattern and its management to derive benefits from a given resource base under specific environmental conditions.	Crop rotation practiced by a majority of farmers in a given area or locality.

S. No.	Cropping Pattern	Cropping System
2	The cropping patterns used on a farm and their interaction with farm resources, other farm enterprises and available technology which determine their make up.	Type and management of crops in time and space.
3	Pattern of crops taken up for a given piece of land, or order in which crops are cultivated on a piece of land over a fixed period, associated with soil, management practices such as tillage manuring and irrigation	Yearly sequence and spatial arrangement of crop or crops and fallow on a given area. The proportion of area under various crops at a point of time in a unit area

Rice-based cropping patterns

Rice is grown in high-rainfall areas or in areas where supplemental irrigation is available to ensure good yields. If the crop has to depend solely on rainfall, it requires not less than 30 cm per month of rainfall over the entire growing period. However, only 9% of the area in the country comes under this category, and it lies in the eastern parts. Nearly 45% of the total rice area in India receives 30 cm per month of rainfall during at least two months (July and August) of the south-westerly monsoon and much less during other months. In contrast to these parts, the eastern and southern regions comprising Assam, West Bengal, coastal Orissa, coastal Andhra Pradesh, Karnataka (most part), Tamil Nadu and Kerala receive rainfall of 10 to 20 cm per month in four to eight consecutive months, starting earlier or going over later than the south-westerly monsoon months. With supplemental irrigation, 2 or 3 crops are taken in these areas. However, it has been observed that on an all-India basis, nearly 80% of rice is sown during June-September and the rest during the rest of the season.

INTERCROPPING

The growing of two or more crops simultaneously on the same piece of land at the same time with a definite row arrangement, called intercropping. There is a crop intensification in both time and space dimensions. For example, growing setaria + red gram in 5:1 ratios *i.e.* after every row of setaria, one row of redgram is sown.

Objectives of intercropping systems

- Insurance against total crop failure under aberrant weather conditions or pest epidemics.
- Increase in total productivity per unit land area.
- Judicious utilization of resources such as land labour and inputs.

Intercropping was originally practiced as an insurance against crop failure under rainfall conditions. At present the main objective of intercropping is higher productivity per unit area in addition to stability in production. Intercropping systems utilize resources sufficiently and their productivity is increased.

When two crops are to be grown together, they are chosen in such a way that there is variation in their growth duration. The peak periods of growth of the two crops species should not coincide. In such arrangements, a quick maturing crop completes its life cycle before the other crop starts. Due to greater differences in maturity and growth demands of the crop components, more opportunity is provided for greater exploitation of growth factors. This is achieved either by generic difference in crop species or manipulation of planting dates. Normally short and long duration crops are grown together.

Principles of Intercropping

- The crops grown in association should have complementary effects rather than competitive effects.
- The subsidiary crop should be of shorter duration and of faster growing habits to utilize the early slow growing period of the main crop and they must be harvested when the main crop starts growing. e.g. autumn planted sugarcane remains dormant after germination until February during which potato, berseem, lucerne, mustard could be taken successfully as companion intercrops. Cotton could be intercropped with onion, green gram or black gram; and chillies and tomato could be intercropped with onion.
- Component crops should have similar agronomic practices.
- Erect growing crops should be intercropped with cover crops like pulses. These check the weeds and reduce soil erosion. The loss of water due to evaporation is also reduced.
- The component crops should have different root depths. This will minimize competition for nutrients, water and root respiration.
- A standard plant population of main crop should be maintained whereas in the case of subsidiary crops, the plant population could increased or decreased as per demand of the situation.
- The planting method and management should be simple, less time consuming, economical and profitable so that it may have wider adaptability.
- Component crops of similar pests and disease pathogens and parasite infestations should not be chosen.

Based on the percent of plant population used for each crop in intercropping system, it is divided into two series:

1. Additive series: It is mostly adopted in India, one crop is sown with 100% of its recommended population in pure stand, which is known as the base crop. Another crop known as intercrop is introduced into the base crop by adjusting or changing the geometry. The population of intercrop is less than its recommended population in pure stand. LER of additive series is greater than replacement series. Additive series is more efficient than replacement series in intercropping system.

2. Replacement series: Both the crops growing together are called component crops. By scarifying certain proportion of population of one component, another component is increased.

Component crop: is used to refer to either individual crop making up the intercropping situation. Intercrop yield is the yield of a component crop when grown in intercropping and expressed over the total inter-cropped area. (*i.e.* area occupied by both the crops). A simple addition of both the intercrop yields a combined intercrop yield.

Base crop: is the one which is planted as its optimum sole crop population in an intercropping situation and the second crop is planted in between rows of base crop for obtaining bonus yield from intercrop without affecting the base crop yield.

Inter-cropping may be divided into the following groups:

1. **Parallel cropping:** Under this cropping two crops are selected which have different growth habits and have a zero competition between each other and both of them express their full yield potential. *e.g.* (i) Green gram or black gram with maize, (ii) Green gram or soybean with cotton.
2. **Companion cropping:** In companion cropping the yield of one crop is not affected by the other, In other words, the yield of both the crops is equal to their pure crops and the standard plant population of both crops is maintained. *i.e.* (i) mustard, wheat, potato, *etc.* with sugarcane, (ii) Wheat, radish, cabbage, sugar beet *etc.*, with potato.
3. **Strip intercropping:** It is growing two or more crops simultaneously in different strips wide enough to permit independent cultivation but narrow enough for the crops to interact agronomically.
4. **Relay intercropping:** Growing two or more crops simultaneously during part of the life cycle of each. A second crop is planted after the first crop has reached its reproductive stage but before it is ready for harvest.
5. **Multistoried cropping or Multi-tier cropping:** Growing plants of different height in the same field at the same time is termed as multistoried cropping. It is mostly practiced in orchards and plantation crops for maximum use of solar energy even under high planting density. *i.e.*
   - Eucalyptus +Papaya + Berseem
   - Sugarcane + Potato + Onion
   - Sugarcane + Mustard + Potato
   - Coconut + Pineapple + Turmeric/Ginger

Intercropping is mostly prevalent in plantation crops like coconut and arecanut. The practice of growing different crops of varying heights, rooting pattern and duration are called multi-tier cropping. The objective of this system of cropping is to utilize the vertical space more effectively. In this system, the tallest components have foliage tolerant of strong light and high evaporative demand and the shorter component(s) with foliage requiring shade and or relatively high humidity. e.g. Coconut + black pepper + cocoa + pineapple.

The multi-tier cropping plan is shown in Fig. 1.1. In this system, coconut is planted with a spacing of 7.5 m. Rooted cutting of black pepper are planted on either side of coconut about 75 cm away from the base. On the coconut trunk at a height of about one meter from the ground level, the vines of pepper are trailed. A single row of cocoa is planted at the center of the space between coconut rows. Pineapple is planted in the interspace. Coconut growing to a height of more than 10 m Cocoa with its pruned canopy of about 2.5 m height and pineapple growing to about 1 m height from the first and ground floors, respectively.

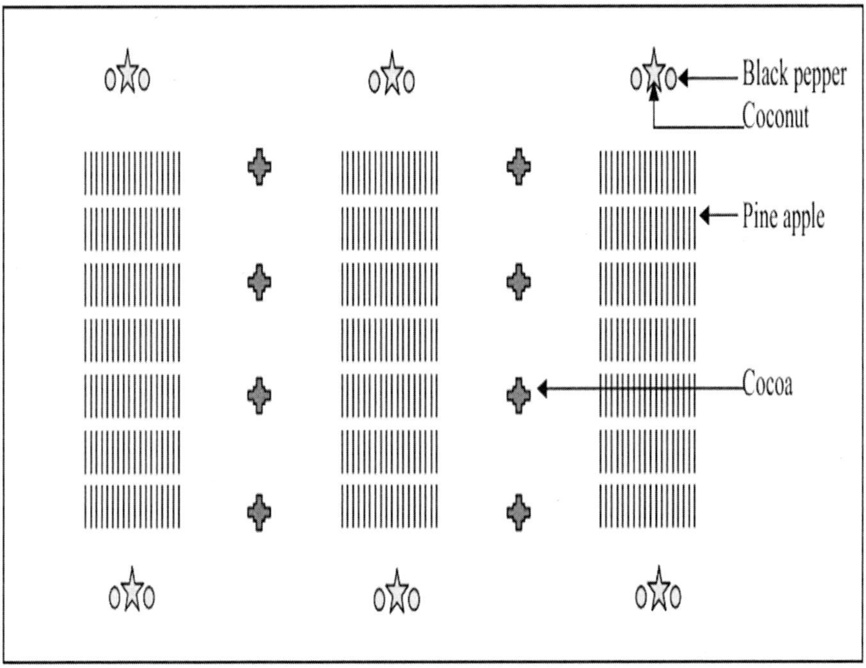

Fig. 1.1. Multistoried cropping

In another multi-tier system in coconut, ginger or turmeric and partial shade loving vegetables form the first tier, banana the second tier, pepper the third tier and coconut or arecanut the final tier. In the arecanut plantation, tuber crops are predominantly intercropped. Elephant yam, tapioca, greater yam and sweet potato are common inter-crops in humid tropics. Banana and pineapple are also cultivated as intercrops in arecanut gardens. In coffee based multi-tier cropping systems, the first tier is with pineapple, second tier with coffee, third tier with cocoa/mandarin orange and final tier with fast growing shade trees necessary for coffee plantation.

Advantage of intercropping

- Intercropping gives higher income per unit area than sole cropping.
- It acts as an insurance against failure of crop in a year.
- Intercrops maintain soil fertility as the nutrient uptake is made from both layers.
- Reduces soil runoff.

- Better control of weeds, pests and diseases.
- One crop provides physical support to the other crop.
- One crop provides shelter to the other crop.
- Erosion control through providing continuous leaf cover over the ground surface, and it is the small farmers of limited means who is most likely to benefit.
- Total biomass production/unit area/time is increased because of fullest use of land as the inter-row space will be utilized which otherwise would have been used for weed growth. The farmer gets all his required agriculture commodities from a limited space. Thus the profit/unit area becomes higher. The fodder value becomes higher when a legume is intercropped with a non-legume *i.e.* Napier + Cowpea – Napier + Berseem. It offers best employment and utilization of labour, machine and power throughout the year.

Disadvantages of intercropping

- Yield decreases as the crops differ in their competitive abilities.
- Management of intercrops having different cultural practices seems to be difficult task.
- Improved implements cannot be used efficiently.
- Higher amount of fertilizer or irrigation water cannot be utilized properly as the component crops vary in their response to these resources.
- Harvesting is difficult.

Intercropping with cereals: Intercropping with cereal is an excellent way of improving the resource utilization because cereals utilize the rainy season resources while late maturing crops exploit the post-rainy season resources such as residual moisture. Sorghum is most commonly intercropped with pigeon peas on a variety of soils. Sorghum is harvested after 3½ to 4½ months and pigeon pea matures in about 6-9 months depending on the genotype. In monocropping of groundnut climate resources like rainfall, temperature and solar radiation are utilized only during the rainy season, however when red gram is introduced as an intercrop, these resources are used up at the start of the winter season and also benefit from late shower rains. Soybean is also a compatible companion crop with maize. Pearl millet is a quick tillering and fast growing crop that attains full canopy development within 20-30 days of seedling establishment. It can be intercropped with groundnut, black-gram or castor.

Intercropping in pulses: For pigeon pea, short duration grain legumes such as blackgram and soybean are the best companion crops in Peninsular India. Groundnut is also a suitable intercrop.

Intercropping in cotton: It is initially a slow growing crop. Any short duration and fast growing crops such as groundnut, blackgram, greengram or cluster bean are compatible companion crops.

Intercropping in sugarcane: Sugarcane is slow growing up to 80-90 days. Since the crop is planted in rows 0.8-1.0 m apart, considerable space is available for intercropping. Short duration crops maturing in 80-90 days can be advocated as intercrops. Black-gram and soybean are found suitable. The green manure crops like dhaincha can be sown in the inter-space and incorporated at about 2 months.

MIXED CROPPING

It is the growing of two or more crops simultaneously on the same piece of land at the same time, without a definite row arrangement, called mixed farming. *e.g.* Pearl millet + Moong bean/Moth bean, Pearl millet + Cluster bean

Principles behind mixed cropping

- All crops cannot fail under adverse climate conditions.
- An epidemic attack of any insect, pest or disease kills only one crop without affecting the rest of the crops.
- The farmers grow different crops which fulfill their daily need or demand for cereals, pulses and oilseeds.
- Mixed cropping checks soil erosion, weeds, etc.
- It improves or maintains the soil fertility.
- Family labour and cattle are employed throughout the whole year.
- Legume and non-legume mixture increases the fodder quantity and quality of both.
- It reduces the cost of cultivation.

Classification of mixed cropping

- **Mixed crops:** In this case, the seeds of different crops are mixed together and then sown either in lines or they are broadcast. The system is not scientific and it causes problems in performing all the agricultural operations and harvesting of the crops.
- **Companion crops:** Under this method the seeds of different crops are not mixed together but different crops are sown in different rows i.e. between two rows of mustard five to eight rows of wheat or between two rows of pigeon pea two to three rows of groundnut are sown. This method of sowing facilitates weeding, interculture, plant protection operation and even harvesting.
- **Guard crops:** Under this system of cropping, the main crop is grown in the center, surrounding by hardy or thorny crops such as safflower around pea or wheat, mesta, around sugarcane, sorghum around maize etc. with a view to provide protection to the main crop.
- **Augmenting crops:** When sub crops are sown to supplement the yield of the main crop, the sub crops are called as augmenting crops, such as mustard with berseem. Here the mustard helps in getting higher tonnage of fodder in spite of the fact that berseem gives poor yield in fist cutting.

Advantages of mixed cropping

- It offers highest return on farm business, as the by-products of farm are properly utilized.
- It provides employment throughout the year.
- Efficient utilization of land, labour, equipment and other resources.
- The crop by products such as straw, bushes, fodder etc. is used for feeding of livestock and in return they provide milk.
- Manures available from livestock maintain soil fertility.
- It helps in supplying all the food requirement of the family members.
- Intensive cultivation is possible.
- If one source of income is lost he can maintain his family from other source of income.
- Milk cattle provide draft animals for crop production and rural transport

Differences between Intercropping and Mixed cropping

S. No.	Intercropping	Mixed cropping
1	The main object is to utilize the space left between two rows of the main crop	To get at least one crop under favourable conditions
2	More emphasis is given to the main crop	All crops arc cared equally
3	There is no competition between both crops	There is competition between all crops growing
4	Inter crops are of short duration and are harvested much earlier than the main	The crops are almost of the same duration
5	Sowing time may be same or different	It is same for all crops
6	Crops are sown in different rows without affecting the population of main crop when sown as sole crop	Either sown in rows or mixed without considering the population

SYNERGETIC CROPPING AND ALLEY CROPPING

- **Synergetic cropping:** Here the yields of both crop grown together are found to be higher than the yield of their pure crops on unit area basis. *i.e.* Sugarcane + Potato
- **Alley cropping:** Food crops grown in alleys formed by hedgerow intercropping. It is recommended for humid tropics. In semiarid regions of India, alley cropping provides fodder during dry period.

Alley cropping is an agroforestry system in which fast growing N-fixing stubs are planted as hedgerows. Food crops are grown in alleys formed between hedges of trees and shrubs. This system is most suitable for marginal and sub-marginal lands. Scientists at International Institute of Tropical Agriculture (IITA), Nigeria developed this system of agroforestry in 1947.

The essential feature of this system is that hedge rows are cut back at about one meter height at planting and are kept pruned during cropping to prevent shading and reduce competition with field crops. In semiarid regions, alley cropping provide fodder during dry period since mulching the crop with hedgerow pruning usually do not contribute to increase crop production.

Advantages:

- Provision of green fodder during lean period of the year.
- Higher biomass production per unit area than arable crops.
- Efficient use off season rainfall in the absences of the crop.
- Additional employment during off-season.
- It serves as a barrier to surface runoff leading to soil and water.
- Conservation based on objectives following three types of alley.
- It improves soil fertility and is more remunerative under rainfed conditions.
- Systems are recognizing:
  (a) Forage alley cropping.
  (b) Forage cum mulch system
  (c) Forage cum pole system.

(a) Forage alley cropping: In this system both yield of crop and forage assume importance. Luciana (Squabble), Sesbania are recommended for hedgerow. pigeon pea or caster crops are suitable for growing in the allies. Crop yield decreases with decrease in the row width.

(b) Forage cum mulch system: In this systems hedgerows are used for both forage and mulch. Lapping is used for mulching during the crop season.

(c) Forage cum pole systems: Luciana alleys are established at 5-meter intervals along the contours. Hedgerows are established by direct seeding and topped every two months at 1.0 m height during crop season and every four months during the off season. A Luciana plant at every 2-meter is allowed to grow in to a pole. Crop yields are usually reduced due to competition from hedgerow.

Classification of alley cropping: According to the purpose for which the alleys are raised, alley cropping may be grouped into:

- Forage alley cropping.
- Green manure-cum-mulch alley cropping.
- Forage-cum-mulch alley cropping.
- Forage-cum-green manure alley cropping. Among them, forage alley cropping is recommended as an alternate land use system for semi-arid regions with main benefit of green fodder production during dry season.

MULTIPLE CROPPING

1. Sequence cropping: Growing of two or more crops in sequence on the same piece of land in a farming year. Crop intensification is only in time dimension and there is no intercrop competition. Depending on the number of crops grown in a year, it is called as double, triple and quadruple cropping involving two, three, and four crops, respectively.

Example:

(a) Double cropping: (i) Rice-Potato/mustard, (ii) Sorghum-Gram.

(b) Triple Cropping: (i) Rice- Potato-Wheat, (ii) Cowpea-Mustard-Jute.

(c) Quadruple cropping: *Kharif* groundnut-Leafy vegetables-Wheat- Summer green gram.

In India, food crop is predominantly grown in the most suitable season and thus particular food crop is basic to the cropping system followed by the farmers.

Cropping systems are usually referred to as:

I. Rice-based cropping system.

II. Sorghum-based cropping system.

III. Pearl millet-based cropping system

IV. Wheat and gram-based cropping system.

Some of the cropping systems based on commercial crops are:

(i) Cotton-based

(ii) Groundnut-based,

(iii) Sugarcane-based,

(iv) Plantation crop-based

(v) Vegetable-based cropping systems.

The grain production potential in different regions of the country ranges from 11-18 t/ha. Maize-Potato or Toria-Wheat-Moong system is followed at IARI, New Delhi; it was possible to produce 14-15 tons of food per hectare per annum without impairing the soil health. The results of multiple cropping demonstrations under irrigated condition showed that production potential can be as high as 19.8 t/ha in cereal- based cropping system of rice-rice-rice. All rice sequence is followed in some pockets of the river command areas of Orissa, Andhra Pradesh, Karnataka and Tamil Nadu.

Systems with more than three crops: Bradfield experimental at IRRI, Philippines, with various cropping pattern that could maximize productivity per unit area. The most successful pattern was a sequence of five crops consisting of rice, sweet potato, soybean, sweet corn and green soybeans. This pattern gave three times higher gross income than normally obtained in a rice-rice pattern.

Experiment conducted at IARI, New Delhi showed that four crop sequence of green gram- maize- potato- wheat gave the highest production of 13.6 t/ha. However, three crop sequence of green gram- maize- wheat was found to be the most ideal for the small and marginal farmers. Crop sequences and their production potential, New Delhi, India are given in Table 1.1

Table 1.1. Crop sequential total production in India

Crop sequences	Total production (Q/ha)
Maize-Wheat	90.1
Green gram-Maize-Wheat Green gram-Pigeon pea-Wheat Cowpea-Pearl millet-Wheat	97.7 68.8 92.9
Green gram-Maize –Potato-Wheat Green gram-Maize-Toria-Wheat	135.9 112.8

Cropping systems in dry lands: The intensity of cropping can be increased through sequential intercropping depending on the rainfall and moisture storage capacity of soils as shown below.

Cropping pattern with varying rainfall and soil moisture storage capacity

Rainfall (mm)	Storage capacity of soil(mm)	Cropping pattern
350-625	100	Single crop in *kharif*
650-720	100	Intercropping can be attempted
780-900	150	Sequential cropping is possible
900- above	200	Sequential cropping is assured.

Ratoon cropping or Rationing: Refers to raising a crop with re-growth coming out of roots or stalks after harvest of the crop or cultivation of crops regrow after harvest, although not necessary for grain. Generally sugarcane and napier-bajra hybrid are kept as ratoon.

CROPPING SCHEME

The plan according to which crops are grown on individual plots of a farm with an object of getting maximum return from cash crop without impairing the fertility of soil is known as cropping scheme.

Objects of Cropping Scheme

Cropping scheme provides an idea beforehand about the following:

(i) What crop is to be cultivated in a farm.

(ii) What amount of area is to be allotted for a particular crop.

(iii) The relative claim for acreage of the competitive crop.

(iv) Selection of crops as per the facilities available for power, irrigation, inputs, labour transport etc.

(v) Utilization of inputs and other resources available on the farm without wastage.

(vi) Preparation of budget for each crop.

Principles or Characteristics of a Good Cropping Scheme

(i) Area under individual plots: The areas of individual plot for each crop should be approximately same year after year unless price variation takes place.

(ii) Number of plots: The number of plots should be equal to the duration or multiple of it. When the total duration of rotation in cropping scheme is 4, then the number of plots may be 4 or multiple of it. *i.e.* 8, 12, 16 and so on.

(iii) Selection of crops: Cropping scheme is related to the profitable use of productive resources and management.

(iv) Profitable crops: the rotation should be planted around most profitable crops.

(v) Meet the requirement: The cropping scheme should be so planned as to provide maintenance of soil fertility and other physical chemical properties.

(vi) The scheme should have one leguminous crop in year in its rotation for the maintenance of soil fertility and other physical chemical properties.

Selection of crops:

Selection of crops depends on the situation of the farm to obtain maximum profit.

(a) Near a city: Farm is near the city, farmer should grown vegetable on at least 60% of the net cultivable area remaining 10-15 % under fodder.

(b) Near a sugar factory: 60% areas under sugarcane or under sugar beet, remaining area under orchards or other field crops..

(c) Near a canning factory: Farmer should grow oilseed crops like groundnut, sunflowers.

(d) Near dairy farm: The farmer should grow fodder crops. 10-15% area under oil seeds and pulses in order to supply concentrate to the dairy cattle along with fodder.

(e) Near cold storage: Farmer should grow potatoes onion, vegetable, etc.

(f) Near highway/ railway station: Farmer can cultivable perishable crops like vegetables.

Zaid crops: Those crops grown from February to May are called Zaid crops. Irrigation needs of these crops are usually higher than *kharif* and *rabi* crops.

Examples of cropping scheme: Cropping scheme for a farm of 10 ha which is situated near a city and have all facilities for intensive cropping, soil- sandy loam is as under.

Net Cultivable area = 9 ha (10% area under building and layout).

Cropping Scheme

Plot No.	*Kharif* crop	Area (ha)	*Rabi* crop	Area (ha)	Zaid Crops	Area (ha)
1	Maize	2	Potato	2	Sesamum	2
2	Groundnut	2	Wheat	2	Moong	2
3	Brinjal	2	Cabbage	2	Chilli	2

Plot No.	*Kharif* crop	Area (ha)	*Rabi* crop	Area (ha)	Zaid Crops	Area (ha)
4	Radish	2	Cauliflower	2	Watermelon	2
5	Bhendi	1	Pea	1	Cucumber	1
	Total	9		9		9

Total cropped area = 9 + 9 + 9 = 27 ha

Cropping intensity% = Total cropped area/Net cultivable area = (27/9) × 100 = 300%.

CROP ROTATION AND ITS CHARACTERISTICS

Crop rotation

It is a process of growing different crops in succession on a piece of land in a specific period of time with an object to get maximum profit from least investment without impairing soil fertility.

Recurrent succession of crop on the same piece of land either in a year or over a longer period of time. Component crops are so chosen so that soil health is not impaired, e.g., Cotton-Gram, Sugarcane-Wheat. Or it means growing a set of crops in a regular succession on a piece of land in a specific period of time, with an object to get maximum profit with least investment without impairing soil fertility. e.g., Sorghum-Gram, Groundnut-Wheat.

Principles of Crop Rotation

- The crops with tap roots should be followed by those which have a fibrous root system
- The leguminous crops should be grown after non-leguminous crops.
- More exhaustive crops should be followed by less exhaustive crop.
- Selection of crops should be demand-based.
- Selection of crops should be problem-based.
- The crops of the same family should not be grown in succession because they act like alternate hosts for insects, pests and disease pathogens.
- An ideal crop rotation is one which provides maximum employment to the family and farm labour, the machines and equipment are efficiently used then all the agriculture operations are done simultaneously

Characteristics of Good Crop Rotation

- It should be adaptable to the existing soil, climatic and economic factors.
- The sequence cropping adopted for any specific area should be based on proper land utilization or it should be so arranged in relation to fields that crop yields can be maintained and also build up organic matter content of the soil.
- Rotation should contain sufficient area under soil improving crops (legumes) to maintain and also build up organic matter content of the soil.

- In areas where legumes can be grown successfully, the rotation should provide sufficient average of legumes to maintain "N" supply of the soil.
- It should provide food grains, pulses, oilseed etc. to family and roughage, fodder to cattle.
- It should help in control of pests, diseases, and weeds.
- It should provide maximum area under the most profitable crops adapted to the area.
- It should be so arranged to make it economical in production and labour utilization.

Advantages of an Ideal Crop Production

- There is overall increase in yield of crops mainly due to maintaining physical-chemical properties of soil. Soil fertility is restored by fixing atmospheric nitrogen, encouraging microbial activity (more OM) and protecting soil from erosion, salinity and acidity.
- It helps in controlling insects, pests and soil borne diseases. It also controls weeds. *i.e.* repeated wheat culture (growing) increases wild oats and *P. minor* infestation. Similarly growing berseem continuously encourages chicory (kasani) infestation, but an alternate cropping of berseem and wheat helps in controlling wild oats and *P. minor*.
- Prevents or limits periods of peak requirements of irrigation water. Crops require high irrigation if followed by crops requiring light irrigation, this will not affect the soil physical condition.
- It facilitates even distribution of labour. Following crop make proper utilization of all resources and inputs. Family and farm labour, power, equipment and machines are well employed thought the year.
- Farmers get a better price for his produce due to higher demand in local market. So there is regular flow of income over year.

Inclusion of crops of different feeding zones (root system) and nutrient requirement could maintain the better balance of nutrient in soil. Growing crops of different root depths avoids continuous depletion of nutrients form same the depth. e.g. deep rooted crops take nutrients from deeper zone and during that period the upper zone gets enriched. Similarly, surface feeding roots take nutrients from the upper zone when the lower zone get enriched. So growings same crop without rotation results in loss of soil productivity utilizes the nutrients from the entire soil mass and cost of cultivation is reduced.

Diversification of crops reduces risk of financial loss due unfavorable conditions. Diversification of crops means a variety of crops can be grown for meeting the domestic needs of farmers and livestock, to reduce risk of market fluctuations, mechanism of farming and growing expensive crops. So, all variety of crops are grown in rotation for more benefit.

It improves soil structure, percolation and reduces changes of creation of hard- pan in the sub soil and also reduces soil erosion.

Some crop plants are found to produce phytoalexins when they get infected by diseases. Repeated cultivation of such crops results in harmful effects over crop plants and lower crop yield is obtained. *i.e.* crop- phytolalexins produced by diseased plants. Groundnut -Resvertrol, Soybean- Glyceollins

The family needs of food, fuel, fiber, spices, sugar etc. are fulfilled and also fulfill needs of livestock.

It is advantageons to raise short duration crops (catch crop/ vegetables) when long season crops cannot be raised due to some reasons.

Factors to be considered while planning crop rotation: Growing different crops is very beneficial, but sometimes the desired crops cannot be grown because of certain governing factors (soil and climate), irrigation, non-availability of bullocks and other powers, market facilities and type of farming.

Factors of Crop Rotation

- Net profit per hectare
- Growth habit and nutrient requirement of different crops.
- Soil type and slope.
- Infestation of weeds, pests and diseases.
- Irrigation facilities.
- Climatic conditions.
- Land, labour, power and other resources.
- Food habit and requirements.
- Market facilities.

In light of the above factors or consideration the mechanics of setting good crop rotation can be summarized as below or there are certain accepted principles, based on which the crops should be selected for crop rotation.

(i) The crops with taproot should be followed by those which have a fibrous root system. This helps in proper and uniform use of nutrients from the soil and root do not compete with each other for uptake of nutrients.

(ii) A shallow rooted grain crop, deep rooted cash crop and restorative crop (legume crop) should be included in the rotation for providing food, fodder, cash and maintaining the fertility and productivity of soil.

(iii) Leguminous crops should be grown after non-leguminous crops because leguminous fix atmospheric "N" into the soil and add more organic matter to the soil, while non-leguminous are exhaustive crops. Apart from this, legumes need more phosphate and less nitrogen while non- legumes need more nitrogen and relatively low phosphorus. So nutrient requirements of these crops are different and such combination helps farmers in reducing cost of cultivation.

(iv) Selection of the crops should be based on soil, climate season and market demand.

(v) More exhaustive and high input requiring crops should be followed by less exhaustive crops and low input requiring crops like potato, sugarcane, maize, etc. need more inputs such as better tillage, more fertilizer, higher number of irrigations, more plant protection measures than other crops like oil seeds, pulses, etc. which need less inputs.

(vi) As per availability of irrigation water, two or three crops are taken in a year on the same land under irrigated conditions. However a dry crop should be included in the rotation to avoid damage to the soil due to continuous irrigation.

(vii) In case of rain fed farming (assured rainfall) on moisture retentive soils, after harvest of *Kharif* crop some drought tolerant crop requiring less moisture like pulses or cereals crops may be grown. *i.e.*, sorghum, pearl millet, green gram, black gram etc during *kharif* and wheat, barley, sarson, gram etc during *rabi* season.

(viii) The selection of crxops should be problem based e.g. on sloppy lands which are prone to soil erosion, an alternate cropping of erosion promoting (erect growing crops like millet etc) and erosion resisting crops like legumes, should be adopted. Selection of crops should suit the farmer's financial conditions.

(ix) Both wide spaced crop and thickly planted crops should be included in rotation for control of weeds. Wide spaced crops like maize, sugarcane etc can be selected and weeds can be controlled due to frequent inter culture and dense (thick) cops like forage or legume which smother weeds.

(x) Select crops with different botanical families in a rotation in order to maintain soil fertility and to avoid crop-weed association.

(xi) Effect of previous crop on succeeding crop should be considered for obtaining maximum yield and harvest quality of produce.

(xii) Enough elasticity may be kept in rotation so that if pest or diseases destroys a crop, another crop can be substituted

(xiii) Fertile and well-drained land should be utilized for important good rotation, less fertile land for soil improving crops (legumes) and salt tolerant crops on acidic, saline or alkali soils.

(xiv) The ideal crop rotation should be built up around a hub crop for which the greatest comparative advantages exist. For example, in areas of dairy industry grow fodder crops as well as oil seeds like groundnut or pulses which supply cattle feed (oil cakes and roughage) or in irrigated areas near cities, growing of vegetables or floriculture will be profitable.

(xv) Selection of crops should be demand-based, i.e. the crops, which are needed by the people or area, so that produce can be sold at a higher price. The area devoted to each crop should be constant from year to year.

Competitive relationships

Competition: Three broad categories of competition are

(i) **Mutual inhibition:** The actual yield of each species is less than expected under sole crop condition.

(ii) **Mutual co-operation:** The yield of each species is greater than expected.

(iii) **Compensation:** One species yields less and the other more than expected.

The more competitive species in a compensation situation can be called **dominant**, and less competitive species **dominated**.

RESEARCH FINDINGS OF DIFFERENT CROPPING SYSTEM

1. Rice-based intercropping

A field experiment was conducted at farmers' field in West Bengal to assess productivity, profitability, resource-use efficiency including energy value under coastal saline zone with 6 different rice (*Oryza sativa* L.)-based cropping systems, viz. rice-greengram, rice-sunflower (*Helianthus annuus* L.), rice-sunflower + greengram, rice-okra, rice-rice and rice-*khasari* (*Lathyrus sativus* L.). The highest rice-equivalent yield was recorded in rice-okra (33.6 t/ha) cropping system. Higher productivity and profitability also recorded in rice-okra system followed by rice-sunflower. Rice-okra system showed better land-utilization efficiency (69.1%) followed by rice-sunflower than the other systems. Rice-okra system also showed higher irrigation water-use efficiency (1,416.3 kg/ha-cm) and employment generation (475 man-days) than the other systems (Table 1.2). The rice-okra cropping system proved the most remunerative cropping system in the coastal saline zone of West Bengal (Ray *et al.*, 2016).

Table 1.2. Productivity and resource-use efficiency of rice-based cropping systems (pooled data of 2 years)

Cropping system	Productivity (kg/ha/day)	Land-use efficiency (%)	Profitability (/ha/day)	Irrigation requirement (cm)	Irrigation water-use efficiency (kg/ha-cm)	Employment generation (man-days/ ha)
Rice-Green gram	24.5	60.6	74	2.8	1318.3	303
Rice-Sunflower	33.1	67.5	123	13.9	881.0	320
Rice-Sunflower + Green gram	28.3	67.5	116	13.8	771.2	311
Rice-okra	91.9	69.1	330	23.6	1416.3	475
Rice-rice	29.1	67.4	112	33.7	315.9	340
Rice-*khasari*	21.7	55.8	95	0	0	223
CD (P=0.05)	2.40	0.65	16.00	0.53	155.89	-

Whereas, in an experiment conducted at Varanasi, Uttar Pradesh Yadav *et al.* (2013) concluded that Rice-Potato-Onion system recorded significantly higher production efficiency (97.5 kg/ha/day), land-utilization efficiency (91.4%), economic efficiency

(738/ha/day), energy input (61.08×103 MJ/ha) and energy output (187.09×103 MJ/ha). The lowest (39.4 × 103 MJ/ha) energy intensiveness was recorded in Rice-Pea-Cowpea cropping system (Table 1.3).

Table 1.3. Effect of cropping sequence on equivalent yield (EY) of crops in system during rainy, winter and summer season.

Treatment	Rice grain yield (t/ha)		Potato EY of *rabi* crops (t/ha)		Onion EY of *zaid* crops (t/ha)		Rice equivalent yield of system (t/ha)	
	2005–06	2006–07	2005–06	2006–07	2005–06	2006–07	2005–06	2006–07
Cropping sequence Rice–potato–onion	3.98	3.92	19.47	19.77	18.94	19.30	35.34	35.81
Rice–green pea–onion	3.99	3.97	5.73	5.90	19.35	19.66	24.47	24.83
Rice–potato–cowpea	4.07	4.01	20.32	20.58	13.81	14.20	31.93	32.40
Rice–green pea–cowpea	4.08	4.03	5.20	5.35	13.30	13.70	19.18	19.59
Rice–rajma–onion	4.16	4.14	11.94	12.26	19.17	19.49	29.56	30.06
Rice–rajma–cowpea	4.18	4.15	11.61	11.96	13.57	13.96	24.73	25.30
Rice–maize–cowpea	4.21	4.16	10.25	10.60	13.82	14.19	23.85	24.39
CD (P=0.05)	NS	NS	1.94	1.92	1.26	1.26	1.43	1.57

Shah *et al* (2015) showed that Hybrid rice 'JRH 5'-*gobhi sarson* (*Brassica napus* L.)-okra (*Abelmoschus esculentus* L.) system was maximum production efficiency, rice-equivalent yield and water productivity followed by scented rice-berseem (*Trifolium alexandrinum* L.) fodder + seed system, and the pre-dominant cropping systems of the region, *i.e.* rice-wheat and rice-chickpea (*Cicer arietinum* L.) (Table 1.4).

Table 1.4. Rice equivalent yield, cumulative cropping days, production efficiency and economic analysis under different crop-sequences (2006-07 to 2010-11)

Crop sequence	Rice equivalent yield (t/ha/year)				Production efficiency (kg/ha/day)	Benefit cost ratio
	Kharif	*Rabi*	Summer	Total		
Rice-wheat	5.72	5.97	-	11.69	45.50	2.85
Rice-chickpea	5.88	5.28	-	11.16	45.54	3.01
Hybrid rice-onion-greengram+ Residue	7.44	2.82	3.77	14.02	42.37	2.46
Rice-berseem+seed	7.00	7.56	-	14.56	47.42	2.97
Hybrid rice-potato-maize cob+ fodder	5.77	5.48	2.14	13.39	40.93	2.16
Hybrid rice-*gobhi sarson*-maize	5.86	3.47	1.87	11.20	34.46	2.36
Hybrid rice-vegetable pea- hybrid. Sesame	5.93	2.92	2.52	11.37	35.20	2.27

Crop sequence	Rice equivalent yield (t/ha/year)				Production efficiency (kg/ha/day)	Benefit cost ratio
	Kharif	*Rabi*	Summer	Total		
Hybrid rice-*gobhi sarson*-groundnut+maize (4:2)	5.85	3.61	2.56	12.02	34.85	2.51
Hybrid rice-*gobhi sarson*-okra	6.00	3.65	9.18	18.84	53.81	3.75
Hybrid rice-garlic-sorghum-chari' fodder + cowpea (Local) 4:2 rows	6.02	1.16	2.05	9.23	28.66	1.54
Hybrid rice-marigold-maize cob+ fodder	7.27	2.91	1.84	12.01	37.77	2.20
CD (P=0.05)	2.47	1.74	0.99	1.48	-	0.76

The net returns were significantly higher with the hybrid rice-*gobhi sarson*-okra, followed by hybrid rice-potato-groundnut and scented rice 'Sugandha 5'-berseem. Similarly, benefit cost ratio was also significantly higher with hybrid rice 'JRH 5'-*gobhi sarson*-okra, followed by both exist cropping systems, *viz.* rice-chickpea and rice-wheat. The hybrid rice 'JRH 5'-garlic-sorghum+cowpea system gave the lowest benefit cost ratio, which indicated less benefit from this sequence. Diversification and intensification significantly resulted in the higher productivity, over existing rice-wheat and rice-chickpea cropping systems in Madhya Pradesh (Table 1.4).

The production potential, nutrient uptake, water expense efficiency and economics of seven rice based cropping systems were evaluated by Mall *et al.*, 2014. Rice-potato + wheat (1:1)-green gram recorded highest rice equivalent yield (215.81 q/ha), system productivity (59.13 kg rice/ha/day), water expense efficiency (35.74 kg/ha/mm) and land use efficiency (95.89%) as compared to other cropping systems. Crop sequences with potato as *Rabi* crop resulted in significantly higher N, P and K uptake. The use of resources and their efficiencies were higher, *i.e.* economic analysis revealed that the maximum net profit (Rs. 120602), benefit: cost ratio (1.26) and monetary efficiency (₹305.41/ha/day) were recorded in rice- potato + wheat (1:1)- green gram crop sequence closely followed by rice-potato-green gram which were significantly superior over other cropping systems (Table 1.5).

Table 1.5. Indices of resource-use efficiency and economics of different rice-based cropping systems (pooled data of 3 years)

Treatments	Water expense Efficiency (kg/ha/ mm)	Land use efficiency (%)	Total variable cost (₹/ha)	Gross return (₹/ha)	Net return (₹/ha)	B:C Ratio	Monetary efficiency (₹/ha/day)
Rice - Wheat	24.14	70.78	45481	84948	43660	0.91	108.13
Rice - Mustard - Green Gram	20.27	85.84	50445	80271	34469	0.61	81.72
Rice - Rajma - Green Gram	27.36	88.58	56662	105404	54129	0.89	133.54
Rice - Potato - Green Gram	31.59	77.63	87052	188376	110110	1.22	277.60
Rice - Wheat + Mustard (5:1) - Green Gram	18.38	89.04	58904	99851	46405	0.72	112.18

Treatments	Water expense Efficiency (kg/ha/ mm)	Land use efficiency (%)	Total variable cost (₹/ha)	Gross return (₹/ha)	Net return (₹/ha)	B:C Ratio	Monetary efficiency (₹/ha/day)
Rice - Wheat + Rajma (5:1) - Green Gram	20.08	89.04	64042	120229	61770	0.91	153.94
Rice - Potato + Wheat (1:1) - Green Gram	35.74	95.89	89536	201010	120602	1.26	305.41
C.D. (P=0.05)	5.35	-	-	9209	13362	0.12	23.08

Hence, rice-potato + wheat (1:1)-green gram and rice-potato-green gram were found to be the most productive, resource- use efficient and remunerative cropping system under irrigated conditions and can be followed in place of rice-wheat systems for higher profitability (Table 1.5).

Prasad *et al.* (2013) evaluated the production potential, nutrient uptake, resource use efficiency and economics of eight rice (*Oryza sativa* L.)-based cropping systems. Eight cropping systems *viz.* rice-fallow, rice-wheat {*Triticum aestivum*, rice-mustard {*Brassica juncea*-greengram {*Vigna radiata*}, rice-rajmash (*Phaseolus vulgaris*)–greengram, rice-potato (*Solanum tuberosum*)-greengram, rice-wheat+ mustard (5:1)–greengram, rice-wheat + rajmash (5:1)-greengram, and rice-potato+ wheat (1:1)-greengram were evaluated in the study. Among the cropping systems, rice-potato + wheat (1:1)-greengram cropping system recorded highest system productivity (20.39 t REGY/ha), land use efficiency (95.89), production efficiency (55.58 kg REGY/ha/day) and employment generation (293 man days/ha) as compared to other cropping systems. Crop sequences with potato as *rabi* crop resulted in significantly higher N, P and K uptake, while S uptake was significantly higher in rice-mustard-greengram sequence (Table 1.6).

Table 1.6. System productivity and nutrient uptake of different rice-based cropping systems

Treatments	System productivity (t REGY/ ha)		Nutrient uptake (kg/ha)							
			Nitrogen		Phosphorus		Potassium		Sulphur	
	2008–09	2009–10	2008–09	2009–10	2008–09	2009–10	2008–09	2009–10	2008–09	2009–10
Rice–Fallow	3.35	3.47	69.2	70.5	13.2	13.1	60.2	57.2	9.6	10.6
Rice–Wheat	8.36	8.81	160.1	170.3	31.4	34.0	150.0	161.6	21.2	22.4
Rice–Mustard–Greengram	8.00	8.48	170.1	184.6	32.9	35.8	140.0	152.0	38.7	42.2
Rice–Rajmash–Greengram	10.58	10.84	170.2	182.1	28.3	31.9	133.4	141.3	25.4	26.9
Rice–Potato–Greengram	19.99	20.47	267.3	282.4	59.0	66.5	323.7	245.0	32.7	32.6

Treatments	System productivity (t REGY/ ha)		Nutrient uptake (kg/ha)							
		Nitrogen		Phosphorus		Potassium		Sulphur		
	2008–09	2009–10	2008–09	2009–10	2008–09	2009–10	2008–09	2009–10	2008–09	2009–10
Rice–Potato + Wheat (1:1)–Greengram	18.97	21.61	272.3	294.0	57.7	78.9	330.4	276.1	32.7	33.6
SEm±	0.49	0.33	10.4	4.5	1.8	1.0	11.6	4.9	1.1	0.7

Economic analysis revealed that the maximum net profit (125 × 103 /ha), benefit:cost ratio (1.61) and monetary efficiency (342.7 /ha/day) were recorded in rice-potato-greengram crop sequence closely followed by rice-potato + wheat (1:1)-greengram, which were significantly superior over other cropping systems. Based on study, rice-potato + wheat (1:1)-greengram and rice-potato-greengram were found to be the most productive, resource-use efficient and remunerative cropping system under irrigated conditions and can be followed in place of rice-wheat or rice-fallow systems for higher profitability (Table 1.7).

Table 1.7. Indices of resource-use efficiency and economics of different rice-based cropping systems (Pooled data of 2 years)

Treatment	Land use efficiency (%)	Production efficiency (kg REGY/ ha/ day)	Irrigation water requirement (cm)	Employment generation (days/ha)	Total Variable cost (×10³ /ha)	Net returns (× 10³ / ha)	B:C ratio	Monetary efficiency / ha/day)
Rice-Fallow	34.3	9.3	102	107	19.9	22.2	1.12	61.0
Rice-Wheat	69.9	23.5	138 (6)	212	42.9	49.4	1.15	135.3
Rice-M -GG	87.7	22.6	144 (7)	283	51.4	37.7	0.73	103.3
R-Rajmash-GG	84.9	29.4	144 (7)	239	54.3	58.1	1.07	159.1
R-Potato-GG	76.7	55.4	162 (10)	275	77.7	125.1	1.61	342.7
R-W + M (5:1)-GG	89.0	27.7	162 (10)	286	55.3	51.4	0.93	140.8
R-W + Rajmash (5:1)-GG	89.0	30.5	162 (10)	282	56.0	59.8	1.07	163.9
R-Potato + W (1:1)-GG	95.9	55.6	174 (12)	293	82.5	120.4	1.46	329.8
CD (P=0.05)	—	2.4	—	—	8.8	—	0.16	24.1

Note: R(Rice) , W(Wheat) , M(Mustard), GG(Green gram)

A field study was conducted to identify sustainable rice (*Oryza sativa*)-based cropping systems with higher productivity, economics and efficient use of water in Kymore Plateau and Satpura Hills zone of Madhya Pradesh. In these 12 cropping systems, rice was sequenced with feasible *rabi* viz., wheat, chickpea, onion, berseem

(*Trifolium alexandrinum*), potato, gobhi sarson (*Brassica napus* var. napus), vegetable pea, (*Pisum sativum*), garlic (*Allium sativum*), marigold (Tagetes erecta) and summer crops viz., maize (*Zea mays*), green gram (*Vigna radiata Wilczek*), sunflower (*Helianthus annuus*), groundnut (*Arachis hypogaea*), okra (*Abelmoschus esculentus Moench*) and cowpea (*Vigna unguiculata*). Hybrid rice 'JRH-5'-garlic-maize+cowpea cropping system recorded the highest productivity (27.35 t/ha/year) in terms of rice equivalent yields (REY) with net monetary returns of Rs. 1,75,980/ha/year, 4.47 benefit eq. cost ratio and water productivity of 137.9 kg/ha/cm. Both the existing cropping systems, viz. rice–wheat and rice–chickpea recorded the least productivity, net monetary returns, benefit cost ratio and water productivity. The highest total uptake of N, P and K nutrients was under rice–gobhi sarson–okra cropping system. After completion of two crop cycles, there was a slight improvement in organic carbon and total N content of soil due to inclusion of legumes (Soni *et al*., 2012).

Continuous adoption of rice-rice cropping system has led to deterioration of soil quality resulting in a serious threat to its sustainability in high rainfall zone of South Gujarat, India. Therefore crop diversification with wider choice in the production of a variety of crops is being promoted to restore the soil quality. A field experiment was conducted by Jha *et al.* (2011) at Navsari, India during 2003-2007 on heavy black soil to evaluate the production potential, sustainability, resource-use efficiency and economics of 10 cropping systems based on rice. System productivity of rice-fenugreek-okra was the highest (25.73 ton/ha) followed by rice-onion-cowpea (24.15 ton/ha). Lowest system productivity was registered with rice-wheat-fallow system (7.85 ton/ha). Sustainable yield index (0.94), production efficiency (102.94 kg/ha/day), and field water use efficiency (32.99 kg/ha-cm) were maximum with rice-fenugreek-okra system. Similarly, net return (Rs. 96286/ha), net return per rupee invested (Rs 2.84), and production efficiency (Rs. 385.14 /ha/day) and field water use efficiency (Rs 1234.43/ha-cm) in economic terms was maximum with rice-fenugreek-okra cropping sequence. Thus rice-fenugreek-okra was the most productive, sustainable, resource-use efficient and remunerative cropping system followed by rice-onion-cowpea system. The effect of crop diversification and intensification in rice wheat cropping system on production potential, profitability and energy use efficiency on sandy clay loam soil at Agricultural Research farm, B.H.U., Varanasi, U.P was evaluated by (Saha *et al.*, 2010). The overall productivity of the system in terms of RGEY was noticed significantly highest in rice-potato-black gram (312.73 q $ha^{-1}yr^{-1}$) over rest of the cropping sequence except rice-maize (cob) + veg. Pea (1:2)-green gram cropping sequence (301.91 q $ha^{-1}yr^{-1}$). The net energy return was maximum under rice maize (cob) + veg. Pea (1:2)-green gram cropping sequence (163.78 '000 MJ q $ha^{-1}yr^{-1}$) followed by rice-potato-black gram. Energy output: input ratio was also highest in rice-maize (cob) + veg. Pea (1:2)-green gram cropping sequence (4.01). Early pigeon pea- wheat-maize (fodder) cropping sequence registered maximum land utilization efficiency (97.81%) followed by rice-rapeseed-onion cropping sequence (90.68%) and the value was lowest in rice-wheat cropping sequence (68.49%).

Brar *et al.* (2007) revealed that rice-gobhi sarson-summer moong was identified to be the most productive and remunerative with rice equivalent yield of 135.6 q/ha and 128.8 q/ha and profitability of 103.2 and 92.4 Rs/ha/day at Ludhiana- Moga and

Faridkot- Ferozepur- Moga region, respectively. However, rice-wheat was found to be the most stable cropping system (stability index 0.92) in the Ludhiana-Moga region, while rice-gobhi sarson-summer moong (stability index 0.85) was more suitable for Faridkot-Ferozepur- Moga region from stability point of view (Table 1.8).

Table 1.8. Productivity, stability and economic viability of various crop sequences at two regions in southwestern Punjab

Sequence	Pooled yield (q/ha)			REY (q/ha)	Stability index				Net return (Rs/ha)	B:C ratio
	K*	R*	S*		K*	R*	S*	Sequence		
Ludhiana- Moga zone										
Rice-wheat	62.3	53.4	-	120.4	0.88	0.92	-	**0.92**	32804	1.89
Rice-gobhi sarson	62.3	17.9	-	108.6	0.89	0.79	-	0.87	29530	1.88
Rice-gobhi sarson- summer moongbean	62.5	**18.1**	**10.2**	**135.6**	0.88	0.80	0.85	0.90	**37667**	**1.92**
Rice-gram-summer moongbean	62.8	9.93	10.2	116.5	0.89	0.73	0.76	0.86	26729	1.65
Rice- gram	62.9	9.96	-	90.4	0.88	0.67	-	0.85	19114	1.57
CD (P=0.05)	-	-	-	11.7	-	-	-	-	-	-
Faridkot-Ferozpur- Moga zone										
Rice-wheat	61.2	53.8	-	119.7	0.79	0.73	-	0.81	32398	**1.87**
Rice-gobhi sarson	61.1	16.1	-	102.9	0.79	0.70	-	0.82	26224	1.78
Rice-gobhi sarson- summer moongbean	**61.2**	**16.3**	**9.79**	**128.8**	0.80	0.72	0.76	**0.85**	**33723**	1.82
Rice-gram-summer moongbean	61.3	10.4	9.56	114.8	0.80	0.48	0.83	0.82	25743	1.63
Rice- gram	61.1	10.2	-	89.2	0.80	0.61	-	0.81	18418	1.55
CD (P=0.05)	-	-	-	7.59	-	-	-	-	-	-

Note: K- *kharif*; R- *Rabi*; S-Summer, Price(Rs/q): Rice-580; Wheat-630; Gobhi sarson-1500; Gram-1600; Moongbean-1500

2. Wheat-based intercropping

Hua *et al* (2017) revealed that wheat-maize intercropping integrated with no-till and stubble mulching yielded 1.6 t ha^{-1} more grains and 7.1 MJ ha^{1} more energy, while emitted 1.2 t ha^{-1} less soil CO_2 than conventional wheat-maize intercropping and the system increased plant biomass carbon by 16%, and enhanced CO_2 emission efficiency per unit of grain yield by 39%, compared with the conventional system. Integrated system decreased the soil evaporation by 11%, increased energy yield per unit of water by 19%, and lowered CO_2 emission per unit of water by 9%. Wheat-maize intercropping integrated with no-till and stubble retention can be used to increase crop yield efficiently while reducing soil CO_2 emission and enhancing crop water use effectively in dry areas. Bargaz *et al.* (2017) noted that total root dry weight (RDW), length, and surface area significantly increased in P deficient intercropped wheat and soybean. Greater root allocation to deeper soil layers was evident for P-deficient intercropped wheat. Shallow roots of intercropped wheat exhibited highly stimulated acid phosphatases (APase) activity under P-deficient conditions while shallow roots of monocropped soybean exhibited higher APase activity in comparison to deeper roots, irrespective of P treatment. Root fungal diversity was significantly higher in intercropped wheat and was significantly correlated with RDW, root APase activity, shoot P, and soil available P. Root bacterial diversity was higher in both intercrops, and was significantly correlated with RDW and shoot N concentration. Narkhede *et al* (2015) revealed that turmeric (*Curcuma longa*) + castor (*Ricinus communis*) (4:1) intercropping system gave significantly highest net profit, economic efficiency of 616.9/day/ha on net monetary return (NMR) basis and soybean equivalent yield (SEY), followed by soybean-onion, and Bt. Cotton (*Gossypium hirsutum*)-wheat (*Triticum aestivum*)-coriander (*Coriandrum sativum*) (SEY 8.5 t/ha) (Table 1.9).

Table 1.9. Soybean-equivalent yield, B:C, net returns and WUE as influenced by different crop sequences (mean of 2008-10)

Crop sequences	SEY (t/ha)	Net returns ($x10^3$ Rs/ ha)	B:C ratio	WUE (kg/ha-cm)
Sorghum-wheat	4.8	65.1	2.0	51.2
Bt cotton-groundnut	6.8	101.8	2.7	53.2
Soybean-onion	8.2	125.2	2.9	79.3
Soybean-wheat-cowpea	5.8	78.8	2.0	43.4
Pigeonpea-groundnut	7.0	112.9	3.7	51.6
Maize-wheat-greengram	7.8	117.1	2.7	56.0
Bt cotton-wheat-coriander	8.5	123.4	2.4	62.1
Maize-chickpea-okra	6.1	78.4	1.7	45.7
Soybean-Chickpea-coriander	5.7	76.5	1.9	49.0
Turmeric+ castor	10.5	171.1	3.8	139.0
Bt cotton-chickpea	5.9	86.0	2.4	66.0
Soybean-safflower-maize fodder	4.9	65.0	1.8	38.3
CD (P=0.05)	0.77	10.90	-	-

Note: Prices Rs./t: sorghum 7963,sorghum fodder 1667, wheat 11533, wheat straw 500, cotton 31000, cotton stalk 1000, Groundnut 25277,groundnut haulm 3000, soybean 20517, onion 6000, greengram 36,467, pigeonpea 31003, pigeonpea stalk 1000, maize 8167 (fodder 1000), cowpea fodder 1250, okra 6333, turmeric 14000, castor 12000,safflower 20087, coriander 8000, green maize fodder 1250. The highest benefit: cost ratio was recorded in turmeric + castor (4:1 ratio) intercropping system (3.8), followed by pigeonpea-groundnut (3.7). The turmeric + castor intercropping system (4:1) recorded the lowest employment generation (136 man-days/ha/year) (Table 2.9).

Walia *et al.* (2011) concluded that in rice-wheat system, the population of grassy weeds was 74 m^{-2}, which decreased drastically in maize-wheat-moongbean, summer groundnut- *toria+ gobhi sarson* and in the other cropping systems. But surprisingly, the *Phalaris minor* infestation was reduced. Summer groundnut-toria+ gobhi sarson recorded lowest broadleaf weed population whereas in rice-wheat system it was maximum (Table 1.10).

Table 1.10. Weed Population, water productivity, Nutrient use productivity and B: C ratio of different cropping systems.

Cropping system	Weed population (no./m^2)		Water productivity (kg grain/m^3 irrigation water)	Nutrient use productivity (kg grain/kg nutrient applied)	B: C ratio
	Grass*	BLW			
Rice-Wheat	74 (36)	25	0.573	39.12	1.453
Maize-Wheat	12 (7)	17	1.770	38.13	1.544
Maize-Wheat-Mungbean	10 (5)	11	1.640	47.13	1.397
Maize-Potato-Mungbean	17 (4)	19	2.178	49.80	1.484
Maize-Potato-Onion	22 (4)	17	2.151	53.30	2.012
Cotton-Wheat	19 (12)	21	1.583	48.57	1.266
Cotton-African Sarson	12 (5)	16	1.742	63.06	1.701
Cotton-Gobhi sarson	14 (4)	13	1.743	58.67	1.512
G.nut-Toria+ *Gobhi sarson*	11 (3)	10	2.298	83.99	2.723
G.nut-Potato-Bajra	10 (4)	17	2.417	68.47	1.718
SE±	1.7 (1.2)	1.0	-	-	
CD (p=0.05)	5 (4)	3	-	-	

The results further revealed that *Phalaris minor* population is increasing in rice-wheat system whereas in all other cropping systems it has reduced. It may be attributed due to the change in the physical conditions of the soil on account of puddling. The groundnut-potato-bajra (fodder) cropping system showed highest water productivity followed by groundnut-*Gobhi sarson*+toria, maize-potato- summer moongbean and maize-potato-onion. The lowest water productivity was observed with rice-wheat cropping system. The groundnut - *Gobhi sarson* + Toria cropping system showed highest nutrient use productivity of 83.99 kg grain/kg nutrient applied followed by groundnut-potato-*bajra* (fodder) cropping system (68.47 kg grain/kg nutrient applied) which may be due to higher yield potential and price of these crops and groundnut being a leguminous crop required less nutrients (15-9-20; N-P-K; kg/ha). Consequently maize-potato-onion; summer groundnut-potato-pearlmillet (fodder) and maize-potato-summer mungbean are biologically efficient, resource conservative, and highly profitable crop sequences (Table 1.10).

3. Maize-based Intercropping

Sharma *et al.* (2017) showed that by growing one row of cowpea in between two rows of maize (90×20 cm), no effect was observed on the productivity of rainfed maize. Productivity of the succeeding wheat crop was enhanced by 13% which resulted in a higher net return (117 US$ ha^{-1}) than in a maize-wheat system. This system also reduced runoff and soil loss by 26% and 43%, respectively, compared to only a maize cropping system. Kermah *et al.* (2017) recorded that sole crops of maize and legumes produced significantly more grain yield per unit area than the respective intercrops of maize and legumes. Land equivalent ratios (LERs) of all intercrop patterns were greater than unity indicating more efficient and productive use of environmental resources by intercrops. Sole legumes intercepted more radiation than sole maize, while the interception by intercrops was in between that of sole legumes and sole maize. The intercrop however converted the intercepted radiation more efficiently into grain yield than the sole crops. Economic returns were greater for intercrops than for either sole crop. Cowpea and groundnut performed better than soybean when intercropped with maize, though the larger absolute grain yields of soybean resulted in larger net benefits. Hu *et al.* (2016) elucidated that maize/pea strip intercropping may serve as a model for effectively boosting a system's productivity. Intercropped pea was the dominant plant, as shown by the highly positive competitive ratio (averaging 1.35) and its aggressivity (averaging 0.31) values compared with intercropped maize. Ameliorating N application in the maize/pea strip intercropping intensified inter specific competition, improved the N_2 fixation of intercropped pea and increased the complementary growth of intercropped maize. Mishra (2014) conducted a field experiment in Kanpur, India and studied the effect of winter maize (*Zea mays*) based intercropping systems on maize yield, associated weeds and economics under irrigated condition. Thirteen maize-based cropping systems such as maize sole, potato (*Solanum tuberosum*) sole, mustard (*Brassica juncea*) sole, toria (*Brassica campestris* var. *toria*) sole, pea (*Pisum sativum*) sole, linseed (*Linum usitatissimum*) sole, wheat (*Triticum aestivum*) sole, maize + potato (1:1), maize + mustard (1:1), maize + toria (1:2), maize + pea (1:2), maize + linseed (1:2) and maize + toria (1:2). Maize + potato system recorded higher yield attributes and grain yield of maize followed by maize + pea than sole stand of maize. Potato was showed most compatible intercrop planted with winter maize as it gave higher maize-equivalent yield, land equivalent ratio, productivity, monetary returns and lowered weed population, weed dry-biomass and highest weed-control effciency under irrigated conditions. Pea was the next best intercrop with winter maize.

Choudhary *et al.* (2014) conducted the experiment at Banaras Hindu University, Varanasi to know the production potential and nutrient status of winter maize intercropped with various vegetable like radish (*Raphanus sativus*), spinach (*Spinacia oleracea* L.) and carrot (*Daucus carota*). Days to 50% tasseling (110.50), days to 50% silking (117.0) and barren on plants $plot^{-1}$ (1.70), maize equivalent yield (282.46 q ha^{-1}) was significantly higher in intercrops than the sole maize crop. Association of carrot with maize crop improved availability of residual nitrogen, phosphorous and potassium. Nigade *et al.* (2012) observed that performance of different pulses grown as intercrop with finger millet (*Eleusine coracana*) in different row proportions at the Zonal Agricultural Research station, Shenda Park, Kolhapur (Maharashtra). The result revealed that the blackgram or mothbean is recommended as an intercrop in 8:2 or 4:1 row proportion in finger millet under sub-montane zone situation in rainfall zone IX with shallow and lighter type of soils for obtaining maximum grain and straw yield as well as net profit. The grain equivalent yield was increased to the tune of 42 to 57 per cent over sole cropping. Shah *et al.* (2011) found that intercropping of maize + greengram recorded significantly the highest maize equivalent yield followed by maize + soybean intercropping system. Significantly higher grain and straw yields of maize as well as maize equivalent yield was recorded under the treatment of soil solarization and it was followed by pre-emergence application of pendimethalin 0.5 kg ha^{-1} + hand weeding (HW) at 45 days after sowing (DAS) and alachlor @ 1.0 kg ha^{-1} + HW (45 DAS).

Walia *et al.* (2009) conducted a long-term experiment during 2002-03 to 2005-06 at Ludhiana, to evaluate the effect often crop sequences on productivity, profitability, weed population and soil fertility. Maximum rice equivalent yield (27.8 q/ha) was recorded in maize (*Zea mays*)- potato (*Solanum tuberosum*)-onion (*Allium cepa*) cropping system followed by summer groundnut-potato-pearlmillet (fodder) and maize-potato-summer mungbean cropping systems. These cropping systems resulted in an additional rice equivalent yield of 6.3 to 8.5 q/ha and irrigation water saving of 97 to 124 cm over rice-wheat system. The highest calorific value was with maize-wheat-mungbean (40,144, 103 kcal), and production efficiency with maize-potato-onion cropping system (72.5 kg/day/ha). The highest net returns (Rs. 86,668/ha/year) was realized from maize-potato-onion cropping system. All the cropping systems recorded lower population of *Phalaris minor* than in rice-wheat system. Banik and Sharma (2009) reported that cereal-legume intercropping systems were superior to monocropping. Maize + French bean gave high maize equivalent yield over sole yield and kernel yield of maize was unaffected in maize + French bean intercropping. Bharati *et al* (2007) found that intercropping of maize reduced the maize yield but significant reduction was recorded only in french bean (*Phaseolus vulgaris*) and toria intercropping system at Pusa in Bihar. Among the intercrops, maximum water-use efficiency (WUE) was obtained with maize + potato (*Solanum tuberosum*) (561.89 kg/ha-cm) and minimum with maize + toria (256.0 kg/ha-cm). All the intercrops with maize recorded significantly higher maize-equivalent yield than sole cropping of maize. Maize + potato recorded the highest (123.48 and 140.07 q/ha) maize-equivalent yield and net return, whereas, sole cropping of maize recorded the maximum net return per rupee of investment. Among intercropping systems, maize+potato generated the highest net return (Rs. 28,781 and 35,661), followed by maize + rajmash.

4. Pulses-based intercropping

Continuous adoption of rice-wheat cropping system (RWCS) has led to depletion of inherent soil fertility resulting in a serious threat to its sustainability in the Indo-Gangetic plain region (IGPR) of India. The inclusion of legumes in RWCS assumes a great significance to restore soil fertility. But farmers in the IGPR rarely grow legumes in the system. We, therefore, carried out farmers' participatory diagnostic survey in the Upper Gangetic plain zone (UGP) to understand farmers' fertilizer management practices for wheat (*Trititicum aestivum*) following rice (*Oryza sativa*) or pigeonpea (*Cajanus cajan*). The survey indicated that most of the farmers in UGP grew pigeonpea in place of rice under RWCS as only a break crop at a 2-3 year interval. The farmers applied, on average, 11 kg N ha^{-1} and 24 kg P ha^{-1} to wheat sown after rice, and 12 kg

N ha^{-1} and 19 kg P ha^{-1} to wheat sown after pigeonpea. Wheat yields, however, were lower (3.3 t ha^{-1}) when sown after pigeonpea than after rice (3.7 t ha^{-1}). The survey was followed by a field experiment at Modipuram, Meerut, India to examine the effect of inclusion of pigeonpea in place of rice on soil fertility, N and P use efficiency and yields of wheat. In 1998-1999, wheat yields after pigeonpea were lower than after rice, but improved significantly ($p < 0.05$) by 11.4-15.1% in pigeonpea plots compared with those in rice plots during 1999-2000 and 2000-2001, respectively. The use efficiency of applied N and P fertilizers in wheat, measured as agronomic efficiency and apparent recovery, was increased with combined use of fertilizer N and P at recommended rate, and also with inclusion of pigeonpea in place of rice (Singh *et al.*, 2005).

Role of mulches in intercropping

Choudhary and Kumar (2014) found that the productivity of sequential crops was 18-35% higher with mulch application than no mulch. Land-use efficiency was found maximum with maize–Indian mustard system (71.9%) followed by maize-groundnut system.

Root length, dry weight, volume number and density were recorded higher with groundnut followed by Indian mustard and lowest with French bean. However, specific root length was highest with pea (12.0 cc/g) (Table 1.11).

Table 1.11. Root growth as influenced by mulch in maize-based cropping systems (mean of 3 years)

Treatment	Root length (cm)	Root dry weight (g/plant)	Root volume (cc)	Root numbers	Root density (mg/cc)
		Cropping system			
Maize–pea	46.2	3.8	42.7	12.7	90.1
Maize–mustard	46.0	4.9	38.0	16.5	130.5
Maize–Frenchbean	34.5	2.9	36.5	10.2	80.8
Maize–groundnut	50.2	6.9	43.2	20.8	160.5
SEm±	1.2	0.2	1.4	0.9	5.8
CD (P=0.05)	3.7	0.5	4.0	2.7	17.3
		Mulch			
Without mulch	40.3	4.1	38.6	16.8	107.7
With mulch	48.1	5.2	41.6	13.2	123.3
SEm±	1.9	0.5	1.3	1.3	10.3
CD (P=0.05)	5.7	NS	NS	NS	NS

Weed-smothering efficiency was the maximum with maize–groundnut (59.4 %), followed by maize-Indian mustard system (58.6%). Mulched plot registered the least said weed parameters with 65% weed-smothering efficiency. Soil-moisture content 30, 60 and 90 days after sowing was higher with maize-groundnut system, followed by maize–Indian mustard system, and between the mulch, no mulched plot had considerably lower soil moisture than the mulched ones (Table 1.12).

Table 1.12. Weed dynamics as influenced by mulch in maize-based cropping systems

Treatment	Weed density (No./m²)	Weed dry weight (g/m²)	Weed-Control efficiency (%)	Weed index (%)	Consumptive water (mm)	Water-use efficiency (kg/ha mm)
Cropping system						
Maize–pea	9.4 (92.3)	5.5 (31.0)	43.5	21.2	1,062.3	7.1
Maize–Indian mustard	8.7 (79.5)	4.7 (22.7)	58.6	25.2	1,074.6	6.7
Maize–Frenchbean	9.9 (102.2)	5.8 (34.7)	46.8	-	1,075.3	8.9
Maize–groundnut	8.2 (70.2)	4.7 (22.2)	59.4	16.6	1,047.9	7.5
Maize–fallow	11.4 (135.2)	6.2 (39.8)	27.4	58.8	1,100.6	3.8
SEm±	0.2	0.2	0.03	–	–	0.5
CD (P=0.05)	0.6	0.5	0.08	–	–	1.4
Mulch						
Without mulch	11.6 (134.9)	6.4 (41.0)	29.3	28.2	1,056.4	5.5
With mulch	7.5 (56.9)	4.4 (19.1)	65.0	20.6	1,070.4	4.2
SEm±	0.3	0.3	0.04	–	–	1.9
CD (P=0.05)	0.9	0.6	0.11	–	–	5.2
C × M	NS	NS	*	–	–	NS

*Figure in parentheses are original value and normal values are square root transformed value with $\sqrt{x+0.5}$

Midega *et al.* (2013) studied the effective control method for the weed that involves intercropping cereals with the fodder legume, silverleaf desmodium, *Desmodium uncinatum* through mechanisms that involve soil shading and nitrogen fixation. Because the parasitic weed *Striga hermonthica* is a major constraint to maize production in sub-Saharan Africa and causes serious food insecurity. Intercropping maize with *D. uncinatum* gave the most significant suppression of *S. hermonthica*, with addition of N only minimally enhancing its effect. Application of N both to sole maize and mulched plots reduced seasonal *S. hermonthica* infestations but the effect was largely insignificant. Efficacy of *D. uncinatum* in *S. hermonthica* suppression resulted in better growth and yields of maize. Shah *et al.* (2011) found that significantly weed smothering efficiency (WSE) (%) calculated at 20 and 45 DAS and at harvest clearly indicated that intercropping of maize with soybean having higher WSE than the maize with greengram.

Weed Control in Intercropping

Chahal *et al.* (2005) reported that dry matter production by *P. minor* recorded at the time of harvest was reduced to a level of 100% in rice-potato-wheat, rice-potato-sunflower and rice-berseem cropping systems compared to rice-wheat and rice-*gobhi sarson*. This might be due to complete elimination of *P. minor* either due to digging operation of potato or due to repeated cuttings of berseem fodder. Further, rice-*gobhi sarson* cropping system caused 99% reduction in *P. minor* dry matter accumulation compared to rice-wheat (unweeded check) cropping system and it was even 82.9% less than rice-wheat (herbicide) cropping system as the growth pattern of *gobhi sarson* provided good smother on *P. minor* and as a result growth and development of *P. minor* was controlled. In rice-wheat (unweeded check) cropping system, seeds/ panicle were significantly higher (average 319) than other treatments and thus resulted in higher seed production potential which was 87.7% higher than *rice-gobhi sarson* system during 2nd year indicating thereby the favourable effect of *Brassica napus* in checking the seed production potential of *P. minor* (Table 1.13).

Table 1.13. Effect of different cropping systems on panicles m^{-2}, dry matter accumulation (q ha^{-1}), seeds panicle^{-1} at harvest and distribution of seeds of *Phalaris minor*

Cropping system	Dry matter (q/ha)		Panicles/m^2		Seed/panicle		Before sowing seeds of *P.minor* kg^{-1} soil		After harvesting seeds of *P.minor* kg^{-1} soil	
	1999-2000	2000-01	1999-2000	2000-01	1999-2000	2000-01	1999-2000	2000-01	1999-2000	2000-01
Rice-wheat (Herbicide)	1.79 (2.25)	1.76 (2.09)	4.48 (19.0)	4.70 (22.0)	17.49 (305.0)	18.25 (332.0)	5.04 (25.0)	4.27 (17.5)	4.84 (22.5)	4.20 (17.5)
Rice-wheat (Control)	6.60 (42.6)	6.80 (45.5)	20.70 (432.0)	21.10 (447.0)	18.28 (341.0)	19.67 (386)	8.02 (65.0)	8.87 (80.0)	9.67 (92.5)	10.15 (102.5)
Rice-potato-wheat	1.00 (0.0)	1.00 (0.0)	1.00 (0.0)	1.00 (0.0)	1.00 (0.0)	1.00 (0.0)	3.05 (10.0)	2.16 (5.0)	2.16 (5.0)	1.00 (0.0)
Rice-potato-sunflower	1.00 (0.0)	1.00 (0.0)	1.00 (0.0)	1.00 (0.0)	1.00 (0.0)	1.00 (0.0)	2.74 (7.5)	1.58 (2.5)	1.58 (2.5)	1.00 (0.0)
Rice-berseem	1.00 (0.0)	1.00 (0.0)	1.00 (0.0)	1.00 (0.0)	1.00 (0.0)	1.00 (0.0)	2.16 (5.0)	1.58 (2.5)	1.58 (2.5)	1.00 (0.0)
Rice- gobhi sarson	1.27 (0.61)	1.06 (0.13)	3.00 (8.3)	2.70 (6.3)	14.57 (211.0)	14.38 (206.0)	3.95 (15.0)	2.74 (7.5)	2.47 (7.5)	2.16 (5.0)
CD (p=0.05)	0.29	0.40	1.38	1.48	0.93	0.76	1.54	1.71	1.85	1.91

Figures in parenthesis are original values

The data on soil seed bank under different cropping systems revealed that rice-wheat (unweeded check) cropping system added significantly more seeds in soil than other cropping systems at 0-7.5 cm soil depth during both the years. It was followed by rice-wheat (herbicide treated) which was at par with *rice-gobhi sarson* system. There was about 50 per cent reduction in seed bank in rice-potato-wheat, rice-potato-sunflower and rice-*gobhi sarson* during 2000-01 compared to 1999-2000. However, in case of rice-berseem system, seed bank reduction was 66.7 per cent. At the end of third year of cropping system, soil seed bank of P. *minor* was completely exhausted by rice-potato-wheat, rice-potato-sunflower and rice-berseem systems. However, in case of *rice-gobhi sarson* system, there was 33.3% reduction in soil seed bank in 2000-01 as compared to 1999-2000. There was 10.8% increase in seed bank of *P. minor* in rice-wheat (unweeded check) system during 2000- 01 over 1999-2000 because of more population pressure of *P. minor.* At farmers' fields, it was found that the soil seed bank of P. *minor* became zero in rice-toria-sunflower, rice-berseem and rice-onion-wheat cropping systems. The soil seed bank of *P. minor* was reduced to 59.8 and 66.7% in rice-potato-sunflower and rice-*gobhi sarson*, respectively in 2000-2001 as compared to 1999-2000 in Kapurthala district.

2 FARMING SYSTEM AND HISTORICAL BACKGROUND

DEFINITION

Farming system is a complex interrelated matrix of soil, plants, animals, implements, power, labour, capital and other inputs controlled in part by farm families and influenced by varying degrees of political, economic, institutional and social forces that operate at many levels. In other words, it is defined as unique and reasonably stable arrangement of farm enterprises that the household manages according to its physical, biological, economic and socio-cultural environment in accordance with the household's goals, preferences and resources. Conceptually it refers to a set of elements or components that are interrelated which interact among themselves. Selection of farm enterprises depends upon the resources available at the farm such as type of soil, availability of water, inputs, machinery, labour etc.

It is a resource management strategy to achieve economic and sustained production to meet diverse requirement of farm households while preserving resource base and maintaining a high level of environmental quality. For example, it represents integration of farm enterprises such as cropping systems, animal husbandry, fisheries, forestry, sericulture, poultry, etc. for optimal utilization of resources bringing prosperity to the farmer. The farm products other than the economic products, for which the crops are grown, can be better utilized for productive purposes in the farming systems approach.

Different scientists have defined, farming system differently. However, most definitions, convey the same meaning, that it is a strategy to achieve profitable and sustained agricultural production to meet the diversified needs of the farming community through efficient use of farm resources without degrading the natural resource base and environmental quality. Relatively recent definitions include:

According to Paul Harris, it is a system which comprises inter-related sets of enterprises with crop activity as base, and provide ways to recycle byproducts and other wastes from one component to become an input for another part of the system. It reduces cost and improves soil health and production and/or income.

- Farming system is a resource management strategy to achieve economic and sustained agricultural production to meet diverse requirements of farm livelihood while preserving resource base and maintaining a high level of environment quality (Lal & Miller 1990).

- Farming system is a set of agro-economic activities that are interrelated and interact with themselves in a particular agrarian setting. It is a mix of farm enterprises to which farm families allocate its resources in order to efficiently utilize the existing enterprises for increasing productivity and profitability of the farm. These farm enterprises are crop, livestock, aquaculture, agro-forestry and agri-horticulture (Sharma *et al.* 1991).
- Farming system is a mix of farm enterprises such as crop, livestock, aquaculture, agro forestry and fruit crops to which farm family allocates its resources in order to efficiently manage the existing environment for the attainment of the family goal (Pandey *et al.* 1992).
- Farming system represents an appropriate combination of farm enterprises (cropping systems horticulture, livestock, fishery, forestry, poultry) and the means available to the farmer to raise them for profitability. It interacts adequately with the environment without dislocating the ecological and socioeconomic balance on the one hand and attempts to meet the national goals on the other (Jayanthi *et al.* 2002).
- Farming system is a decision-making unit comprising the farm household, and cropping and livestock system that transform land, capital and labour into useful products that can be consumed or sold (Fresco & Westphal 1988).

HISTORICAL BACKGROUND

In the history of human settlements on the planet earth, agriculture has transformed in tune with the growing population and its challenging needs. The transformation has been quite remarkable since the end of World War II. Food and fibre productivity spurred up due to adoption of new technologies viz., high yielding varieties, farm mechanization, increased fertilizer and pesticide use, specialized farming practices, water resource development and improved irrigation practices and government policies that favored maximizing production. It was in the early 1960s, the Green Revolution took shape in developing countries, especially India. It led to the attainment of self- sufficiency in food grain production. This has been described by Plunkett (1993), scientific adviser to the CGIAR, as the greatest agricultural transformation in the history of humankind, and most of it has taken place during our lifetime. The change was brought about the rise of science-based agriculture which permitted higher and more stable food production, ensuring food stability and security for a constantly growing world population. A major problem was that these benefits have been poorly distributed. Many people have missed out and hunger still persists in many parts of the world. Estimates by the FAO and WHO (1992) suggest that around 1 billion people in the world have diets that are 'too poor to abstain the energy required for healthy growth of children and minimal activity of adults'. The causes are complex and it is not entirely the fault of overall availability of food. Moreover, the process of agricultural modernization has been an important contributing factor, in that the technologies have been more readily available for the betterment of farming.

Modern agriculture begins in the research station, where researchers have access to all necessary inputs i.e., fertilizers, pesticides and labour at all the appropriate times. But when the package is extended to farmers, even the best performing farms cannot match the yields which the researchers get. For high productivity per hectare, farmers, need access to the whole package of inputs i.e modern seeds, water, labour, capital or credit, fertilizers and pesticides. Many poorer farming households simply cannot adopt the whole package. If one element is missing, the seed delivery system fails or the fertilizer arrives late, or there is insufficient irrigation water, then yields may not be much better than those for traditional varieties. Even if farmers want to use external resources, very often delivery systems are unable to supply them on time. Where production has been improved through these modern technologies, there have been adverse environmental and social impacts in both the advanced and developing countries including India. These include the following:

- Overuse of natural resources, causing depletion of groundwater, and loss of forests, wild habitats etc. Sometimes more exploitation of natural resources may cause water-logging and increased salinity.
- Contamination of the atmosphere by ammonia, nitrous oxide, methane and other products of burning, play a role in ozone depletion, global warming and atmospheric pollution.
- Contamination of food and fodder by residues of pesticides, nitrates and antibiotics.
- Contamination of water by pesticides, nitrates, soil and livestock water, causing harm to wildlife, disruption of ecosystems and possible health hazards in drinking water.
- Build up of resistance to pesticides in pests and diseases including herbicide resistance in weeds.
- Damage of farm and natural resources by pesticides, causing harm to farm workers and public, disruption of ecosystems and harm to wildlife.
- Erosion of genetic diversity – the tendency in agriculture to standardize and specialize by focusing on modern varieties, causing the displacement of traditional varieties and breeds
- New health hazards for workers in the agrochemical and food processing industries

In addition to the above adverse effects, the increasing human as well as cattle population is imposing intense pressure on the available natural resources. Accordingly, a challenge has emerged that required a new vision, holistic approach for ecosystem management and renewed partnership between science and society.

In December 1983, the UN General Assembly established the World Commission on Environment and Development. On 27th of April, 1987 at the Queen Elizabeth Hall in London, the then Prime Minister of Norway, Mrs. Brundtland, who was also the Chairman of the World Commission of Environment and Development, released the publication *Our Common Future* by the World Commission on Environment and

Development (WCED) and said "Securing our common future will require new energy and openness, fresh insights, and an ability to look beyond the narrow bounds of national frontiers and separate scientific disciplines. The young are better at such vision than we, who are too often constrained by the traditions of former, more fragmented world. We must tap their energy, their openness, their ability to see the interdependence of issue." She suggests that we must adopt a new paradigm based on a completely new value system. "Our generation has too often been willing to use the resources of the future to meet our own short-term goals. It is a debt we can never repay. If we fail to change our ways, these young men and women will suffer more than we, and they and their children will be denied their fundamental right to a healthy productive, life-enhancing environment." Her speech made it clear that we are consuming resources which belong to the next generation and generation to come.

Sustainable Agriculture

We must recognize that because resources are limited, we need a sustainable way of life. Almost at the same time, the realization of the prime importance of staple food production for achieving food security for future generations has brought the concept of "Sustainable Agriculture" to the forefront and began to take shape in the following three points:

(i) The interrelatedness of all the farming systems including the farmer and the family.

(ii) The importance of ecological balance in the system.

(iii) The need to maximize desired biological relationships in the system and minimize the use of materials and practices that disrupt these relations. Sustainability of agricultural systems has become a global concern today.

OBJECTIVES

Sustainability is the objective utilization of inputs without impairing the quality of the environment with which it interacts. Therefore, it is clear that farming system is a process in which sustainability of production is the objective. The overall objective is to evolve technically feasible and economically viable farming system models by integrating cropping with allied enterprises for irrigated, rainy, hilly and coastal areas with a view to generate income and employment from the farm. The specific objectives are:

- To identify existing farming systems in specific areas and access their relative viability.
- To formulate a farming system model involving main and allied enterprises for different farming situations.
- To ensure optional utilization and conservation of available resources and effective recycling of farm residues within the system and
- To maintain sustainable production system without damaging resources/ environment.

- To raise overall profitability of the family farm by complementing main and allied enterprises with each other.

Integrated Farming System (IFS)

IFS may be defined as linking together two or more normally separate components or enterprises which then become subsystems of a whole farming system. Two main features of IFS are:

- Waste or by-products of one subsystem becomes an input of a second subsystem;
- Improved space utilization in which the two subsystems essentially occupy part or all of the space required for an individual subsystem

CHARACTERISTICS OF FARMING SYSTEM RESEARCH (FSR)

Farming system research activities must be farmer-oriented, system-oriented, have a problem-solving approach, interdisciplinary, complements mainstream disciplinary research, test the technology in on-farm trials and provide feedback from the farmers. The strategy of FSR should emphasize that the research agenda should be determine by farmers needs through an understanding of the existing farming systems rather than its perception by the researchers. The farming systems research and extension should be dealt within a holistic manner on farmers participatory mode with problem solving approach, keeping a gender activity, interdisciplinary and interactive approach. It should emphasize extensive on-farm activities and complement the experimental on-station research and acknowledges the location specificity of technical solutions and document the inter-dependencies among multiple clients. Greater importance is placed on feedback to modify the content of subsequent on-farm trials, if necessary, by changing research priorities focusing on policy shifts based upon micro level analysis. Thus farming system represents an appropriate combination of farm enterprises, viz., cropping systems, horticulture, livestock, fishery, forestry, poultry and the means available to the farmer to raise them for profitability. It interacts adequately with the environment without disclosing the ecological and socio-economic balance on one hand and attempts to meet national goal on the other. Hence there is a need to do research on farming systems for sustainable income along with the maintenance of natural resources. It is designed to understand farmers' priorities, strategies and resource allocation decisions. The main characteristics of Integrated Farming System research are:

- It is holistic or system approach
- It is farmer participatory
- It envisages location specific technology solutions
- It is problem solving
- It adopts a bottom up approach
- It is gender sensitive

- It is specific for client group- small/marginal farmer
- Its ultimate objective is sustainability
- It focuses on actual adoption
- It recognizes interdependence among multiple clients
- It encompasses extensively on farm activities and collaboration between farmers and scientists.

To **summarize, Farming** systems research has three core characteristics:

(i) It uses systems thinking. Situations deemed 'problematic' are understood as emergent phenomena of systems, which cannot be comprehensively addressed by using only a reductionist, analytical approach. It requires thinking about the interconnections between a system's elements, its dynamics, and its relation with the environment. It studies boundaries, linkages, synergies and emergent properties. The aim is to understand and take into account interdependencies and dynamics. It means keeping the 'bigger picture' in mind, even when a study focuses on a specific aspect or sub-system.

(ii) It relies on **interdisciplinarity.** Agronomic sciences (crop production, animal husbandry) are working closely with social sciences (sociology, economics, political sciences) and 'interdisciplinary' sciences (e.g. human geography, landscape planning). This interdisciplinary approach is essential to understand farming in a systematic way. Farming Systems Research is thus distinct from disciplinary research, which can provide complementary insights (e.g. informing the development of new production methods).

(iii) It builds on a **participatory approach.** Integrating societal actors in research is critical to understand 'real world' situations, to include the goals of various actors, and to appreciate their perception of constraints and opportunities. A broad range of societal actors (farmers, extension agents, civil society organizations, associations, etc.) can be involved in research, and may actively shape the research process. The participatory approach also allows integrating local and farmers' knowledge with scientific knowledge, thus fuelling reciprocal learning processes.

Comprehensively implementing all three characteristics in a single research project is a steep challenge, both conceptually and in practice. Doing so may not be feasible in many settings and often not effective. Scarce resources may make it more efficient to focus on those aspects that are most relevant to the specific situation in which the researcher is engaged. However, a farming systems researcher should be aware of the choices made in a specific situation, and is expected to make these choices in an informed and reflexive manner. In the coming sections we will discuss various aspects underlying these three core characteristics, highlighting both the theoretical discourse and the practical challenges in implementation. Although a number of issues are fairly well understood at the theoretical level, their implementation is not standardized, fuelling the diversity of farming systems research in practice.

The characteristics of farming system also depends upon the water supply system of that farming i.e., rainfed farming or irrigated farming. Also different continents have different characteristics of farming system even different countries within a continent have different farming system characteristics. For e.g., Asian farming system, European farming system, Australian farming system, etc.

1. Characteristics of Rainfed Farming

- Crops and varieties which can withstand moisture stress should be cultivated.
- *Kharif* crops are sown after monsoons.
- Not possible to adopt improved methods of cultivation. Only one or two crops are grown.
- Crops rotation is not followed.
- Soils of these areas are deficient in nutrients.
- Mixed cropping should be practiced and also deep-rooted crops.
- Short duration varieties fit well in rainfed areas.
- The crops that are tolerant to drought should be cultivated.
- Soil moisture should be preserved by mulching or by FYM application.
- Soil erosion which may be called "Creeping death" of the soil is a worldwide problem, so necessary measures should be taken to keep the soil productive.

Principles of relevant components of environmentally sustainable farming systems should include:

- Reduction of soil erosion and improving soil conservations.
- Inclusion of legumes and cover crops in crop rotations.
- Agro-forestry as an alternate land use system and
- Judicious use of organic wastes.

2. Characteristics of Irrigated Farming

Crop can be grown throughout the year; moisture is not a limited factor.

- Round the year cropping pattern becomes possible.
- Intensive cropping is possible.
- Production can be increased by proper utilization of productive resources.
- Crop rotation can be executed properly due to adequate irrigation facility.
- Manuring is safely done in irrigated soil.
- The field experiment is possible because of timely irrigation facility.

3. Characteristics of Asian Farming System

The emphasis here is therefore on Asian systems with the expectation that better understanding of crop-livestock-fish integration in Asia will point the way to similar developments in Africa. The characteristics of Asian Farming System are outlined below as an example. A typical Asian farmer lives with his wife and four children on a farm of approx. 1.5 ha and raises mainly rice. The family owns one or two draught animals and raises several chickens and ducks: land tenure and ownership/tenancy arrangements are variable. There may be several enterprises of varying importance.

- Wet-rice, almost always present on farms in tropical Asian and often the principal feature of the farm;
- Multiple cropping of other annual crops with rice in the paddy fields;
- Permanent cultivation of annual or perennial crops, including staple roots and tubers on dry land (upland);
- A mixed garden around the farmstead where fruits, vegetables and root crops are grown; and
- Livestock, cattle or buffaloes kept mainly for draught, and scavenging poultry and/ or pig.

In Asia, as in Europe there is a tendency for mixed gardening to occupy an increasing proportion of the total arable land on the farm and evolve into market gardening, the growing of vegetables for sale. Farming throughout the humid tropics is characterized by the neglect of livestock productivity. Most of the livestock in SE Asia are raised on small-scale farms rather than in commercial operations. Buffaloes and cattle are raised primarily as draught animals although they are slaughtered when their working days are over. During the rice growing season the buffaloes and cattle subsist on rough grazing of poor quality and in the off-season for rice they graze on poor quality feed in paddy fields. Cattle raising for dairy purposes is traditional in the Indian subcontinent but not in SE Asia, although there is now increasing interest in dairying among small-scale farmers in the latter region. Pisci culture contrary to popular belief, is not at all widespread in tropical Asia. Farmers have captured wild fish from rice fields since time immemorial but it has been estimated that less than 1% of the irrigated rice fields in SE Asia are used to culture fish and little has changed since this estimate. There are a few traditional aquaculture systems in tropical Asia; for example, the Indian major carp polyculture system which until recently was not an example of integrated farming because it was extensive with no fertilizer or feed inputs. Overseas Chinese were largely responsible for importing integrated fish farming technology to Malaysia, Singapore and Thailand, and perhaps Indonesia. There has been a significant increase in the development of integrated crop-livestock-fish farming systems in tropical Asia over the past two or three decades but it still probably involves < 1% of the small-scale farmers in the region. Overseas Chinese farmers in Malaysia and elsewhere are involved in more intensive cultivation of crops and livestock and sometimes fish. The livestock kept by Chinese farmers, mainly pigs, are well-integrated into the waste vegetables and rice by products and their manure is applied either to crops or to fishponds. However, most small-scale farmers in Asia may be characterized as Settled Agriculture Phase I with crops dominant and not integrated with livestock or fish.

4. Characteristic features of Integrated Farming in China

I. Integrated farming is a system oriented multi-component agriculture. It views farming in a holistic manner emphasizing the interaction between the components. Giving the multi-biological components as its central part, integrated farming is a complex biological-social-economic system. The final goal of management is to seek for the integrated effects of the whole system, not the effects of its individual compartment.

II. Integrated farming goes beyond narrow sectorial limitations. Whenever possible it tries to combine agriculture, forestry, horticulture, animal husbandry, aquaculture, as well as other biological production, into an interconnected system. Some of the village cottage industries which have an immediate link with the system are also becoming a component of the system.

III. Increase in primary and corresponding secondary productivity is a key indicator in judging the successfulness of the system. This goal is to be reached mainly by multiplying the composition and structure of the system in space and time; by raising the converting effects, by increasing the recycling of nutrients within the system and, through intensive management, using the surplus labour in the rural areas rather than relying on high input from outside the system.

IV. The system is expected to meet the economic needs of the farming community by providing multiple products. It is expected to overcome or mitigate the risks of monoculture, particularly those of irregular rainfall, market fluctuations, outbreaks, insufficient weeding and high fertilizer costs. The IFS is expected to have more flexibility in distributing workload over the course of a year, allowing farmers to earn additional income in their village's small processing industries. The IFS should combine short-term with mid-term and long-term goals, introducing better varieties and crops with higher economic value.

The common structure of Integrated Farming System is given in Fig. 2.1.

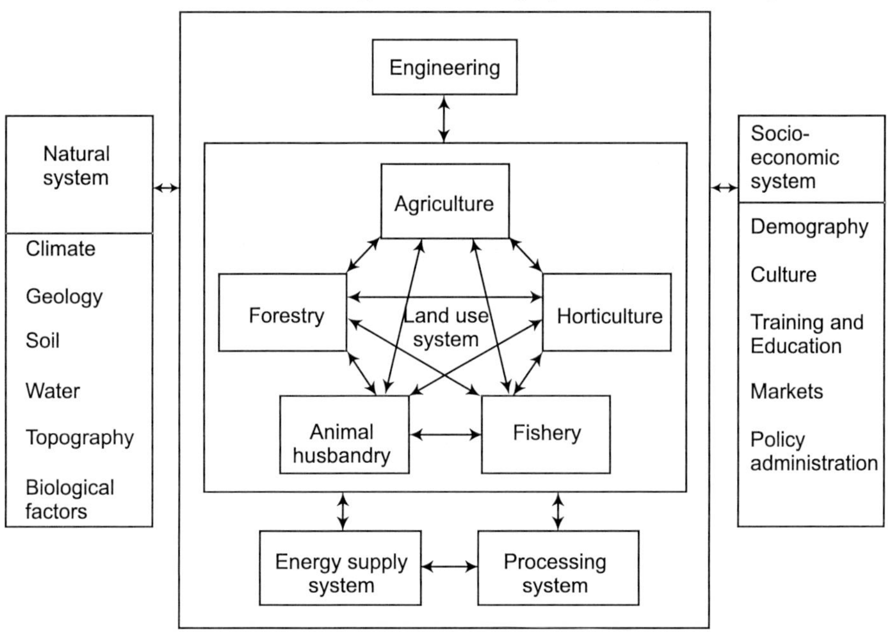

Fig. 2.1. Structure of IFS system

V. Integrated farming has high ecological value, allowing multiple use, achieved through recycling organic materials under the philosophy that by-products (waste) from the use of one resource must, whenever possible, become an input into another use of resources. Integrated farming is trying to use renewable energy and apply energy-saving techniques to complement the inadequate conventional energy. Integrated farming pays great attention to using organic fertilizers and biological pest control. Chemical input, while not excluded, is limited to a certain extent.

VI. Although incentive is necessary, integrated farming tends to develop a mechanism which can help the farmers to develop their economy on a self-reliant basis.

VII. The integrated farming concept can be used in different hierarchies, from the home garden to farmlands and watersheds of the regional and national levels if these areas are designated as integrated units for implementation.

VIII. The ultimate goal for the development of integrated farming is to achieve integrated ecological, economic and social effects, or in other words, to implement the concept of sustainable development in the rural area.

IX. Research for integrated farming should be of a cross disciplinary character with interdisciplinary experts in participation.

X. For the implementation of integrated farming, particularly in medium and macro-scale integrated farming, it is necessary to have the coopcration not only of the scientists and technicians but also of the farmers and related decision makers.

Case Study

Singh et al. (2017) studied the socio-economic characteristics and different farming systems used by 60 Punjab Agricultural University Award Farmers. The results of the study revealed that 96.67% of respondents (Table 2.1) were males and 3.33 per cent respondents were females. About half of the respondents were having the educational level up to matric, 45 per cent were graduates, over 3 per cent were post graduates and remaining were illiterate. Over 63 per cent lived in nuclear families and over 36 per cent respondents belonged to joint families with 54 per cent owned land. Majority of the members, i.e 88.33 per cent had agriculture as a main source of income.

Table 2.1: Distribution of respondents according to their socio-personal characteristics N=60

Sr. No.	Socio-personal characteristics of the respondents	Category range	Frequency	Percentage
1.	Age	30–46 46–62 Above	26 30 4	43.33 50.00 6.66
2.	Gender	Male Female	58 2	96.67 3.33

3.	Education	Illiterate Matric Graduation Post-Graduation	1 30 27 2	1.67 50.00 45.00 3.33
4.	Family type	Nuclear Joint	38 22	63.33 36.67
5.	Family size (No. of family members)	Small (upto 4) Medium (5-8) Large (>8)	9 33 18	15.00 55.00 30.00
6.	Source of Income	Agriculture Agriculture+Service Agriculture+Bussiness	60 4 0	100.00 6.67 0.00
7.	Social participation	Low (0-1) Medium (2-3) High (Above 3)	0 16 44	0.00 26.67 73.33
8.	Mass media exposure	0Low (0-1) Medium (2-3) High (Above 3)	0 7 53	0.00 11.67 88.33

All farmers adopt different integrated farming systems like crop + dairy, crop + floriculture, crop + poultry, crop + beekeeping etc. It was observed that majority of respondents i.e. 63.0% had crop + dairy farming system (Table 2.2). About 4.1 per cent had crops + poultry, 9.6% had crops + beekeeping, 2.7% had crop + forestry, 13.7% had crops + fruits, and 6.9% had crops + floriculture farming system.

Table 2.2. Distribution of the respondents according to different farming systems

Farming system	Frequency	Percentage
Crops + Dairying	46	63.0
Crops + Poultry	3	4.1
Crops + Bee-keeping	7	9.6
Keeping Crops + Forestry	2	2.7
Crops + Fruits	10	13.7
Crops + Floriculture	5	6.9
Total	73	100

ADVANTAGES OF INTEGRATED FARMING SYSTEM

The advantages of unlegrated farming system include pooling and sharing of resources/ inputs, efficient use of family labour, conservation, preservation and utilization of farm biomass including non-conventional feed and fodder resources, effective use of manure/animal waste, regulation of soil fertility and health, income and employment generation for many people and increase in economic resources. It improves space utilization and provides diversified products. The farming system is part of the strategy to ensure sustainable use of the natural resources for the benefit of present and future generations. Important advantages are as under

1. **Productivity:** Integrated farming system provides an opportunity to increase economic yield per unit area per unit time by virtue of intensification of crop and allied enterprises. Time concept by crop intensification and space concept by building up of vertical dimension through crops and allied enterprises.
2. **Profitability:** The system as a whole provides an opportunity to make use of produce/waste material of one enterprise as an input in another enterprise at low/no cost. Thus it helps to reduce the cost of production and increases the profitability and benefit cost ratio.

   Integration of fish in rice system though decrease the grain yield due to presence of fish trenches occupying 10% of the rice area, however, additional income is increased. The profit can be increased more when fish, vegetable system and livestock is included in rice - rice farming system.

   Case Study

   Various components of IFS like cropping, vegetables, fishery, poultry and goat rearing were observed to study the productivity, profitability, energy flow, employment generation and water requirement of IFS over conventional rice-rice system in Tungabhadra project area of Karnataka. Integrated farming system approach recorded 26.3 and 32.3% higher productivity and profitability, respectively over conventional rice-rice system. Among the components evaluated, the highest net returns was obtained from crops (63.8 %), followed by goat (30.9 %), fish (4.0 %) and poultry (1.3 %), respectively. Employment generation and water requirement was 275 man days/ha/year and 1247 mm, respectively under the integrated farming system. Specific energy was low in IFS (3.09 MJ/kg) (Channabasavanna *et al.*, 2009).
3. **Potentiality or Sustainability:** The use organic supplementation through the effective utilization of byproducts of the linked components of Integrated Farming System provides an opportunity to sustain the potentiality of production base for much longer periods.

   Two farming systems for ensuring sustainability are Low Input Sustainable Agriculture (LISA) and Organic farming and waste recycling. Low inputs can be made by recycling of wastages from different integrated components of IFS. For example, nutritional requirement of crops can be met by adopting the crops + dairy farming system. Here in this the waste of dairy farming such as cow dung and urine that is rich in Nitrogen which is the major nutrient required in the plant system can be added in the field. This helps to cut down the fertilizer cost of the farmers. This farm waste not only provides the nutrient but also improves the soil health status for sustainable production. Therefore, the addition of different farming systems along with the crop helps to provide low input strategy.

4. **Balanced food:**Soil health, a key factor for sustainability, is deteriorating and getting polluted due to faulty agricultural management practices viz., excessive use of inorganic fertilizers, pesticides, herbicides, high intensity of irrigation etc. In integrated farming system, organic supplementation through effective use of manures and waste recycling is done, thus providing an opportunity to sustain potentiality of production base for a much longer time. Thus the components of various nature are linked to produce different sources of nutrition.

   Integrated Farming system would not only meet the food requirements but also fulfill the need of protein, fat, vitamins and minerals required for good health. Conjunction of horticulture and agroforestry with cropping would ensure the seasonal access to fruits, fuel, fodder and fibre. Integration of allied enterprises with cropping increases the nutritive value of the products. Integration of cropping with pigeon + fish + mushroom resulted in 31-52 % higher protein yield than cropping alone.

5. **Environmental safety:** The very nature of integrated farming system is to make use or conserve the byproduct/waste product of one component as the input in another component and use of bio-control measures for pest and disease control. It is recognized that single enterprise based farming endangers ecological security. The eco-friendly practices of IFS help to bring down the application of huge quantities of fertilizers, pesticides and herbicides, which pollute the soil water and environment to an alarming level. Therefore, IFS will greatly reduce environmental pollution

6. **Recycling of resources:** Integrated farming system helps to recycle waste material (crop residues and livestock wastes) effectively. Therefore, there is less reliance on outside inputs – fertilizers, agrochemicals, feeds, energy, etc. This helps to maintain the ecological balance. For example crop by-product is utilized as fodder for animals, and animal by-product i.e. milk, and dung may be utilized for increasing income and soil fertility, respectively.

   Case Study

   Among different IFS models crop + cow + goat + poultry birds + fish (F7) emerged as the best integrated farming system in terms of resource recycling, resource management and nutrient budgeting(Table 2.3). The waste material/ by products of crops and animals were recycled and used as inputs for other components of integrated farming system. The F7 system added 19,122 and 20,623 kg/ha during 2012-13 and 2013-14, respectively of organic residues in both the years which is closely followed by crops + goat + dairy farming system (F5) system (18,368 and 19,614 kg ha^{-1} during 2012-13 and 2013-14, respectively). Quantity of N, P and K added and nutrient budgeting varied between IFS models (Vinodkumar et al., 2017).

Case Study

Table 2.3. Total organic residues/manures (kg/ha) and NPK addition (kg/ha) through various farming systems

Treat-ments		Crops	Goat	Poultry birds	Cow	Rabbit	Fish pond silt	Total residues/ manures (kg/ha)	N	P	K
									(kg/ha)		
First Year (2012-13)	F1	1375	–	–	–	–	–	1375	9.9	3.3	15.4
	F2	2805	–	–	–	–	–	2805	20.5	5.9	30.3
	F3	4180	2350	–	–	–	–	6530	74.9	40.8	67.2
	F4	4532	2360	–	–	–	–	7157	85.6	46.9	78.0
	F5	4675	2390	–	11303	–	–	18368	237.5	86.4	123.8
	F6	3053	2530	–	–	850	–	6433	94.9	58.7	71.3
	F7	3355	2550	290	12027	–	900	19122	251.5	94.3	130.8
Second Year (2013-14)	F1	2475	–	–	–	–	–	2475	17.6	5.5	26.4
	F2	3850	–	–	–	–	–	3850	28.2	8.1	41.6
	F3	5060	2470	–	–	–	–	7530	83.6	44.3	77.9
	F4	5720	2480	270	–	–	–	8470	96.7	51.0	92.1
	F5	5896	2510	–	11208	–	–	19614	247.5	90.3	137.7
	F6	3925	2660	–	–	890	–	7475	104.8	61.7	82.6
	F7	4576	2680	299	12088	–	980	20623	262.1	99.3	144.7

7. **Improve nutrient cycling:** Excreta contains several nutrients (including nitrogen, phosphorus and potassium) and organic matter, which are important for maintaining the soil structure and fertility. Animal excreta contains the major inorganic nutrient components (N, P and K). By integrating the animal production along with crops nutrients can be recycled in the soil.

Case Study

Channabasavanna *et al.* (2002) reported the changes in chemical properties of soil under rice-based farming systems (Table 2.4). The available N, P_2O_5 and K_2O contents of soil varied significantly. Maximum available N was recorded in rice (fish)-rice (fish) with poultry and minimum in rice-rice system. The maximum P_2O_5 was recorded in rice (GLM)-rice (GM) systems followed by rice (fish)-rice (fish) with poultry and rice-rice with dairy. The rice (fish) - rice (fish) with or without poultry recorded the highest available K_2O in soil.

Table 2.4: Soil nutrients as influenced by rice based farming systems

Treatment	N (kg/ha)	P_2O_5(kg/ha)	K_2O (kg/ha)
Rice-rice	256	41.2	556
Rice-sesame	336	42.9	725
Rice-chilli	396	43.1	676
Rice-wheat-GM	355	44.2	730
Rice+Fish(Ri)-Rice+Fish(dung)	410	43.9	704

Treatment	N (kg/ha)	P_2O_5(kg/ha)	K_2O (kg/ha)
Rice+Fish(Ri)-Rice+Fish(fish feed)	373	44.0	701
Rice-rice-banana on bunds	286	42.1	626
Rice-rice-drumstick and curry leaf on bunds	270	41.9	634
Rice-rice-paragrass and cowpea	385	44.7	705
Rice GLM-rice	397	45.0	695
Rice-rice-GM	385	44.2	680

Source: Channabasavanna *et al.*, 2002

Channabasavanna and Biradar (2007) reported that nutritional status of soil N show increased trend from 187 kg/ ha to 362 kg/ha (40%) in rice-fish poultry system (Table 2.5). The increase was to the tune of 11.5% over conventional systems. Similarly, P and K content showed increased trend with IFS.

Table 2.5: Soil nutrients as influenced by rice based farming system.

Treatment	*Employment generation (mandays)	*Water requirement (ha.cm)	N (kg/ha) (187**)	P (kg/ha) (187**)	K (kg/ ha) (503)
Integrated farming system					
Rice-fish(pit at one side) and poultry (shed on fish pit)	600	375	250	32.8	490
Rice-fish(pit at one side connected by trenches) and poultry (shed on fish pit)	612	375	255	33.1	485
Rice-fish(pit at the centre) and poultry (reared separately)	625	352	275	33.5	530
Rice-fish(pit at one centre connected by trenches) and poultry (reared separately)	625	347	262	33.5	525
Rice-fish(pit at four corners connected by trenches) and poultry (reared separately)	628	360	269	33.6	510
Mean	618	362	362	33.3	508
Conventional cropping system					
Control(rice-rice system)	6,667	266	235	31.8	495

8. **Income rounds the year:** Unlike conventional single enterprise crop activity where the income is expected only at the time of disposal of economic

produce after several months depending upon the duration of the crop, the IFS enables cash flow round the year by way of sale of products from different enterprises viz., eggs from poultry, milk from dairy, fish from fisheries, silkworm cocoons from sericulture, honey from apiculture etc. This not only enhances the purchasing power of the farmer but also provides an opportunity to invest in improved technologies for enhanced production. There is higher net return to land and labour resources of the farming family. Therefore the flow of regular stable income can be got through the products like egg, milk, mushroom, vegetables, honey and silkworm cocoons from the linked activities in integrated farming

9. **Adoption of new technology:** Big farmers fully utilize the modern technologies. Money flow round the year in the IFS (dairy/mushroom/ sericulture/ vegetable) gives an inducement to the small/ original farmers to go for the adoption of such technologies.
10. **Saving energy:** Availability of fossil fuel has been declining at a rapid rate leading to a situation wherein the whole world may suffer for want of fossil fuel by 2030 AD. In farming system, effective recycling of organic wastes to generate energy from biogas plants and other such techniques can mitigate this energy crisis to certain extent or postponed a later period.
11. **Meeting fodder crisis:** In IFS every inch of land area is effectively utilized. Alley cropping or growing fodder legume along the border or water courses, intensification of cropping including fodder legumes in cropping systems help to produce the required fodder and greatly relieve the problem of non-availability of fodder to livestock component of the farming system.
12. **Solving fuel and timber crisis:** The current production level of 20 million m^3 of fuel wood and 11 million m^3 of timber wood is no match for the demand estimated or 360 m^3 of fuel and 64,4 million m^3 of timber wood in 2000 AD. Hence the current production needs to be stepped up several-fold. A forestation programmes besides introduction of agro-forestry component in integrated farming system without detrimental effect on crop yield will greatly reduce deforestation, preserving our natural ecosystem.
13. **Employment generation:** Since crop-based agriculture is highly season specific and time bound, the intensity of labour requirement increases during sowing and harvesting time of crops. For the rest of the time farmers sit idle if they do not have off-farm activities. All farm resources could be utilized effectively by adoption of this farming system, which keeps the whole family busy throughout the year. Various farm enterprises viz., crop + livestock or any other allied enterprise in the farming system can be adopted which would increase labour requirement significantly and would help solve the problem of underemployment. IFS provides enough scope to employ family labour round the year. Combing crop with livestock enterprises would increase the labour requirement significantly and would help in reducing the problems

of underemployment to a great extent. IFS provide enough scope to employ family labour round the year.

14. **Scope for establishment of agro - industries:** When one of the produce from different components in IFS is increased to commercial level there is surplus for value addition leading to establishment of agro-industries.
15. **Increasing input efficiency:** IFS provides good scope for resource utilization in different components leading to greater input use efficiency and benefit-cost ratio. IFS provides good scope to use inputs in different components with greater efficiency and benefit cost ratio.
16. **Pollution abatement:** In crop-based activity, some of the organics are left as waste material which in turn pollute the environment on decomposition. Application of huge quantities of fertilizers and pesticides pollute the air, soil and water. Much of the wastes could be converted/ recycled to some other forms of economic/ ecological/ social value, under the Integrated Farming System.

    In Punjab, Haryana and Western U.P., burning of rice residue is common practice, which increased the concentration of greenhouse gases in the atmosphere, in addition to huge amount of nutrient loss. Such situation could be avoided by introduction of some more enterprises like animal husbandry on the farm. Rice straw may be used as animal feed.
17. Integrated Farming System provide opportunities as crop insurance cover as money round the year is obtained from different farm produces.
18. Climate change programme from the perspective of adaplation & mitigation
19. Avoids degradation of land resources.
20. Mitigating rural-urban exodus.

3

INTEGRATED FARMING SYSTEM AND ALLIED ENTERPRISES

INTRODUCTION

Agriculture is not only the main source of employment, income and food for over 70% of the population but also the main culture of the rural families. But the linear growth in population and unplanned colonization lead to rapid fragmentation of land holdings and shrinkage in fertile cultivated land. In India, more than 70% of the total population of the country still live in villages and mainly depend on agriculture and/or related enterprises. Marginal and small farmers constitute more than 84% of the 115 million operational holdings in India which are cultivating only 29% of the arable land (Singh *et al.*, 2011). The livelihoods of the small and marginal household families are the major concern. The small land holders are better contributors to the total production (78%) but weak in terms of generating adequate income and sustaining their own livelihood. Small holding (below 0.8 ha) does not generate enough income to keep a farm family out of poverty despite high productivity (Chand *et al.*, 2011). Growth in human and livestock populations has led to an expansion of cultivated land and shortened fallow periods. High population growth has also resulted in some area which was used for agricultural purpose being converted to residential area. This therefore has resulted in great pressure on agricultural land, leading to intensification of land use. Widespread occurrence of ill-effects of green revolution technologies in all intensively cultivated areas is threatening the sustainability of the important agricultural production systems and national food security. A mismatch between the national food grain production and requirement has already crept into the system, which is further widening.

According to the reports of Indian Council of Medical Research (ICMR), the per capita daily requirement for an adult is worked out to be 420 gm cereals, 40 gm pulses, 50 gm leafy vegetables, 60 gm other vegetables, 150 ml milk and 40 gm fat and oil to get 2738 calories of energy and 65 gm of protein to perform voluntary as well as involuntary functions of body (Ray, 2009). In the context of burgeoning human population, increase in demand for food grains and natural resources, shrinking land mass, conversion of cultivable land into non-agricultural purpose, change in ecological environment, less availability and high cost of labour etc., now the livelihood of the small and marginal farmers' is on stake. Marginal and small farmers in general are literally illiterate, financially handicapped, their holdings are small and scattered not suited for high tech agricultural machinery, work in resource poor and risk prone diverse conditions. A rural family having six members including two children requires approximately Rs 64, 000/- per annum for livelihood security. This has to be derived from the income from farming and allied enterprises (Ray, 2009). No single farm enterprise is likely to support the small and marginal farmers for generation of adequate income and gainful employment year round.

The prosperity of any country depends upon the prosperity of its people. Food, clothing, shelter, health, education, security, roads, electricity and drinking water are the today's basic needs. Farmers produce food. About 70% of Indian population is engaged in farming. Therefore, prosperity of India would depend upon the prosperity of its farmers. This in turn depends upon the adoption of improved technology and judicious allocation of resources (land, labour, capital, machinery etc). Human race depends more on farm products for their existence than anything else since food and clothing – the prime necessaries are products of farming. Even for industrial prosperity, farming forms the basic raw material.

This current farming situation in India calls for an integrated effort to address the emerging issues. The integrated farming system approach is considered to be the most powerful tool for enhancing profitability of farming systems, especially for small and marginal farmers to make them copious. In fact, our past experience has clearly evinced that the income from cropping alone is hardly sufficient to sustain the small and marginal farmer's needs. With enhanced consumerism in rural areas, farmer's requirements for cash have also increased to improve their standard of living which is especially true in case of small and marginal farmers. Therefore, farmer's income and food security would have to be increased and supplemented by adoption of efficient allied enterprises like dairy, horticulture (vegetables/ fruits/ flowers/ medicinal and aromatic plants), apiary, mushroom cultivation, fisheries, apiculture, etc. These integrated farming systems required to be planned, designed, implemented and analyzed for productivity and cost economics. These systems also need to be socially acceptable, environment friendly and economically viable. A judicious mix of agricultural enterprises like dairy, poultry, piggery, fishery, sericulture etc. suited to the given agro-climatic conditions and socio-economic status of the farmers would bring prosperity in farming. Agricultural research in farming system made by thoughtfully integrating more than one enterprise with crop leads to greater dividends than single enterprise based farming, especially for small and marginal farmers. It also leads to improvement in nutritional quality of daily diet of farmers. Crop diversification will possibly be an important mechanism for employment generation, income growth, poverty alleviation, food and nutritional security, risk aversion and sustainability of the system from judicious use of scarce natural resources (Kar *et al.*, 2003). In other words "farming system" is a resource management strategy to achieve economic and sustain agricultural production to meet diverse requirement of the farm household while preserving the resource base and maintaining high environmental quality. The farming system in its real sense will help to lift the economy of agriculture and standard of living of the farmers.

Farming system specially refers to a group/combination of enterprises in which the products and or the byproducts of one enterprise serve as the inputs for production of other enterprise. The waste of dairying like dung, urine, refuse, etc. is used for preparation of FYM, which is an input in cropping systems. The straw obtained from the crops is used as fodder for cattle are used for different field operations for growing crops. Thus, the different enterprises of farming systems are highly interrelated.

Farming system takes into account the combined needs of the family- the economic factors like relative profitability of the technically feasible enterprises, availability of farm resources, infrastructure and institutions such as irrigation, marketing facilities including storage and transportation and credit besides the agro-biological consideration-namely-interdependence, if any among various technically feasible enterprises and the performance of individual farmers.

Farming is defined as the way in which the farm resources are allocated to the needs and priorities of the farmers in his local circumstances which include

- Agro climatic condition such as the quantity, distribution and reliability of rainfall, temperature, etc.
- Soil type and topography and
- Economic and institutional circumstances like market opportunities, prices, institutional and infrastructure facilities and technology.

Integrated farming is becoming popular in India as part of sustainable agriculture system. It is being promoted and is gaining acceptance all over the world as part of the latest efforts to encourage agriculture systems that are socially, ecologically and economically sustainable. As it is a judicious mix of agricultural enterprises like dairy, poultry, piggery, fishery, sericulture etc. suited to the given agro-climatic conditions and socio-economic status of the farmers would bring prosperity in the farming. Integrated farming system (IFS) is considered as one of the best option towards intensification of small holder farm income to ensure sustainable livelihood. Integration of resources is made through a combination of land, water and animal resources of a farm through careful planning including recycling of bio-resources. Governments and development agencies have designed projects/programmes in promoting IFS through demonstration of successful models and other means. Integration of resources is a process which involves great thought in investing in one of the resource combined and gaining benefits. Integrated farming system approach not only fulfils the household needs but enrich diet of human being and animals both for nutritional security. Further, diversified nature of the model provides employment opportunity for unemployed rural youth. Economic and livelihood analysis of the system revealed that beside household food, feed, fodder and fuel security, the system generates a sizable amount of savings which will assist to meet other liabilities of the family including education, health and social obligations and overall improvement in livelihood of small farm holders. Over two decades extension agencies have been encouraging farmers to adopt ways of integrating resources for better efficiency and to reduce dependency on adopting practices with high input cost. Any planning in this regard to be ecologically sound, economically viable, adaptable, socially acceptable and humane should be based on the needs of the targeted population and take into account the "6-M Kits" which consists of Manpower, Money, Material, Market, Motivation and Management aspects with Knowledge, Information, Technology and Skill of both extension worker and beneficiaries for its' successful promotion and propagation.

IFS is a judicious mix of one or more enterprises along with cropping having complementary effect through effective recycling of wastes and crop residues and encompasses additional source of income to the farmer. It could be crop-fish integration, livestock-fish integration, crop-fish-livestock integration or combinations of crop, livestock, fish and other enterprises. IFS activity is inter-dependent, inter-related and inter-linked production system. Farming systems consist of several enterprises like cropping system, dairying, piggery, poultry, fishery, beekeeping etc. Integrated Farming System is a combination of one or more enterprises with cropping when carefully chosen planned and executed gives greater dividends than a single enterprise, especially for small and marginal farmers.

Key Principles

- **Cyclic:** The farming system is essentially cyclic (organic resources – livestock – land – crops). Therefore, management decisions related to one component may affect the others.
- **Rational:** Using crop residues more rationally is an important route out of poverty. For resource-poor farmers, the correct management of crop residues, together with an optimal allocation of scarce resources, leads to sustainable production.
- **Ecologically sustainable:** Combining ecological sustainability and economic viability, the integrated livestock-farming system maintains and improves agricultural productivity while also reducing negative environmental impacts.

Components of Integrated Farming Systems

(i) Crops, livestock, birds and trees are the major components of any Integrated Farming System.

(ii) Crops may have subsystem like monocrop, mixed/intercrop, multi-tier crops of cereals, pulses, oilseeds, fodder, vegetables, etc.

(iii) Livestock component may be milch cows, buffaloes, goats and sheep.

(iv) Tree component may include timber, fuel, fodder and fruit trees.

(v) Other enterprises like apiculture, sericulture, poultry, etc.

IMPORTANCE OF INTEGRATED FARMING SYSTEMS

1. **Deteriorating resource base:** During post-green revolution period, our attempt to solve food problem and attain self-sufficiency in food production through excess use of agrochemicals, inevitable dependence on irrigation and high cropping intensity has led to contamination of food with harmful chemicals, pollution of ground water, degradation of soil quality and damage to agriculturally beneficial microorganisms. In many regions both surface and ground water are already becoming unfit for human and animal consumption due to high concentration of pesticide residues. Intensified agriculture, coupled with indiscriminate use of irrigation water and non-judicious

fertilizer application, especially in irrigated areas of the country has led to various kinds of physical and chemical degradation of the soil.

2. **Climate change:** The increasing green-house gases has resulted in global warming. The Intergovernmental Panel for Climate Change (IPCC) projections on temperature predicts an increase of 1.8 to 4.0 °C by the end of this century. Temperature and sea level changes will affect agriculture through their direct and indirect effects on crops, soils, livestock, fisheries and pests. The brunt of environmental changes is expected to be very high in India due to greater dependence on agriculture, limited natural resources, alarming increase in human and livestock population, changing pattern in land use and socioeconomic factors that pose a great threat in meeting the food, fiber, fuel and fodder requirement. Recent studies done at the Indian Agricultural Research Institute indicated the possibility of loss of 4-5 million tonnes in wheat production in future with every rise of 1°C temperature throughout the growing period.
3. **Narrowed biodiversity:** The narrowing of genetic biodiversity occurs as traditional crop varieties and local animal breeds are being replaced by modern ones. These new varieties/breeds are certainly better matched to modern intensive agriculture, but rarely any consideration is given to preserving the bio-diversity of an agricultural ecosystem. In addition, the increased farming density tends to erode the biodiversity of flora and fauna in the agricultural ecosystems. For example, extensive adoption of rice-wheat monoculture in the Indo-Gangetic plains has replaced the other traditional crops. Soil micro-flora is also adversely influenced on account of large-scale use of agro-chemicals and lack of recycling of crop residues in the region.
4. **Multiplicity of Integrated Farming Systems:** Very often, almost all Indian farmers, in pursuit of supplementing their needs of food, fodder, fuel, fiber and finance resort to adopt integrated farming systems, majority of them revolving around the crops + livestock components. Livelihood of small and marginal farmers, comprising more than 80.00% of total farmers, depends mainly on crops and livestock, which is often affected by weather aberrations. Under present scenario, in the absence of scientifically designed, economically profitable and socially acceptable appropriate integrated farming systems models, they are unable to harness the benefits of integration. An important consequence of this has been that their farming activities remain, by and large, subsistent in nature rather than commercial and many a times uneconomical.
5. **Low rate of farm resource recycling:** In the absence of adequate knowledge among farmers about techniques and benefits of recycling of farm, industrial and municipal organic wastes in agriculture, these remain unutilized. A vast untapped potential exists to recycle these solid and liquid organic wastes of farm origin. Recycling of crop residues may be a potential organic source to sustain the soil health. Incorporation of crop residues of either rice or wheat increases the yield of rice and nutrient uptake and also improves the physico-

chemical properties of the soil, ensuring better soil environment for crop growth.

6. **Technology adoption gaps:** In our effort to develop and improve upon existing technologies, involvement of people in conceptualization and extension of technologies would appear very important. The farm family had never been the focal point of our investigations. This top down approach had given a poor perception of the problems that they tried to solve. Due to poor extension mechanisms at national as well as state levels, many farmers, especially those at lower strata of social structure, remain uninformed about many of the development schemes and the desired impact of such schemes is not obtained. One of the reasons for poor rate of transfer of agricultural technologies is poor linkages between the different clientele groups of agriculture. Practically linkages among farmers, service providers, technological and financial institutions are either weak or nonexistent (NAAS, 2009).

The basic aim of IFS is to derive a set of resource development and utilization practices, which lead to substantial and sustained increase in agricultural production (Kumar and Jain, 2005). Hence, integrated farming systems are often viewed as a sustainable alternative to commercial farming systems particularly on marginal lands with the objective of reversing resource degradation and stabilizing farm incomes. Survey on farming systems in the country as a whole also revealed that milch animals (cows and buffaloes), irrespective of breed and productivity, is the first choice of the farmers as an integral part of their farming system. However, from economic point of view, vegetables and fruits (mango and banana in many parts of the country) followed by fish cultivation were the most enterprising components of any of the farming systems prevailed in the country. A number of success stories on integrated farming system models in different parts of the country suggested that farmers' income can be increased manifold by way of integration of enterprises in a farming system mode.

BENEFITS OF INTEGRATED FARMING SYSTEMS

1. **Productivity:** Integrated farming system provides an opportunity to increase economic yield per unit area per unit time by virtue of intensification of crop and allied enterprises.
2. **Profitability:** Use waste material of one enterprise at the least cost as input for other enterprise. Thus, reduction in cost of production, form the utilization linkage of waste material and elimination of middleman interference in most input used.
3. **Sustainability:** Organic supplementation through effective utilization of byproducts of linked component is done thus providing an opportunity to sustain the potentiality of production base for much longer period.
4. **Balanced food:** Integrated Farming System links components of varied nature enabling to produce different sources of nutrition for farm families.
5. **Environmental safety:** In Integrated Farming System, waste materials are effectively recycled by linking appropriate components, thus minimize environment pollution.

6. **Income round the year:** Interaction of enterprises with crops, eggs, milk, mushroom, honey, fish, cocoons, etc. provides flow of money to the farmers round the year.
7. **Adoption of new technologies:** Money flow round the year due to Integrated Farming System gives an inducement to the small and marginal farmers to go for the adoption of new technologies.
8. **Saving energy:** Effective recycling technique the organic wastes available in the system can be utilized to generate biogas. Energy crisis can be postponed to the later period.
9. **Meeting fodder crisis:** Every piece of land is effectively utilized. Plantation of perennial fodder trees on field borders. These practices will greatly relieve the problem of non – availability of quality fodder to the animal component linked.
10. **Solving fuel and timber crisis:** Linking agro- forestry appropriately the production level of fuel and industrial wood can be enhanced without determining effect on crop. This will also greatly reduce deforestation, preserving our natural ecosystem.
11. **Employment generation:** Combing crop with livestock enterprises would increase the labour requirement significantly and would help in reducing the problems of under employment to a great extent. Integrated Farming System provide enough scope to employ family members round the year.
12. **Agro-industries:** When one of produce linked in Integrated Farming System are increased to commercial level there is surplus value addition leading to development of allied agro–industries.
13. **Increasing input efficiency:** Integrated farming system provide good scope to use inputs in different components efficiently and effectively.

FARMING SYSTEM CONCEPT

Farming system is a complex inter related matrix of soil, plants, animals, implements, power, labour, capital and other inputs controlled in parts by farming families and influenced to varying degree by political, economic, institutional and socials forces that operate at many levels. Thus farming system is the result of a complex interaction among a number of interdependent components. Farm activities interact with market forces (socio-economic) and ecosystem (biophysical) for purchasing inputs and disposing outputs by utilizing and degrading natural resources (land, water, air, sunshine etc.).

A farm is a system which has INPUTS, PROCESSES and OUTPUTS

INPUTS - these are things that go into the farm and may be split into physical inputs (e.g. amount of rain, soil) and human inputs (e.g. labour, money etc.)

PROCESSES -these are things which take place on the farm in order to convert the inputs to outputs (e.g. sowing, weeding, harvesting etc.)

OUTPUTS - these are the products from the farm (i.e. wheat, barley, cattle)

Depending on the type of farming e.g. arable/pastoral, commercial/subsistence, the type and amount of inputs, processes and outputs will vary. You need to make sure you are able to define and give examples of inputs, processes and outputs in farming systems.

Income through arable farming alone is insufficient for bulk of the marginal farmers. The other activities such as dairying, poultry, sericulture, apiculture, fisheries etc. assume critical importance in supplementing their farm income.

The modern agriculture emphasize too more dimensions viz., time and space concept. Time concept relates to increasing crop intensification in situation where there is no constraint for inputs. In rainfed areas where there is no possibility of increasing the intensity of cropping, the other modern concept (space concept) can be applied. In space concept, crops are arranged in tier system combining two or more crops with varying field duration as intercrops by suitably modifying the planting method. Income through arable cropping alone is insufficient for bulk of the marginal farmers. Activities such as dairy, poultry, fish culture, sericulture, bio-gas production, edible mushroom cultivation, agro-forestry and agri-horticulture, etc., assumes critical importance in supplementing farm income. It should fit well with farm level infrastructure and ensures full utilization of by-products. Integrated Farming System is only the answer to the problem of increasing food production for increasing income and for improving the nutrition of small scale farmers with limited resources.

Livestock raising along with crop production is the traditional mixture of activities of the farmer all over the country, only the nature and extent varies from region to region. It fits well with farm level infrastructure, small land base and abundant labour of small man and ensures full utilization of by products.

Sustainability is the objective utilization of inputs without impairing the quality of environment with which it interacts. Therefore, it is clear that farming system is a process in which sustainability of production is the objective.

The overall objective is to evolve technically feasible and economically viable farming system models by integrating cropping with allied enterprises for irrigated, rained, hilly and coastal areas with a view to generate income and employment from the farm.

The Goals of Integrated Farming Systems (IFS) are to:

- **Provide a steady and stable income** through rejuvenation/amelioration of the system's productivity and
- **Achieve agro-ecological equilibrium** through the reduction in the build-up of pests and diseases, through natural cropping system management and the reduction in the use of chemicals (inorganic fertilizers and pesticides).

SYSTEMS APPROACH

System could be defined as an organized unitary whole composed of two or more inter dependent and interacting parts, components or subsystems delineated by identifiable boundary or its environmental super system (Singh, 2001). A system consists of several components or subsystems which depend on each other. A system is defined as a unified whole or set of elements/components that are interrelated and interacting among themselves. A system processes input into outputs. Therefore, each system consists of boundaries, components, interactions between components, inputs and outputs.

The term "systems" is derived from the Greek word "synistanai," which means "to bring together or combine." The term has been used for centuries. Components of the organizational concepts referred to as the "systems approach" have been used to manage armies and governments for millennia. The systems approach emerged as scientists and philosophers identified common themes in the approach to manage and organize complex systems. Four major concepts underlie the systems approach:

- **Specialization:** A system is divided into smaller components allowing more specialized concentration on each component.
- **Grouping:** To avoid generating greater complexity with increasing specialization, it becomes necessary to group related disciplines or sub-disciplines.
- **Coordination:** As the components and subcomponents of a system are grouped, it is necessary to coordinate the interactions among groups.
- **Emergent properties:** Dividing a system into subsystems (groups of component parts within the system), requires recognizing and understanding the "emergent properties" of a system; that is, recognizing why the system as a whole is greater than the sum of its parts. For example, two forest stands may contain the same tree species, but the spatial arrangement and size structure of the individual trees will create different habitats for wildlife species. In this case, an emergent property of each stand is the wildlife habitat.The systems approach considers two basic components: elements and processes. Elements are measurable things that can be linked together. They are also called objects, events, patterns, or structures. Processes change elements from one form to another. They may also be called activities, relations, or functions. In a system the elements or processes are grouped in order to reduce the complexity of the system for conceptual or applied purposes. Depending on the system's design, groups and the interfaces between groups can be either elements or processes. Because elements or processes are grouped, there is variation within each group. Understanding the nature of this variation is central to the application of systems theory to problem-solving.

Farming system approach addresses itself to each of the farmer enterprises, inter relationship among enterprises and between the farm and environment. Thus farming system research has the objective of increasing productivity of various enterprises in the farm. Farming system approach introduces a change in farming technique for high production from a farm as a whole with the integration of all the enterprises. The farm produce other than the economic products for which the crop is grown can be better utilized for productive purposes in the farming system approach. A judicious mix of cropping system with associated enterprises like dairy, poultry, piggery, fishery, sericulture etc. suited to the given agro-climatic conditions and socio economic status of farmers would bring prosperity to the farmer.

In crop production, management practices are developed for individual crops and recommendation are made for individual crops. The residual effect of individual crops is not considered in crop based recommendation. In this resources are not utilized efficiently. To a farmer instead of crop, land is a unit and management practices should be for all crops that are to be grown one after the other on a piece of land. Therefore system approach is applied to agriculture for efficient utilization of all resources,maintains sustainability in production and obtaining higher net returns.

Farming is a dynamic biological and open system with human or social involvement. Being primarily biological with a high degree of dependence on weather variables and changing socio-political environments. Farming is inherently more risky than any other system. Farming system can also mean different things to different people. To avoid ambiguity and confusion both terms farming and system should be clearly understood. Farming is the process of harnessing solar energy in the form of economic plant and animal products and system implies a set of inter related practices/processes organized into a functional entity.

Farming systems research (FSR) originates from recognizing the interdependence and interrelationships of natural environment within the farming system. In FSR, the farmers by participating in the research process help in the identification of the research problems as well as take part in testing the possible solution.

Principles involved

The FSR approach includes:

- Viewing the farm as a whole
- Identifying the farming system, the interacting component and delineating boundaries
- Systematic investigation of the nature and extent of interdependence among the enterprises and identifying constraint
- Applying modern technical know-how to the system so as to make it yield optimum results
- Studying the equity gender income, employment and resources use efficiency, and
- Dealing with the issue at integration level through analysis and solution of problems towards sustainable farming system development.

Farming system research is a 'research method' designated to understand farmer's priorities, strategies and resources allocation decisions. It is most often used in conjunction with on-farm highly location specific research with multi and interdisciplinary in nature and uses a whole farm approach for improved technologies to enhance and stabilize agriculture production.

Characteristics of Farming System Research

- It is holistic or system-oriented,
- It is problems solving: involvement of farmers in problem identification and solving process,
- It is farmer participatory,
- It envisages location specific technology solutions,
- It is for specific client group–small/ marginal farmer,
- It adopts bottom up approach,
- It compasses extensive on farm activities, collaboration between farmer and scientist,
- It is gender sensitive,
- It ultimate objective is sustainability,
- It focuses on actual adoption,
- It recognizes interdependence among multiple clients.

Components of IFS

- Agriculture
- Fish farming
- Horticulture
- Duck rearing
- Forestry
- Pigeon rearing
- Apiary
- Mushroom cultivation
- Sericulture
- Azolla farming
- Dairy
- Kitchen gardening
- Poultry
- Fodder production
- Goat rearing
- Nursery
- Sheep rearing
- Seed production
- Piggery

- Vermiculture
- Rabbitory
- Value addition

Elements of Integrated Farming System

Following elements may be included in Integrated Farming System demonstrations depending upon the individual farmers' resources, interest and opportunities.

- Watershed
- Farm ponds
- Bio-pesticides
- Bio-fertilizers
- Plant products as pesticides
- Bio-gas
- Solar energy
- Compost making (Vermi, Japanese, Improved etc.)
- Green manuring
- Rain water harvesting.

Cropping systems based on climate, soil and water availability have to be evolved for realizing the potential production levels through efficient use of available resources. The cropping system should provide enough food for the family, fodder to the cattle and generate sufficient cash income for domestic and cultivation expenses.

Plant interactions:

- **Interactions between component crops:** In intensive cropping, crops are grown in association (intercropping) or in sequence (sequential cropping). In such situations there is possibility of interaction between the component crops. The interaction is mainly due to response of one species to the environment as modified by the presence of other species. Interaction may be competitive or non-competitive or complementary.
- **Interactions in intercropping:** Factors such as light, water, nutrients, oxygen and CO_2 are required for plant growth. In mixed or intercropping situations, the component species compete for the growth factors. The close proximity of the species causes sub-optimal utilization of the growth factors and hence there is inequitable distribution of resources among the plants. Generally competition will develop between two components or within the components.
- **Light:** Intercropping can increase light inception by as much as 30-40%. When one component is taller than the other in an intercropping system, the taller component intercepts most of the solar radiation. In intercropping situation where the component crops have different growth durations, the peak demand for light would occur at different times. In such combinations, competition

for light is less among the component crops and there is greater light use in intercropping than in pure stands. In general the component crops under intercropping situations are grown in such a way that competition for light is minimized. Proper choice of crops and varieties, adjustment of planting density and pattern are the techniques to reduce competition and increase the light use efficiency.

- **Moisture and nutrients:** Competition for water and nutrients results in two main types of effects on the less successful or suppressed component. First, the roots of dominated crop may grow less on the sides of aggressive component. The suppressed components adapt to such conditions by increased capacity for uptake. Also, if one part of the root system is on the depleted side, the remaining part shows compensatory activity and vigour. Secondly, plants affected by competition for soil factors are likely to have increased root/shoot ratio.
- **Allelopathy:** Allelopathy is any direct or indirect harmful effect that one plant has on another through the release of chemical substances or toxins into the root environment. Some crops may be unsuitable to be grown as intercrops because they may produce and excrete toxins into the soil which are harmful to the other components.
- **Annidation:** Annidation refers to complementary interaction which occurs both in space and time.
- **Annidation in space:** The canopies of component crops may occupy different vertical layers with taller component tolerant to strong light and high evaporative demand and shorter component favouring shade and high relative humidity. Thus, one component crop helps the other. Multi-storeyed cropping in coconut gardens and planting of shade trees in coffee, tea and cocoa plantations use this principle. Similarly, root systems of component crops exploit nutrients from different layers thus utilizing the resources efficiently. Generally, one component with shallow root system and another with deep root system are selected for intercropping as in Setaria (shallow) + red gram (deep) intercropping system.
- **Annidation in time:** When two crops of widely varying duration are planted, their peak demands for light and nutrients are likely to occur at different periods, thus reducing competition. When the early maturing crop is harvested, conditions become favourable for the late maturing crop. This has been observed to occur in sorghum + red gram, groundnut + red gram and maize + green gram intercropping system.
- **Other complementary effects:** In an intercropping system, involving a legume and a non-legume, part of the nitrogen fixed in the root nodule of the legume may become available to the non-legume component. The presence of rhizosphere microflora and mycorrhiza on one species may lead to mobilization and greater availability of nutrients not only to the species

concerned, but also to the associated species. Another example is the provision of physical support by one species to the other in intercropping system. Erect crop plants may improve the yield of a climber as in-the case of coconut + pepper, maize + beans. The taller component acts as a wind barrier protecting the short crop as in maize + groundnut, onion + castor and turmeric + castor.

ALLIED ENTERPRISES

Cattle rearing in India is carried out under a variety of climatic and environmental conditions. Draught cattle broadly classified into following groups.

- **Draught breeds:** The bullocks of these breeds are good draught animals, but the cows are poor milkers. e.g. Nagore, Hallikar, Kangeyan, Mali.
- **Dairy breeds:** The cows are high milk yielders but the bullocks are of poor draught quality e.g., Sahiwal, Sindhi, Gir.
- **Dual purpose:** The cows are fairly good milkers and the bullocks are with good draught work capacity e.g. Hariana, Ongole and Kankrej.
- **Exotic breeds:** The exotic breeds are high milk yielders. e.g. Jersey, Holstein Friesian, Aryshire, Brown Swiss and Guernsey.
- **Buffaloes:** Important dairy breeds of buffalo are Murrah, Nili Ravi (which has its home tract in Pakistan), Mehsana, Suti, Zafrabadi, Godavari and Bhadawari. Of these Godavari has been evolved through crossing local buffaloes in coastal regions of AP with Murrah.
- **Housing:** Each cow requires 12-18 sq m space and buffaloes need 12-15 sq m. it is important to provide good ventilation and an open shed of housing is always preferable. Dairy building should be located at an elevated place to facilitate easy drainage. The floor should be rough and have gradient of 2.5 cm for every 25 cm length.

Breeding and maintenance: The cows remain in milk for 9-10 months, the average calving interval being 16-18 months. A cow does not require more than 6-8 weeks of dry period. From the economic point of view, cows should ordinarily be bred during the second and third months after calving. In weak animals and heavy milkers, breeding may be delayed by 1 or 2 months. Cattle come to heat in more or less regular cycles of about 21 days which lasts for about a day. The best time to serve a cow is during the last stage of heat. If artificially inseminated, a second insemination may be given within 6-8 hours after the first insemination. The gestation period varies with individual cows and normally it is about 280 days. In the case of buffaloes, the lactation period last for 7-9 months. She buffaloes come to heat every 21 to 23 days. The gestation period is 310 days. Calf rearing is very important in the case of buffalo maintenance. Since they require abundant water, wallowing is required. Regular deworming is needed for buffalo maintenance. Under Indian conditions, cattle commonly mature at the age of about three years. This period can, however, be reduced by six months under well managed herd.

Feeding: Cattle feed generally contains fibrous, coarse, low nutrient straw material called roughage and concentrates.

Roughages: Dairy cattle are efficient use of roughages and convert large quantities of relatively inexpensive roughage into milk. Roughages are basic for cattle ration and include legumes, non legume hays, straw and silage of legume and grasses.

Concentrates: grains and byproducts of grains and oilseeds constitute the concentrates. They are extensively used in dairy cattle ration. These include cereals (maize, sorghum, oats, barley), cotton seeds, industrial wastes (bran of wheat, rice, gram husk) and cakes of oilseeds (groundnut, sesame, rapeseed, soybean, linseed)

- **Vitamins and mineral mixtures:** It is advisable to feed a supplement containing vitamins A and D. Mineral mixtures containing salt, Ca and P should also be provided in the ration. The ration per animal per day normally includes concentrates at 1 kg for 2 litres of milk yield, green fodder at 20-30 kg, straw 5-7 kg and water 32 litres.

GOAT REARING

In India, the activity of goat rearing is sustained in different kinds of environments, including dry, hot, wet and cold, high mountains or low lying plains. The activity is also associated with different systems such as crop or animal based, pastoral or sedentary, single animal or mixed herd, small or large scale. Goats are mainly reared for meat, milk and hide. Goat meat is the most preferred meat in India. A goat on hooves fetches a better price than a sheep on hooves.

Breeds: There are nearly 102 breeds of goats in the world, of which 20 breeds are alone found in India.

Following are the important breeds in India.

Milk and meat (dual purpose) –Osmanabadi, Barbari, Jamnapari, Sangamneri, Mehasa and Zalwadi, Beetal, Ajmeri and Kachhi.

Meat–Assam, Kali Bengali, Brown Bengali and Marwari.

Wool–Angora, Gaddhi and Pashmina.

Housing: Goats can be maintained under stall fed conditions. Successful goat rearing depends on the selection of site. Goats do not thrive on marshy or swampy ground. Goats are to be provided with a dry, comfortable, safe and secure place, free from worms and affording protection from excessive heat and inclement weather. The kids are kept under large inverted baskets until they are old enough to run along with their mothers. Males and females are generally kept together. The space requirement for a goat is 4.5 to 5.4 sq m.

Breeding and maintenance: Goats mature in about 6-7 months. Breeding is allowed for buck at one year and doe after 10 months of age. Gestation period is 145-155 days. It gives birth to 1-3 kids at a time. Number of evings is three per 2 year. The kitts can be weaned after 30-45 days. Mother can be allowed for mating 45-60 days after eving. Once in five years, the buck can be changed to avoid deterioration due to inbreeding. When the young ones attain a body weight of about 25-30 kg in about nine months, they can be sold.

Feeding: The per head requirement of nutrients in respect of goats is relatively low. Hence, they are suitable for resource poor farmers with marginal grazing lands. They are essentially browsers and eat which any other animals won't touch. Goats eat 4-5 times that of their body weight. Since the profit depends on weight addition, adequate protein and calorie should be given to goats. They eat more of tree fodder and the rest with other grass species. Goats should be fed with concentrates of maize, wheat, horse gram, groundnut cake, fish meal and wheat bran. Common salt and vitamin mixtures should also be added. Abundant clean fresh water should be made available to the goats. Water should be changed every morning and evening. Fresh water is required for digestion, blood circulation and removal of waste from the body. Water is also required for regulation of the body temperature.

SHEEP REARING

Sheep are well adapted to many areas. They are excellent gleaners and make use of much of the waste feed. They consume large quantities of roughage, converting a relatively cheap food into a good cash product. Housing need not be elaborate or expensive. However, to protect the flock from predatory animals, the height of the fencing should be raised to 2 m.

Breeds of Indian sheep

There are three types of sheep in India based on the geographical division of the country.

(i) **The dry temperate Himalayan region:** Gurez, Karanah, Bhakarwal, Gaddi, Rampur-Bushiar.

(ii) **Dry western region:** Lohi, Bikaneri, Marwahi, Kutchi, Kathiawari.

(iii) **Southern region:** Deccani, Nellore, Bellary, Mandya, Bandur.

- **Breeding and maintenance:** One ram can be maintained for 40-50 ewes. Rams are liable to fight when two or more of them are put in a pen. Unlike other farm animals, ewes in general do not come in heat at regular intervals throughout the year but are seasonal in this respect. The duration of heat period will range from 1-3 days and 75% of ewes remain in heat for 21-39 hours. The optimum time of service is towards the end of heat period. Average heat interval is 18 days during the breeding season. The gestation period will vary from 142-152 days with an average of 147 days. A normal ram is in full vigour for breeding from the age of 2.5-5 years. Sheep grow fully at two years of age when the ewe is ready for breeding, under average range condition; ewes may be expected to produce about five crops of lambs.
- **Feeding:** A sheep requires about 1-2 kg of leguminous hay per day depending on the age of the sheep and its body weight. Proteins may be supplied through concentrates such as groundnut cake, sesame cake or safflower cake when the pastures are poor in legumes or when scarcity conditions prevail. Normally 110-225 g of cake is sufficient to maintain an average sheep in good condition. Feeding a mixture of common salt, ground limestone and sterilized bone meal in equal parts is required to alleviate deficiency of minerals in the feed.

POULTRY

Poultry is one of the fastest growing food industries in the world. Poultry meat accounts for about 27% of the total meat consumed worldwide and its consumption is growing at an average of 5% annually. Poultry industry in India is relatively a new agricultural industry. Till 1950, it was considered a backyard profession in India.

Breeds: Specific poultry stocks for egg and broilers production are available.

A majority of the stocks used for egg production are crosses involving the strains or inbred lines of white Leghorn. To a limited extent, other breeds like Rhode Island Red, California Grey and Australop are used.

Heavy breeds such as white Plymoth Rock, White Cornish and New Hampshire are used for cross-bred broiler chickens. Hence, it is essential to consider the strain within the breed at the time of purchase. Several commercial poultry breeders are selling day old chicks in India. It is best to start with the day old chicks.

- **Housing:** Adequate space should be provided for the birds. Floor area of about 0.2 m^2 per adult bird is adequate for light breeds such as white Leghorn. About 0.3-0.4 m^2 per bird is required for heavy breeds. The house should have good ventilation and be reasonably cool in summer and warm during winter, it should be located on well-drained ground from flood waters.
- **Feed:** The feed conversion efficiency of the bird is far superior to other animals. About 60-70% of the total expenditure on poultry farming is spent on the poultry feed. Hence, use of cheap and efficient ration will give maximum profit. Ration should be balanced containing carbohydrates, fats, minerals and vitamins. Some of the common feed stuffs used for making poultry ration in India are:
- **Cereals:** Maize, barley, oats, wheat, pearl millet, sorghum, rice-broken.
- **Cakes/meal:** oil cakes, maize-gluten-meal, fish meal, meat meal, blood meal.
- **Minerals/salt:** Limestone, oyster shell, salt, manganese. From the day old to 4 weeks of age, birds are fed on starter ration and thereafter finisher ration, which contain more energy and 18-20% protein. Feed may be given 2-3 times. In addition antibiotics and drugs may also be added to the poultry ration. Laying hens are provided with oyster shell or ground limestone. Riboflavin is particularly needed.
- **Maintenance:** The chicks must be vaccinated against Ranikhet diseases with F1 Strain vaccine within the first 6-7 days of age. One drop of vaccine may be administered in the eye and nostril. When chicks get the optimum body weight of 1.0-1.5 kg around six weeks, they can be marketed for broiler. Hens may be retained for one year for production i.e. upto the age of 1.5 years. After that they are disposed off for table purpose. It may not be economical to keep the hens beyond 1.5 years since egg production would get reduced. One hen is capable of laying 180-230 eggs in a year starting from the sixth month. In addition, a laying hen produces about 230 g of fresh droppings (75% moisture) daily.

PIGGERY

Pigs are maintained for production of pork. They are fed with inedible feeds, forages, certain grain by-products obtained from mills, meat byproducts, damage feed and garbage. Most of these feeds are either not edible or not very palatable to human beings. Pigs grows fast and is a prolific breeder, farrowing 10 to 12 pigs at a time. It is capable of producing two litters per year under good management conditions. The carcass return is high at 65-70% of the live weight.

- **Breeds:** Imported breeds of large white Yorkshire and Landrace are being used widely. Yorkshire is the most extensively used exotic bred in India. It is a prolific breed having carcass quality, growth rate and feed conversion ability. For a small breeding farm of a unit, the selection of the herd boar is extremely important. A good boar weighs 90 kg in about 5-6 months and is strong on feet and legs. The mother of pig is to be selected should have large litter of eight piglets or more.
- **Housing:** Housing should provide maximum comfort to pigs so that their growth is optimum. There should not be dampness, draft and overheating. Locally available materials can be used for housing. One pig requires 2.7 sq. m with a wall of 1.2 m height. Eight boars can be kept in 2.7-4.5 sq. m area with 2.4-6.0 sq. m open space.
- **Feeding:** Feed plays an important role in successful pig production. Pigs are the moist rapidly growing livestock and suffer more than nutritional deficiencies than the ruminants. Proteins, fats, carbohydrates, minerals, vitamins and ample good water form a complete diet for pig. Pigs have a simple stomach; therefore they must be fed with maximum of concentrates and minimum of roughage. The main ingredients of swine ration are cereals and millets and their byproducts. The fiber contents in swine ration should be very low (around 5-6%) for better feed utilization efficiency. Mixed ration should be also containing 0.5% of added salt. Swine requires comparatively higher %age of Ca and P than do cattle or sheep. When pigs are maintained with agricultural/kitchen wastes or fish and slaughter house waste, the cost of production remains low. On an average, the consumption of feed is 3.5% of total weight. Feed allowance is calculated as 2.5-3.0 kg/100 kg body weight +@ 0.25 kg per piglet with the lactating mothers.
- **Management:** As a general rule, well developed gilts weighing about 100 kg, when 12-14 months old, may be used for breeding. The body weight is more important than age at breeding. Sow with low body weight show higher rate of fetal and pre-weaning mortality and have been proved with mothers with poor nursing ability. The gestation period is on an average 114 days. Litter size at birth may be 1-16 numbers with the body weight of 1-25 kg. Normal period between birth of piglets is 10-20 minutes. Time taken for the whole process of farrowing ranges from 1.5 to 4.0 hours. Sows are weaned after 40 days. It is advantageous to cull the sow after the fifth or sixth litter in a commercial

herd. Weaned sows come into heat in 3-10 days after the weaning and may be allowed to breed. The boar-sow ratio should be 1:15. It is profitable to raise two litters from each sow each year.

Mortality in piglets is an important cause of heavy economic loss leading to failure of pig industry. In general, one fourth of piglets farrowed die before they are weaned. Another one-tenth are also categorized into stunt or unprofitable group due to disease or parasite infection. Thus about 60-65% of the piglets farrowed perform as healthy piglets at slaughter age. Death rate is high during farrowing and the first week after farrowing. The farrowing season also determines mortality rate. Mortality in newborn piglets is maximum when farrowing takes place in acute cold or hot climate. Therefore, mating should be planned in such a way that farrowing could be avoided in such seasons.

RABBIT REARING

Rabbit is reared all over the world in various climatic conditions. Rabbit rearing in India is of recent origin, though hunting of wild rabbits for meat is not uncommon. Rabbit can easily be reared with relatively less concentrate feed and with very high reproduction rates. Rabbit meat is pure white and is low in fat and cholesterol content. Further rabbit meat does not have any religious taboo, unlike pork and meat from cow.

- **Breeds:** Rabbit can be classified into two categories as meat and wool producing ones. Important meat breeds are New Zealand White, Soviet Chinchilla and Kandinrex. Breeds for wool are Angora, Polish Bolamina, Dan and Flirorida.
- **Housing:** Rabbits can be kept in hutches outside or in cages inside. Cages should have doors (30 × 30 cm) with welded wire mesh at floor and on sides. The area required by a rabbit depends on the body weight. For one kg weight, 1.8 sq m area is needed. Angora rabbits cannot be kept at high temperature and should be restricted to hills. Breeding bucks and adult does are housed single. Castrated males can be kept in groups. Young rabbits also live in groups till the age of 4 months and thereafter they are separated. Rabbit keeping require calm and quiet areas. Sudden noise can cause death of rabbits. Disturbance by dogs, cats and humans should be kept to the minimum.
- **Feeding:** Rabbit is a non-ruminant herbivorous animal having a simple stomach. Like any other animal, rabbit requires protein, carbohydrates, fats, minerals, and vitamins in its feed. Rabbits digest the fibre fraction very poorly and diet having 5-7% crude fibre is adequate. Alfalfa and other forages are well utilized by rabbits. The feed ingredients may constitute maize (20%), alfalfa (25%), wheat bran (40%), groundnut cake (12%), mineral mixture (2%) and salt (1%). The feed mixture should be given twice a day, in the morning and evening. Other grasses and vegetables can be given intermittently in small quantities. Water should always be provided.
- **Maintenance:** Rabbits of 5-6 months age are ready for breeding and the body weight should not be less than 2.0 kg. The female rabbit is taken to the cage

of male for mating. At the time of mating, the male produces a particular noise. After that the female rabbit is taken out of the cage of the male. Early morning is the best time for mating. Rabbits mostly breed from March to July. Gestation period is 28-32 days. The eyes of the young ones (pups) open around two weeks and at 3 weeks, they feed themselves. During this period, the pups should not be often handled. Around four weeks, the young ones are separated from the mother. After two weeks of weaning, the female can be allowed for mating. In a year, it gives birth five times, each time producing about six numbers. Mortality in the kindlings is the most important component that determines the economics of rabbit rearing. In general, mortality in adult rabbits is very low ranging from 3-5%. However, it ranges from 10-15% in kindlings up to the age of 30-42 days. Mating during extreme hot and cold season causes high mortality of kindlings.

BEEKEEPING

Beekeeping is one of the most important agro-based industries which does not require raw material from the artisan like other industries. Nectar and pollen from flowers are the raw materials which are available in plenty in nature. Beekeeping can even be started with a single colony.

- **Species:** There are two species, *A. piscerana* and *A. mellifera* are complementary to each other but have different adaptations. *A. cerana* is better acclimatized to higher altitudes of the Himalayan region. *A. mellifera* is more profitable in the plains.
- **Management:** The beekeeper should be familiar with the source of nectar and pollen within his locality. The most important source of nectar and pollen are maize, great millet, bulrush, sunflower and palm. The beginner should start with 2 and not more than 5 colonies. A minimum of two colonies is recommended because in the event of some mishap, such as the loss of the queen occurring in one, advantage may be taken with the other. The hive consists of a floor board, brood chamber, top cover, frames and entrance rod. These parts can easily be separated. The hive may be double walled or single walled. The single walled hive is light and cheap. The most suitable time for commencing beekeeping in a locality is the arrival of the swarming season. Swarming is a natural tendency of bees to divide their colonies under conditions that are generally favourable for the survival of both parent colony and the swarm. This occurs during the late spring or early summer.
- **Honey collection:** Honey should be of good quality to meet the national and international standards. Qualities such as aroma, colour, consistency and floral sources are important. Proper honey straining and processing are needed to improve the quality of the produce. Honey varies in the proportion of its constituents owing to the differences in the nectar produced by different plants. The nectar collected by bees is processed and placed in comb cells for

ripening. During the ripening, sucrose is converted into glucose and fructose by an enzyme called invertase which is added to it by the bees. Honey is an excellent energy food with an average of about 3500 calories per kg. It is directly absorbed into the human blood stream, requiring no digestion.

AZOLLA CULTIVATION

Presently, due to low rainfall and decline in cultivable land if alternate fodder was used as feed it can reduce fodder scarcity and cost. Azolla is a chemical fertilizer-free alternate fodder, which is a balanced feed for all varieties of cattles. Ingredients added in livestock feed grain forages such as maize, cumbu, sorghum and ingredients for the preparation of concentrates like oilseeds price hike caused increasing livestock and poultry feed cost increasing day by day. The protein rich Azolla can be used as regular alternate feed for livestock and poultry we can significantly reduce the production. It has been used as best natural fertilizer for rice.

Nutrients in Azolla: Azolla contains 25-30% protein, 14-15% fiber, 3-4% fat, 45-50% carbohydrates, minerals and many micronutrients.

Production Methods

- Shallow standing water ponds, paddy / rice nursery
- Cement tanks.
- Silpanulinsheet spreaded cavity method

In paddy field, if Azolla added @ one kg per cent and stagnate the water to the level of 5 cm, within two weeks it grows in the entire place.

Cement tank method: 10 feet long, 2 feet wide and 1 feet deep concrete tank spread with 25-30 kg of soil with this mix 5 kg of decomposed dung, with this soils of gravel quarry or well soil @ 100g could be added. Keep the water level up to 5 cm. in this add 5 kg of Azolla within two weeks we can collect 35-40 kg Azolla. Once in ten days pour the cow dung slurry.

Azollaas bio-fertilizer: It has ability to fix atmospheric nitrogen. It contains 4.5% nitrogen. For paddy it can be added @ 200 kg per acre. It gives nitrogen source to the crop and 15-20% yield also increased.

Feeding cattle: Production of 1 kg Azolla costs about 0.75 paise only. When Azolla is used as feed it could be mixed 1:1 proportion with concentrates and fed to the cattles. After familiar with this, mulching cows will started to consume Azolla alone. One kg of Azolla equal to 1 kg of oil cake. For mulching cows 1-1.5 kg, white pigs 1-1.5 kg, Goat 300-500 grams, Rabbits 100 grams and 25-30 grams of Azolla for chickens might be given to these animals.

Approximately 15-20% increase in milk production in dairy cows, in addition to this increase in cholesterol and also cholesterol free solids enhanced the milk quality. Azolla feeding chickens showed increase in body weight and size of the egg. Further, egg yolk seems to be dark in colour. So, production of Azolla under Integrated Farming System not only gives income and also protects the environment.

AQUACULTURE

Ponds serve various useful purposes, viz. domestic requirement of water, supplementary irrigation source to adjoining crop fields and pisciculture. With the traditional management, farmers obtain hardly 300-400 kg of wild and culture fish per ha annually. However, poly-fish culture with the stocking density of 7500 fingerlings and supplementary feeding will boost the total biomass production.

Pond: the depth of pond should be 1.5 to 2.0 m. this depth will help for effective photosynthesis and temperature maintenance for the growth of zoo and phytoplanktons. Clay soils have high water retention capacity and hence are best suited for fish rearing. Pond water should have appropriate proportion of nutrients, phosphate (0.2 -0.4 ppm), nitrate (0.06-0.1 ppm) and dissolved oxygen (5.0 - 7.0). Water should be slightly alkaline (pH 7.5-8.5). if the pH is less than 6.5, it can be adjusted with the addition of lime at an interval of 2-3 days. Higher pH (>8.5) can be reduced with addition of gypsum. Application of fresh dung may also reduce high pH in the water. Soil of the pond should be tested for N and P content. If the nutrient content is less, nitrogenous fertilizers like ammonium sulphate and urea and phosphate fertilizers like super phosphate can be added. Organic manures such as FYM and poultry droppings may also be applied to promote the growth of phyto and zooplanktons.

Species of fish:

- Among the Indian major carps, Catla (*Catlacatla*) is a fast growing fish. It consumes a lot of vegetation and decomposing higher plants. It is mainly a surface and column feeder.
- Rohu (*Labeorohita*) is a column feeder and feeds on growing plants, decomposing vegetation, large colonial algae, zooplanktons and detritus to a small extent.
- Calbasu (*Labiocalbasu*) is a common feeder on detritus.
- Mrigal (*Cirrhimusmrigale*) is a bottom feeder, taking detritus to a large content, diatoms, filamentous and other algae and higher plants.
- Common carp (*Cyprinuscario*) is a bottom feeder and omnivorous.
- Silver carp (*Hypophthalmichlthysmolitrix*) is mainly a surface and phytoplankton-feeder and also feeds on microplants.
- Grass carp (*Cyernuscarpia*) is a specialized feeder on aquatic plants, cut grass and vegetable matter. It is also a fast growing exotic fish.

Poly fish culture: The phytophagous fish (Catla, rohu and mrigal) can be combined with omnivorous (Common Carp), plankton- feeder (Silver Carp) and mud eaters (Mrigal and Calbasu) in a composite fish culture system.

Management: The fish are to be nourished with supplementary feeding with rice brands and oilseed cakes. This will enable faster growth and better yield. Each variety of carps could be stocked to 500 fingerlings with the total of 5000-8000 per ha. This stocking density will enable to get a maximum yield of 2000-5000 kg/ha of fish annually.

SERICULTURE

Sericulture is defined as the practice of combining mulberry cultivation, silkworm rearing and silk reeling. Sericulture is a recognized practice in India. It plays an important role in socio-economic development of rural poor in some areas. The climatic conditions in India are favourable for luxuriant growth of mulberry and rearing of silkworms throughout the year. The temperature in Karnataka state, major silk producing site in India, ranges from 21.2 to 30°C. climatic conditions in Kashmir are favourable to rear silkworm from May to October.

Moriculture: Cultivation of mulberry plants is called moriculture. There are about 20 species of mulberry, of which four are commonly cultivated. They are *Morusalba, M indica, M. serrata* and *M. latifolia*. The crop can yield well for 12 years, after which they are pulled out and fresh planting is done. Yield of mulberry leaves is 30-40 t/ha/ year.

Silkworm rearing: There are four types of silkworms:

- Mulberry silkworm (*Bombaxmori*)
- Eri silkworm (*Philosamiaricini*)
- Tussar silkworm (*Antheraeamylitta*)
- Muga silkworm (*Antheraeaassami*)

Rearing and maintenance: The fertilized moth is covered with an inverted funnel or cellule and eggs are allowed to be laid over a cardboard. Parasites may be removed by brushing the egg masses with a fine brush. This will also enable to obtain an uniform hatch. In a bamboo tray rice husk is spread. Tender chopped mulberry leaves are added to the tray. The hatched out larvae are transferred to the leaves. It is important to change the leaves every 2-3 hours during the first 2-3 days. The cocoon is constructed with a single reelable thread of silk. If the moths are allowed to emerge from the cocoons, the silk thread is cut into pieces. Hence, the pupa are killed 2-3 days before the emergence of the moth and processed. The cocoons required for further rearing are kept separately and moths are allowed to emerge from them. A family of three can manage 0.75 acre without engaging extra labour. Under irrigated conditions one hectare of mulberry garden gives employment to 13 people throughout the year. It requires 607 man days and 827 woman days.

MUSHROOM CULTIVATION

Mushroom is an edible fungi with great diversity in shape, size and color. Essentially mushroom is a vegetable that is cultivated in protected farms in a high sanitized atmosphere. Just like other vegetables, mushroom contains 90% moisture with high quality protein. Mushrooms are fairly good sources of vitamin C and B complex. The protein have 60-70% digestibility and contain all essential amino acids. It is also rich source of minerals like Ca, P, K, Cu. They contain less of fat (0.3%) and CHO and are considered good for diabetic and blood pressure patients. Fibre content in mushrooms is very high which helps in excretion of waste and prevention of constipation.

Types of Mushroom

- Oyster mushroom-*Pleurotus spp.*
- Milky mushroom-*Calsybeindica*
- Button mushroom-*Agricusgisporus*

Oyster mushroom basically occurs in dead woods in nature and resembles the shape of oyster shell. It can be grown on a wide range of agricultural wastes. It can be grown on a wide range on a wide range of temperature and with low investment. The production of this mushroom is being done on a small scale in different parts of the country and is most popular one. Milky mushroom is new of its kind and was introduced for commercial cultivation in the year 1998 from TNAU. The mushrooms are umbrella shaped, robust, attractive and milky white in nature. This variety is tropical in nature and can be grown when button and oyster mushroom cannot be grown due to high temperature. Button mushroom is particularly suitable for winter season, in the northern states. Cultivation of this mushroom requires more investment. This mushroom cannot be cultivated in paddy or wheat straw.

Method of production

Oyster mushroom: Take fresh paddy straw and cut into small pieces of 3-5 cm length. Soak them in water for 4-6 hours and then boil for half an hour. Drain the water and dry the straw in shade till it is neither too dry nor wet. Take polythene bags of 60 x 30 cm of size and make two holes of one cm diameter in center of the bag such that they face opposite sides. Tie the bottom of the bag with a thread to make a flat bottom. Fill the bag with paddy straw to 10 cm height. Then inoculate with the spawn. Likewise prepare 4-5 layers of straw and spawn alternately. The last layer ends up in the straw of 10 cm height. Keep this in a spawn running room maintained a temperature of about 20-25 °C. and with RH 85 - 90%. After 15-20 days when the spawn running is completed, cut open the polythene bag and take it to the cropping room and allow it to grow for 7 days and harvest the mushroom. Yield: 0.5–1.0kg/bed.

DUCK REARING

Ducks account for about 7% of the poultry population in India. They are popular in coastal and water logged states like West Bengal, Orissa, Andhra Pradesh, Tamil Nadu, Kerala, Tripura and Jammu & Kashmir. Ducks are predominantly of indigenous type and reared for egg production on natural foraging. They have a production potential of about 130-140 eggs/bird/year. Ducks are quite hardy, more easily brooded and resistant to common avian diseases. In places like marshy river side, wetland and barren moors where chicken or any other type of stock do not flourish, duck farming can be a better alternative.

- **Breeds:** The important Indian breeds are Sylhet Mete and Nageswari which are mostly found in the eastern region of the country. Their annual production 150 eggs/bird/year. Improved breeds for eggs and meat production are available. Khaki Campbell and Indian Runner are the most popular breeds for egg laying. Khaki Campbell has a production potential of 300 eggs/bird/

year. Indian Runner is the second best producer. White Pekin, Muscovy and Aylesbury are known for meat production. White Pekin is the most popular duck in the world. It is fast growing and has low feed consumption with fine quality of meat. It attains about 3 kg of body weight in 40 days. Indigenous types, however, still continue to dominate in duck farming. Desi ducks are robust, well adapted to local conditions and free of diseases.

- **Housing:** Ducks prefer to stay outside day and night even during winter or rains. In mild climate, it is possible to raise ducks without artificial shelter. A free fence of atleast 1.2 m high enclosing the yard is enough to stop any predator. One nest of size 0.3 x 0.3 x 0.45 m to every 3 ducks is sufficient. In case of laying birds, a mating ratio of 1 drake: 6-7 ducks and in meat type 1: 4-5 is allowed. The duck house should be well ventilated, dry, leak and rat proof. The roof may be thatched or asbestos sheeted. A water channel of 0.5 m wide and 0.2 m deep is constructed at the far end of both sides parallel to the night shelter in the rearing or layer house.
- **Feeding:** Ducks normally require lesser attention. They supplement their feed by foraging, eating fallen grains in harvested paddy fields, small fishes and other aquatic materials in lakes and ponds. However, for intensive rearing, pellet feeding may be given. Ducks prefer wet mash due to difficulties in swallowing the dry mash. Hence, ducks should never have access to feed without water. During the first 8 weeks, birds should always have an access to feed. Later on they may be fed twice a day in the morning and late afternoon.
- **Maintenance:**The general management of ducks is similar to that of chickens. The incubation period is 28 days. A broody duck or hen may be used for small-scale hatching and incubator for large-scale hatching. During the early part of the life, newly hatched ducklings require warm temperature under the natural or village conditions. A duck or broody hen can take care of 10-15 ducklings. Artificial brooding may be resorted for large number of ducklings. High egg-laying strains of duck come into production at 16-18 weeks of age. Ducks are resistant to common avian diseases. Some of the common diseases in ducks are duck plague, ducks virus hepatitis, duck cholera and aflatoxicosis.

TURKEY REARING

Turkey is a robust bird and can be reared in humid tropics. It actively feeds on a variety of crop residues and insects in the farm and hence can be raised with much ease.

- **Physical features:**There are a few peculiar features in the physical appearance of a turkey. Male has black beard attached to the upper breast regions. This beard is small in females of white turkeys and absent in those of coloured varieties. A fleshy protuberance called snood is present near the base of the beak in both males and females. The head bears caruncles and a large flap of skin below the chin is known as dewlap. Broad Breasted Bronze breed are ordinarily seen in India. Norfolks and the Cambridge are the other important breeds.

- **Housing:**Housing for turkeys may be of range system, confinement or a combination of both. There is a growing trend towards confinement housing. Breeds require a floor space of 0.4 m^2/bird in confinement and 32.5 m^2/bird in the range rearing.
- **Feeding:** Turkeys have much higher growth rate compared to chickens. Hence they have relatively higher feed and nutrient requirement. Turkey can be fed with all types of dry feed comprising pellets, wetmesh, grain and green feed in well balanced combination. They also need plenty of clean water.
- **Maintenance:** Mating ratio in natural flock is 14 hens: 1 tom for small and 12:1 for medium sized turkeys. The birds should be carefully segregated when they reach the mating stage. Mating takes place during summer. Particular attention should be bestowed in regulating mating. The males should be kept away from their mates in the morning and selectively mated with other birds by turn. Mating in the evening increases the number of fertile eggs. Generally turkeys lay eggs on alternate days from six onwards. In a year, they lay about 150 eggs. The bird attains a weight of 5-8 kg in a year. The birds should not be allowed to fatten too much as it would make them vulnerable for diseases. Changing the breeding stock periodically would be gainful for the farmer.

PIGEONS FOR MEAT

Domestic pigeons (*Columba livia*) belonging to the sub-family Columbia can be reared for its meat. Females are smaller than the males and finer in the head. They hold their tails higher than males and their pelvic pones are spaced wide apart. The males are large, aggressive and make a loud cooing noise. They strut around the females during the breeding season. There is difference in the growth rate among the males and females. The former grows faster than the latter. Thus, even at an early stage, the gender difference can be easily made based on their body weight.

Feeding: Adults feed the young ones (squabs) with pigeon milk, a fluid that is secreted in their crop. Pigeon milk looks like thick crusted and has high protein content. Pigeons require water for drinking and for bathing to remove external parasites. A pair of breeds drinks about 500 ml of water per day. Pigeons are not fussy eaters, but do best on diets having whole grains. Nutrient requirement of pigeons is crude protein (13.5%), carbohydrates (65%), crude fibre (3.5%) and fat (3.0%). The daily feed intake of an adult pair is about 120 g.

Maintenance: Each pair normally produces 14 squabs in a year. The incubation period of the eggs is 18 days. Both partners build their nest and take turns to sit on the eggs. The female starts laying eggs again when the squabs are two weeks old. The young ones become independent in 35 days. Squabs of 30-35 days are best suited for meat purpose since recovery is high with the body weight of 350 g. In addition to the meat, pigeon droppings (25 g per bird day) can be used as fish feed or enriched manure.

JAPANESE QUAIL

Quail belongs to class Aves, family Phasianidae and genus *Coturnir*. Quails, introduced initially as experimental birds, have become popular for their eggs and meat. Efforts are being made by Central Avian Research Institute, Izatnagar, to develop the guinea fowl for meat production. Males are lighter than females and are identified by the cinnamon coloured feathers on the upper throat and lower breast region. Females are similar to males in colour except that the feathers on the throat and upper breast are pointed and much lighter cinnamon in colour.

Maintenance: Quails need small space and are highly suitable for cage rearing. Therefore, it is possible to rear them even in urban areas where space is a major constraint. They are socially and commercially more adaptable due to their small body size, shorter generation interval, highly prolific nature and simplicity in rearing and are more economical. Females lay eggs as early as at the age of 5 weeks. The average age of the first egg production is 40 days. They are in full production by 50 days. They have a capacity to lay 300 or more eggs each per annum. The shell of a typical quail's egg is a mixture of greenish black and brownish speckles or patches of varying sizes. An egg weighs about 10-12 g on an average and is comparatively thin and delicate in nature. Albumen content is 56 to 59%, yolk 28-32% and shell 12-14%. The yolk content in relation to albumen is more in quail's egg than in chicken's egg. Special care should be taken for collecting and handling the egg. The rapid growth rate of the Japanese quail renders it an ideal broiler. At six weeks, quails attain a body weight of 125-150 g with a feed efficiency of 3.0. Japanese quails are fairly resistant to most of the common diseases affecting chicken. However, they are sensitive to abrupt environmental changes during first and second week of their life.

VERMICOMPOST

Vermicompost is a one of the important natural fertilizers. In the past, agriculture was done using only natural fertilizers. Because of this increased production and also soil fertility was preserved. But the current situation due to growing population, because of Green Revolution, by using chemical fertilizers high yielding varieties is created. Excessive use of chemical fertilizers and pesticides caused reduction in soil fertility and also reduced the number of beneficial microorganisms. Therefore, conversion of this degraded land into fertile lands, organic manures play a major role for the transformation. Naturally available organic residues are ingested by worms. It was digested by the intestine microorganisms and enzymes and then tiny pellets form forced out through anal of the earthworm is called vermi composting. Earthwarm is called as "farmer's friend". According to living place, earthworm is categorized as top, intermediate and grassroots worms. Top and intermediate type of earthworm's are best suited to produce compost. Earthworms takes food as 2 to 5 times organic matter of its body weight, out of this 5 -10% of them spend as food and the rest of the organic matter excreted as fertilizer.

Selection of earthworm: Despite, nearly 3,000 varieties of earthworms, available worms such as Eudrilus and Eisenia of African species are suitable for commercial production of manure, as they are the top feeders in the soil. The species should have higher reproduction rate, adjusting to climatic conditions and economic importance has to be selected. And also reproduce quickly, suitable for temperatures and with the economic importance of the species should be selected. *Perionyxexcavatus, Eudriluseugeniae*, and *Eiseniafoetida* are such kind of earthworm, much of it has been used to produce vermicompost.

Collecting earthworms: Using local earthworm helps to improve the fertility of the soil and environment. Choose a shady place in the garden. The place of tree shade where chemical fertilizers has not been applied or select a place where organic waste from the house pass out would be right choice. Choose a small place and spread cow dung and place on straw or sprinkle dust and cover with old gunny bags and frequently sprinkle water based on need. After two weeks we can see top and middle layer earthworms. While collecting earthworms soil also to be taken together.Culturing of earthworms and manufacturing fertilizers, worms are fed continuously. Nitrogenrich cowdung, goat dung and pig dung may be used as feed.

Earthworm production systems: Vermicompost production methods may change according to prevalent situation in the place of worm culturing area. Some of the methods which include are cavity, basin system and bed methods etc. Cavity method is a very simple method for preparation of the earthworm manure. In this method large-scale vermicompost production is possible on commercial scale. In this system, long beds are prepared and grow worms and produced fertilizer.

In this method, dig a pit in the dimension of 10" × 3" × 2" on a shaded and elevated surface and even extend the length up to 20 feet. Release 2000 – 2500 earthworms per square metre.

Thus, inner side of the excavated pit is lined with bricks. Bottom of the pit has to be filled with coir waste or agriculture waste. Finally add cowdung or decomposed agriculture waste or leaves has to be spread over there. Finally cover the pit with biogas slurry or cowdung mixture and release the earthworms. After this pits are to be covered with straw or stakes to preserve the moisture.

Separation method of earthworm excreta

Once decided to collect the excreta, stop sprinkling water two days in advance. Earthworms move to moisturized portion in the bottom. This is directly used for cultivated land or longterm usage, bagged it without loss in moisture and it was used later. If maintained the compost without loss in moisture, it can be used for six to eight months.

Uses:

- Increasing the microorganism populations in the soil.
- Increase the soil aeration.
- Play a role in increasing the growth of roots.
- Environmental protection.

BIOGAS PLANT

Biogas is a clean, unpolluted and cheap source of energy, which can be obtained by a simple mechanism and little investment. The gas is generated from cowdung during anaerobic decomposition. Biogas generation is a complex biochemical process. The celluloitic materials are broken down to methane and carbon dioxide by different groups of microorganisms. It can be used for cooking purpose, burning lamps, running pumps, etc.

Types of biogas plant

(i) **Float dome type:** Different models are available in this category, e.g. KVIC vertical and horizontal, Pragati model and Ganesh model.

(ii) **Fixed dome type:** The gas plant is a dome shaped underground construction. The masonry gas holder is an integral part of the digester called dome. The gas produced in a digester is collected in the dome at vertical pressure by displacement of slurry in inlet or outlet. The entire construction is made of bricks and cement. The models available in this category are Janata and Deen-bandhu.

The selection of particular type depends on technical, climatologically, geographical and economic factors.

Size: The size of a biogas plant is decided by the number of family members and the availability of cattle. One cubic meter capacity plant will need 2-3 animals and 25 kg of dung. The gas produced will meet the requirement of the family of 4-6 members. It would suffice to have a 2 cubic meter plant to cater to the needs of a family of 6-10 members.

Site selection and management: The site should be close to the kitchen or the place of the use. It will reduce the cost of gas distribution system. It should also be nearer to the cattle shed to reduce the cost of transportation of cattle dung. Land should be levelled and slightly above the ground level to avoid inflow or run off of water. Plant should get clear sunshine during most part of the day. Generation of dung has directed bearing on the quantity of gas generated. The amount of gas production is considerably higher in summer followed by rainy and winter seasons. Gas production would be maximum at a temperature between 30 and 35° C. if the ambient temperature falls below 10°C, gas production is reduced drastically. Biogas slurry is obtained after the production of biogas. It is enriched manure. Another positive aspect of this manure is that even after weeks of exposure to the atmosphere, the slurry does not attract flies and worms.

Resource(s) use efficiencies and optimization techniques

Yield gaps and resource use efficiencies (RUEs) are typically assessed at field level. For regional assessments these are generally directly up-scaled, whilst in the real world they are explained by a range of factors that intersect and are integrated at farming systems level.

Farmers take decisions as to field, crop and livestock management given their access to knowledge and information, personal circumstances, and in the context of the broader socioeconomic, institutional and political environment. Comparative analysis of best, average and worst performing farms using frontier analysis may be combined with theory and information on production ecology to explore options for sustainable intensification. The analysis may be enriched using case studies which are under different phases of agricultural development and, where possible, both the past and possible future dynamics will be analyzed.

The innovativeness of the research is that a mix of agronomic and economic methods may be used, which allows to make use of the richness of individual farm level data in addition to regional data. Production ecological concepts may be integrated into the economic methods, allowing assessment of yield gaps and resource use efficiencies from an agronomic perspective. We may assess the influence of regional conditions, farm and farmer characteristics as well as the resulting management. Case studies may include arable farming in Europe, irrigated rice-based farming in the Philippines and Vietnam, and mixed crop-livestock farming in West Africa.

We may develop a generic method and test this method in contrasting case studies, with the specific objectives to:

(i) Establish yield gaps and RUEs of different farm types using crop simulation models and approaches of frontier analysis;

(ii) Estimate and explain current variability in yield and RUE of land, water, nutrients, labour and capital across different farm types;

(iii) Use a design and re-design cycle to explore options for sustainable intensification by alleviating constraints to improving production and resource use efficiency through comparative studies across farming systems at different stages of intensification.

Resource use efficiency in Indian agriculture

Resource use efficiency in agriculture is defined to include the concepts of technical efficiency, allocative efficiency and environmental efficiency. Public investment, subsidies and credit for agriculture are used in an efficient manner. There are large-scale interregional as well as interfarm variations in factor productivity due to varying influence of different factors in different regions. A number of management factors such as timeliness and method of sowing, transplanting,irrigation and application of right doses of inputs and input mix play an important role in influencing inter-farm variation in crop productivity. Growing marginalization and fragmentation of land holdings coupled with rising incidence of informal tenancies and poor rural infrastructure such as road, electricity, markets and education affect factor productivity. The availability of good quality irrigation water coupled with flexibility of irrigation and drainage system and appropriate methods of application as well as pricing of irrigation water is crucial for sustainable use of land and water resources.

Three types of efficiency identified in the literature viz. technical efficiency, allocative efficiency and overall or economic efficiency. Technical efficiency is the ability of a firm to produce a given level of output with minimum quantity of inputs under a given technology. Allocative efficiency is a measure of the degree of success in achieving the best combination of different inputs in producing a specific level of output considering the relative prices of these inputs.Economic efficiency is a product of technical and allocative efficiency. In one sense, the efficiency of a firm is its success in producing as large an amount of output as possible from a given set of inputs. Maximum efficiency of a firm is attained when it becomes impossible to reshuffle a given resource combination without decreasing the total output.

4 INTEGRATED FARMING SYSTEM MODELS IN DIFFERENT AGROCLIMATIC ZONES

Integrated Farming System (IFS) is a component of farming system research which introduced a change in farming techniques for maximum production in a cropping pattern and take care of optimal utilization of resource. IFS may be defined as linking together two or more normally separate components or enterprises which then become subsystems of a whole farming system.

MAJOR FEATURES OF IFS

(i) Waste or by-product utilization in which the wastes or by-products of one subsystem become an input to a second subsystem

(ii) Improved space utilization in which the two subsystems essentially occupy part or all of the space required for an individual subsystem.

Farming system research is the scientific integration of different interdependent and interacting farm enterprises for the efficient use of land, labour and other resources of a farm family which provide year round income to the farmers specially located in handicapped zones. Diversification of existing farming system and adopting scientific approach of farming brought out significant change in total farm production, profits and besides generating more employment opportunities economized cost of production and thus making farming more profitable and sustainable.

In order to plan agricultural activities more accurately each region into sub-regions based on soil, climate (temperature), rainfall and other agrometeorological characteristics. Planning Commission had demarcated the geographical area of India into 15 agro-climatic regions. These are further divided into more homogeneous 72 sub-zones. The main objectives of agro-climatic regions are:

- To optimize agricultural production
- To increase farm income
- To generate more rural employment
- To make a judicious use of the available irrigation water
- To reduce the regional inequalities in the development of agriculture.

The average size of holding is decreasing and number of marginal farmers are increasing.(Fig. 4.1)

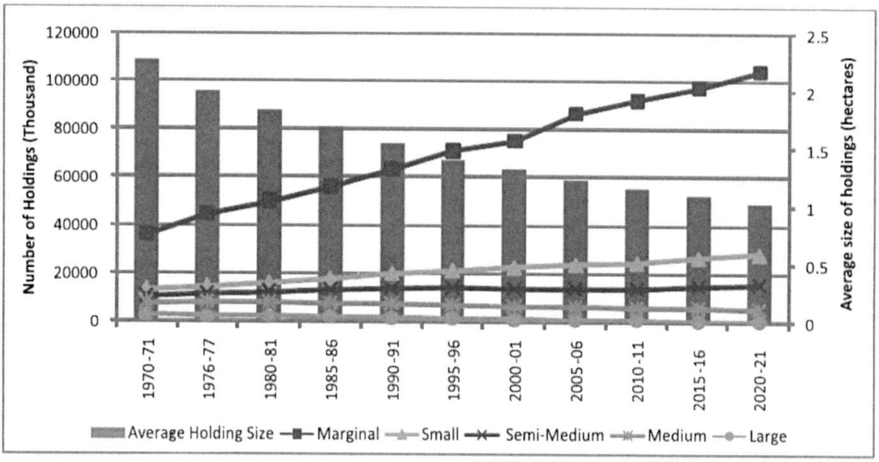

Fig. 4.1: Number of operational holdings and average holding size in India

The 15 agro-climatic zones are:

- **Zone 1:** Western Himalayan Region: Jammu and Kashmir, Uttar Pradesh
- **Zone 2:** Eastern Himalayan Region: Assam, Sikkim, West Bengal and all North Eastern states
- **Zone 3:** Lower Gangetic Plains Region: West Bengal
- **Zone 4:** Middle Gangetic Plains Region: Uttar Pradesh, Bihar
- **Zone 5:** Upper Gangetic Plains Region: Uttar Pradesh
- **Zone 6:** Trans-Gangetic Plains Region: Punjab, Haryana, Delhi and Rajasthan
- **Zone 7:** Eastern Plateau and Hills Region: Maharashtra, Uttar Pradesh, Orissa and West Bengal
- **Zone 8:** Central Plateau and Hills Region: MP, Rajasthan, Uttar Pradesh
- **Zone 9:** Western Plateau and Hills Region: Maharashtra, Madhya Pradesh and Rajasthan
- **Zone 10:** Southern Plateau and Hills Region: Andhra Pradesh, Karnataka, Tamil Nadu
- **Zone 11:** East Coast Plains and Hills Region: Orissa, Andhra Pradesh, Tamil Nadu and Pondicherry
- **Zone 12:** West Coast Plains and Ghat Region: Tamil Nadu, Kerala, Goa, Karnataka, Maharashtra
- **Zone 13:** Gujarat Plains and Hills Region: Gujarat
- **Zone 14:** Western Dry Region: Rajasthan
- **Zone 15:** The Islands Region: Andaman and Nicobar, Lakshadweep

The CA practices need to be adopted selectively in different rainfall, soil and agro-ecological situations. Besides reduced tillage with mulching, other CA practices such as rain water harvesting and recycling, appropriate soil-water conservation measures, INM, use of compost, recycling of crop residues, green manuring, crop rotation with legumes and cover crops have proved effective in enhancing the sustainability of the yields and improving soil quality across the rainfed agro-ecologies. Thus, the advantage of CA practices can be availed in rainfed agro-ecology, if practiced over a long period. However, the initial investment in rainfed regions on such practices may be arranged through different on going government schemes. Higher carbon sequestration through CA practices may also give additional benefits in terms of CERs, if made tradable in future. For promotion of CA practices across diverse agro-ecologies, appropriate technology, policy and institutional support would be a prerequisite. Strengthening delivery system at the village level is important for the adoption of any innovation. Convergence of various schemes, viz. NREGA, NHM, RKVY, NFSM, watershed programme, etc. at the local level involving all major stakeholders would surely contribute towards promotion of CA in rainfed regions. To further enrich our understanding, there is need to undertake in-depth micro-level studies on the impact of CA in different agro-climatic zones.

IFS MODELS OF DIFFERENT AGRO CLIMATIC ZONES

Agro-climatic zone	Specific characteristics	Integrated farming System
Western Himalayan Region	Sub-zones consist of Jammu and Kashmir, Himachal Pradesh and Uttar Pradesh. Skeletal soils of cold region, podsolic soils prone to erosion hazards. There is alluvial soil in the valleys of Kashmir, Kullu and Doon, and brown soil in the hills.Main problems of this region are poor accessibility, soil erosion, landslides, inclement weather, inadequacy of marketing and storage facilities.	Crop, Dairy, Horticulture crops, Vegetables, Apiary, Fish, Poultry, Vermicomposting, Agro-forestry and Plantation crops Horticultural crops as a major component and agriculture mainly on the hill terraces and slopes with maize, rice, wheat, pulses and fodder crops
Eastern Himalayan Region	Sikkim, Arunachal Pradesh, Meghalaya, Nagaland, Manipur, Tripura, Mizoram, Assam. Shifting cultivation is practiced in nearly one-third of the cultivated area with heavy runoff, massive soil erosion.The red-brown soil is not highly productive. Jhum (shifting cultivation) prevails in the hilly areas.	Primitive crop husbandry with rice, pulses etc. Agroforestry systems are also common. Piggery and poultry are the chief livestock activity. There are orchards of pineapple, litchi, oranges and lime.

Agro-climatic zone	Specific characteristics	Integrated farming System
Trans-Gangetic Plains Region	Zone consists of Punjab, Haryana, Union Territories of Delhi and Sriganganagar district of Rajasthan(also called the Satluj-Yamuna Plains); highest net sown area, highest irrigated area, highest cropping intensity and high groundwater utilization.	Crop + Dairy, recycling of farm wastes, crop residues, green manuring and adopting several resource conservational practices Intensive crop husbandry like rice-wheat, maize/mustard/pulses and livestock, dairy, cattle and buffaloes.
Upper Gangetic Plains Region	Zone consists of 32 districts of Uttar Pradesh , irrigation is through canals and tube wells. Besides modernizing traditional agriculture the region needs special focus on dairy development and horticulture.	Crop + Dairy with crop residues, green manuring and adopting several resource conservational practices provide more benefits. This is an intensive agricultural region wherein wheat, rice, sugarcane, millets, maize, gram, barley, oilseeds, pulses and cotton are the main crops.
Middle Gangetic Plains Region	Zone consists of 12 districts of eastern Uttar Pradesh and 27 districts of Bihar plains.	In regard to saving in market inputs, as a result of proper management of farm resources and recycling of farm wastes, all type of crop residues, GM and use of bio-fertilizers etc.
Lower Gangetic Plains Region	Zone consists of districts in West Bengal in the lower gangetic plains region. These soils are mostly alluvial and prone to floods	Crop + dairy (70 %) followed by (ii) Crop + Vegetable (20%). Integration of fishery in to existing farming system along with on farm processing and value addition of farm produces and proper land distribution among the farm enterprises. Further recycling of farm wastes, crop residues, green manuring and use of bio fertilizers.
Eastern Plateau and Hills	This zone consists of eastern parts of Madhya Pradesh, southern part of West Bengal and most of inland Orissa. The soils are shallow and medium in depth. Irrigation is through tanks and tube-wells	Crop + Dairy (cow and buffaloes) and Crops + Goat/ Poultry (60%) or Crops +Dairy + Poultry (20%) are most used system.

Agro-climatic zone	Specific characteristics	Integrated farming System
Central Plateau and Hills	The region is spread over Bundelkhand, Baghelkhand, Bhander Plateau, Malwa Plateau, and Vindhyachal Hills.Soils are mixed red, yellow and black.There is scarcity of water.	Crop and Dairy with adoption by 80% farmers followed by Crop + Dairy + Horticuture / Fishery Cotton and sorghum-millets/pulses with dairy cattle, sheep and goats and poultry are the secondary livestock and animal husbandry enterprises. Crops grown are millets, wheat, gram, oilseeds, cotton and sunflower. In order to improve agricultural returns, measures to be adopted are water conservation through water saving devices like sprinklers and drip system; dairy development, crop diversification, ground water development, reclamation of ravine lands.
Western Plateau and Hills	Zone comprises the major part of Maharashtra, Madhya Pradesh and one district of Rajasthan. Average rainfall of the zone is 904 mm, The net sown area is 65 % and forest occupy 11%. The irrigated area is only 12.4% with canal being the main source	The lower value crops of jowar, bajra and rainfed wheat should give way to high value oilseeds. Five per cent area under rain-fed cotton and jowar could be substituted with fruits like ber, pomegranate, mango and guava. Improvement of milk production of cattle and buffalo through cross-breeding along with poultry development should be encouraged.
Southern Plateau and Hills	This zone comprises 35 districts of Andhra Pradesh, Karnataka and Tamil Nadu. Dryland farming is adopted in 81% of the area and the cropping intensity is 111%	It is an area of dry-zone agriculture where millets, oilseeds, and pulses are grown. Coffee, tea, cardamom and spices are grown along the hilly slopes of Karnataka plateau. Some of the area under coarse cereals may be diverted to pulses and oilseeds. Horticulture, dairy development and poultry farming should be encouraged.

Agro-climatic zone	Specific characteristics	Integrated farming System
East Coast Plain & Hills	The zone comprises of east coast of tamil nadu, Andhra pradesh and Orissa. Soils are mainly alluvial and coastal sands. Irrigation is through canals and tanks	IFS model at 1.25 ha area consisting of crops (0.38 ha), dairy animal (2 cows), Poultry (80 birds), Duckery (20 ducks), Horticulture (0.09ha), Fishery (0.33 ha) and other supplementary enterprises like mushroom. This IFS approach improved the nutritional status of family members by way of producing sufficient amount of milk , fruits & vegetables, mushroom, meet in form of fish/ poultry/duckry birds and thereby reduced market dependency for household domestic needs
Western Coast Plains & Hills	This zone comprises west coast of Tamil Nadu, Kerala, Karnataka, Maharashtra and Goa.The soils are laterite and coastal alluvial. Development of infra-structural facilities and promotion of prawn culture in brackish water should be encouraged.	Rice cultivation with other enterprises like fishery, poultry, and piggery etc. capture fisheries of marine ecosystem is a specialized enterprise. The agricultural development must focus attention on raising of high value crops (pulses, spices, and coconut).
Gujarat Plains & Hills	Zone consists of 19 districts of Gujarat. This zone is arid with low rainfall in most parts and only 32.5% of the area is irrigated largely through wells and tubewells	The economy of the region is based on animals and mainly dairy based farming system with crops as necessary components. Animal husbandry with the camels, sheep and goats, growing with fodder and field crops
Western Dry Region	This zone comprises nine districts of Rajasthan and is characterized by hot sandy desert, erratic rainfall, high evaporation. Famine and drought are common features of the region.	IFS model (1.45ha) with diversified nature of farm components including crops (1.0 ha), dairy animals (2 cows), goatary (11 goats), Horticulture (0.25ha) and other supporting activities including vermicomposting, kitchen gardening and boundary plantations etc.
Island	This zone covers the island territories of Andaman and Nicobar and Lakshdeep which are typically equatorial with rainfall of 3000 mm over eight to nine month. It is largely a forest zone with undulated lands	Plantation crops + Pig (50 %) followed by Crop + Dairy animal cows + Fish (45 %). Pigs are important animal of the region and a symbol of social status too. IFS model of 1.0 ha area developed at the center comprise crops (0.61ha), dairy (2 cows), horticulture (0.20 ha) and piggery.

Conclusion

The profitability of Integrated Farming Systems in well known to the world and can be considered for its widespread adoption by small and marginal farmers. Declining size of landholdings without any alternative income augmenting opportunity is resulting in fall of farm income, and causing agrarian distress. A large number of smallholders have to move to non-farm activities to augment their income (NCAER, 2009). IFS approach not only fulfils the household needs but enriches diet of human being and animals both. Further, diversified nature of the model provides huge employment opportunity for unemployed rural youths. Economic and livelihood analysis of the system revealed that beside household food and fodder security, the system generated a sizable amount of savings which will assist to meet other liabilities of the family including education, health and social obligations and overall improvement in livelihood of small farm holders. The farmers having sufficient land and other farm resources can prefer integration of horticultural crops viz. fruits, vegetables and floriculture as an additional enterprise along with prevailing ones. Whereas, marginal farmers or land less farmers living nearby fruit orchards can integrate apiary and mushroom into their existing farming systems. Farmers having sufficient irrigation water or living in low lying riverbed areas can choose fishery as an additional enterprise. Similarly, farmers living in vicinity of towns and cities can grow vegetables and green fodders based on market demand and availability. Farming systems under small farm holders can only be made profitable if farmers adopt a conservative approach at all stages of farming. For this he has to utilize each and every inch of land for raising suitable field and plantation crops, select low cost viable enterprises for diversification, recycle all farm wastes and crop residues within the system itself and make productive use of farm boundaries and waste lands if any. Further, farmers are also adviced to make use of renewable sources of energy such as solar and biogas etc. Potential improvements and increased productivity from the various enterprises can only come from a better understanding of the nature and extent of the interactions of various enterprises and natural resources, economic benefits, as well as the impact on the livelihoods of small farmers and the environment. Research on these aspects provides major challenges for sustainable agricultural development through integrated farming systems in the future.

IFS MODELS AS PER AGRO-CLIMATIC CONDITIONS

In an effort for a holistic integration of farming enterprises with cropping with the objectives of increasing income and recycling of farm wastes and byproducts to sustain the soil productivity under varied agro-ecological situations, studies on IFS involving various components like livestock, fishery, agroforestry etc, were carried out.

IFS model for upland irrigated areas

According to Jayanthi *et al.* 2002, during 1998-2001, the study involved cropping, poultry, pigeon, goat and fishery in all possible combinations, with a view to recycle the residue and byproducts of one component over the other. In one hectare farm an area of 0.75 was assigned for crop activity, 0.10 ha for growing fodder grass to feed the goat unit (20+1), 0.03 ha allotted to goat shed and the remaining 0.12 ha was allotted to three fish ponds. Three integrated farming systems viz., crop + fish + poultry (20 Bapkok layer birds), crop + fish + pigeon (40 pairs) and crop + fish + goat (Tellichery breed of 20 female and 1 male maintained in 0.03 ha deep litter system) were tried for three years. Polyculture fingerlings of 400 numbers (catla, rohu, mirgal/common carp and grass carp) in the ratio of 40:20:30:10, respectively, reared in 3 ponds of size 0.04 ha (depth 1.5 m) each. Fishes were fed with poultry, (poultry/pigeon droppings obtained from 20 Bapkok layers/40 productive pairs of pigeon) sheltered over two fish ponds and goat droppings (3 animals – 800 kg droppings) to assess the feasibility of rearing fish by using different manures as feed.

Under integrated farming system, cropping sequences includes (i) sugarcane (planted) – sugarcane (ratoon) banana (3 years), (ii) banana-turmeric-rice-banana (3 years) and (iii) maize-rice-sesame-sunhemp (annual) each in 0.25 ha and bajra–napier grass + desmanthus (perennial) in 0.10 ha. Conventional cropping system comprising (i) rice-rice-blackgram (ii) maize –rice-blackgram (iii) maize-rice-sunhemp and rice-rice-sunhemp each in 0.25 ha as practiced by the farmers was taken up for comparison. To sustain the productivity of soil through integrated nutrient supply, recycled poultry, pigeon and goat manures and composted crop residue (banana waste and sugarcane trash) as vermicompost each 6.25 t ha were tested along with 100, 80 and 60% of the recommended fertilizer for the sequences of cropping.

Integrated Farming System provides an opportunity to increase economic yield per unit area per unit time by virtue of intensification of crop and allied enterprises (Jayanthi et.al. 2002). Research results on integrated farming system for three years revealed that integration of crop with fish, poultry, pigeon and goat resulted in higher productivity than cropping alone under lowland. Crop + fish + goat integration recorded higher grain equivalent yield of 39610 kg/ha than other systems (Table 4.1). Similarly as individual animal component, the goat unit (20+1) gave the highest productivity of 8818 kg (mean over three years). This could also provide 11.0 t of valuable manure apart from supplementing the feed requirement of 400 number of fish. While assessing the feasibility of fish by using poultry, pigeon and goat droppings as feed, the fish fed with poultry droppings resulted in higher fish yield (825 kg/0.04 ha ponded water) than the two other sources of feed.

Table 4.1. Productivity of Integrated Farming System and its linked components

Farming system	*Component productivity (kig)					System productivity (kg/ha)	% increase over cropping
	Crop	Poultry	Pigeon	Fish	Goat		
Cropping alone	12995					12995	—
Crop+fish+poultry	26352 (89.0)	1205 (4.1)		2052 (6.9)		29609	128
Crop+fish+pigeon	24854 (85.2)		2545 (8.7)	1774 (6.1)		29173	124
Crop+fish+goat	25725 (68.3)			1975 (5.2)	9979 (36.6)	37679	190

Productivity is expressed in rice grain equivalent yields. Figures in parenthesis indicate % contribution of each component, (jayanthi et al 2002).

Jayanthi 2002 reported that the highest net return of Rs 131118 and per day return of Rs 511/ha were obtained by integrating goat + fish + cropping applied with recycled fish pond silt enriched with goat droppings. Higher net return of Rs 3.36 for every rupee invested was obtained by integration of pigeon + fish + cropping applied with recycled fish pond silt with pigeon droppings. Integration of crop with fish and poultry resulted in higher fish productivity under lowlands. The poultry pigeon and goat droppings were utilized as feed initially and at the end of a year after the fish harvest, about 4500 kg of settled silt from each pond were collected. The pond silt was utilized as organic sources to supply sufficient quantity of nutrients to the crops.

Integrated Farming System for Wetland

Coimbatore (T. N). Components are

1. Cropping (Rice-fish–poultry-mushroom), 2. Fish culture, 3. Poultry, 4. Mushroom production, 5. Cropping 0.36 ha, 6. Fish pond 0.04 ha.

This farming system was compared with conventional cropping normally followed in the region. Rice-Rice-green gram and rice-rice-green manure (0.20 ha).

Component description:

(i) Fish pond area of 0.04 ha with 1.5 m depth.

(ii) Different fingerlings of different fish species with total of 7500/ ha.

(iii) Harvesting of fish commenced from trench month. A poultry shed at corner of fish pond (shed 2.2 m^2).

(iv) Free fall of poultry dropping into the fish pond (Twenty Bapcock chicks of 18 weeks old reared).

(v) The feed components were purchase only in 1st year of cropping.

(vi) The birds started laying eggs around 22nd weeks up to 72nd when they were culled out.

(vii) A mushroom shed of 5x3 m was constructed with local materials. Oyster mushroom was produced utilizing rice straw as the base material.

On an average net profit of Rs. 11, 755 was obtained in rice-poultry-mushroom system as compared to Rs. 6,335 only from conventional system of cropping. Additional employment of 174 man days was generated due to IFS. (Jayanthi *et al*, 2002).

Fish culture

Ponds near to poultry shed. Different fingerlings of fish in ponds, with density of population maintained were 10000 fingerlings / ha. The fish were harvested after six and a half months. A net return of Rs. 17,200 was obtained by integrating different enterprises by introducing poultry-cum-fish culture with cropping a total employment of 385 man-days was generated.

Crop + dairy + biogas + silviculture (Irrigated upland, Tamilnadu)

IFS experiments were conducted at irrigated uplands of TNAU, Coimbatore. The major source of water was met from deep borewell. the major cropping system followed under irrigated uplands are sorghum–cotton–maize, ragi–cotton–maize and ragi -cotton–sorghum. IFS involving crop, dairy, biogas and silviculture was taken from 1987-92. The normal cropping pattern followed in Coimbatore is ragi – cotton – sorghum. In the integrated approach, the same cropping pattern was slightly modified by inclusion of intercropping such as green gram in cotton, cowpea in sorghum and sunflower as border crop in ragi (0.75 ha). The perennial grass fodder (0.15 ha) and legume fodder. Lucerne (0.05 ha) were also raised in the holding. Three jersey cross bred milch cows with 2 calves were included under dairy component. For effective recycling of farm and animal waste, a biogas unit of 2 m^3 capacity was installed for the production of fuel, light and enriched manure. Sixty kg of cowdung expected out of 5 animals is sufficient to produce 2 m^3 of gas every day which is equivalent to 1.5 litres of kerosene. Two hundred numbers of subabul trees were planted all along the boundary of the field for fodder and timber production.

Integrated Farming System(IFS) for irrigated situations

A model integrated farming system to suit the small and marginal farmers of garden land condition was studied at TNAU,Coimbatore, during 1988-1993 (Rangaswami *et al.* 1995). An area of one ha was selected for IFS and compared with conventional cropping system (CCS).

The above integrated system was compared with the conventional cropping system of cotton, sorghum, finger millet in 0.20 ha area. Economic returns from the system: Maize flour, cotton seed and wheat bran obtained from the crop components were recycled for preparing dairy feed from the second year. About 45.5 t of dung was recycled for the biogas plant. Mean revenue of Rs. 34600/ ha was realized in IFS as compared to Rs. 13950 obtained in CCS. Employment opportunity was also enhanced to the tune of 770 man days per year under IFS as against conventional cropping.

Integrated Farming System (IFS) for dry land

Integrated farming system for dry lands suggested for Coimbatore and Aruppukottai, Tamil Nadu, are described below: Model for Coimbatore, Tamil Nadu (mean Annual rainfall; 640 mm).

Crop components for one hectare:

S.No	Cropping	Area (ha)
1	Sorghum+Cowpea both grain purpose	0.20
2	Sorghum+Cowpea both fodder purpose	0.20
3	Leucaena (Tree fodder) + Cenchrus (Grass fodder)	0.20
4	*Aracia senegal* (tree fodder) + grass	0.20
5	*Prosonic cineraris* (tree fodder) + grass	0.20

Animal Components: Telicherry goat: 6 (5 female + 1 male – stall fed) Conventional cropping system = 0.20 ha (sorghum+ cowpea- grain purpose). The animal components, tell cherry goat, numbering 6 (5 female+ 1 male) were kept in a shed (6x4 m). The goats were stall fed form the cropping components. Two kg each of green and dry fodder and 100 g of concentrate were given to each animal. At the end of second year, all the 4 male goats were disposed retaining one number. From the end of third year onwards, 20 female and one male were retained. The litter form the goat shed was used for composting and recycled as manure. A farm pond was dug in an area of 300 m2 and the rain water harvested was utilized for pot watering the tree saplings.

Economic returns from the system: Additional revenue of Rs. 3750 ha/year was obtained from IFS over CCS. The employment generation under IFS was 153- man days/ ha / year whereas it was only 40-man days/ ha/ year in the CCS (Sivasankaram et al. 1995).

Crop + goat (Tamilnadu): An integrated farming system study involving grain crop, fodder crop, fodder trees, perennial grasses and goat rearing in an area of one hectare of rainfed land was carried out at Coimbatore from 1987 to 1992. Generally in dryland farmers raise only fodder sorghum yielding 3 to 10 tonnes of fodder/ha depending upon the rainfall. In the integrated approach, the cropping pattern was modified by including both sorghum grain crop (0.2 ha) and sorghum fodder crop (0.2 ha). To meet the fodder requirements of goats, 0.2 ha of land was raised with Subabul and *Cenchurus ciliaris*, a pasture grass as an intercrop. Trees like Acacia, Senegal and *Prosopis cineraria* were raised in 0.20 ha. The goat unit comprised Tellichery goats 20 eves and 1 buck. The results revealed that through short duration field crops and perennial crops the feed requirement for one productive unit consisting of 20 eves and one buck for all the 365 days can be met. After 5 years, the perennial fodder trees could bear the shock and yield sufficient quantity of loppings to supplement the feed requirement of 21 adults along with millets, legumes and perennial grass linked. The Tellichery goat will be a good breed to build up body weight for every unit of feed secured through different sources under rainfed conditions. It is a dual purpose animal, where it gives 800 to 1000 ml of milk after satisfying the full requirement of dependent kids on it every day. Twenty productive females could give 45 kids per annum. Each kid at the time of weaning will weigh around 12 kg. Moreover the unit of 21 animals with different stages of kids under deep litter system would give 11.2 to of valuable manure. This when applied to the soil, will not only be an excellent source of primary, secondary and micronutrients for the crops but also absorb more moisture, retained in the soil and releases to the crop approximately for better yield. The net income from the farming system was Rs. 5671/ha/year and that of control (Rs 1919/ha/year). Out of the total income from the IFS, 59% was from goat rearing. The additional net income, realized from IFS was Rs 3752/ha/year as compared to cropping alone. The additional employment gained through IFS over cropping was 314 man days/ha/year.

Hill regions

The north-eastern hill regions of India are unique in their physiographic characteristics of land, varied agro-climatic conditions, land tenure system and cultivation practices from the rest of the country. The region falls under high rainfall zone and is characterized by hilly terrain with wide variation in elevation from 100 m to 5500 m. Majority of the farmers are maintaining dairy cattle and major portion of green fodder comes from lopping of the fodder trees and locally available grasses. Crops and cropping patterns of such land uses will differ depending on the type of enterprise such as milk, beef, mutton, wool, pork and poultry production. Combination of cultivated of perennial legumes, grasses, shrubs and trees can extend availability of green fodder upto February at low altitude, thereby shortening the requirement of conserved fodder for lean season feeding. Land upto 100% slope having soil depth greater than 1m can be used for agri-horti-silvipastoral system. This system has the potentiality of maintaining 1.18 livestock unit/ha of land use. In Himachal Pradesh studies undertaken by AICRP-IFS Palampur centre revealed that Livestock based farming systems followed by cereals based farming systems were the dominating farming systems in Zone I and II of Himachal Pradesh. In these zones 63.9 and 59.7% of the farmers were dependent on Livestock based farming systems and 33.3 and 36.8% farmers, respectively, were dependent on cereals based farming systems. In Zone III fruit growing was main activity and 72.2% of the households were dependent on fruit based farming systems. This was followed by vegetable based farming systems from which 23.6% of the households earn their livelihood. On an average, livestock based farming system was the major activity for more than 65% marginal farmers, 36% small farmers, 37% medium farmers and 23% large farmers. Cereals based farming systems was the main activity of the small (39.4%), medium (42.6%) and large farmers (44.3%).

Agroforestry-based land management systems of Indian Himalayas

The forest cover in India is a 19.27% of the total geographical area and is grossly inadequate to meet the nation's requirements for forest products, maintenance of ecological processes and environmental stability. Agroforestry, which is a combination of trees, agriculture and other land use technologies, has the potential for meeting the above requirements. The three basic types of agroforestry systems are agrosilvicultural, silvi pastoral and agro-silvi pastoral. One of the essential characteristics of agroforestry system is its site-specificity. Five major agro-ecological zones for agroforestry have been identified in India. In the mid-western Himalayas, of all the systems, the agro-horti-silvicultural system is highly diverse in vegetation composition and has the highest productivity of 25.8 mt/ha/year. In the Shivaliks, the well- known tree species are Acacias, *Leucacena leucocephala, Dalbergia sissoo, Melia azedarach, Eucalyptus hybrida* and *Emblica officinalis*. These trees are grown with local fodder grasses such as *Chrysopogon fulvus* and *Heteropogon contortus*. Bhabbar (*Eulaliopsis binata*), which is a commercial grass, used in the manufacture of paper, has also done very well in association with most of the above tree species. These two- layered agroforestry systems involving trees and grasses have demonstrated their superiority over traditional rainfed crops on poor quality soils. Moreover, on sloping non-arable lands in the Shivaliks, World Bank assisted projects implemented from 1980 to 1999 with people's participation have helped in conserving natural resources and in substantially augmenting production and income from these agroforestry systems.

In the alluvial plains of Punjab, Haryana and Uttar Pradesh, poplar has done exceedingly well in association with a large number of annual crops particularly wheat, sugarcane and vegetables. Eucalyptus can be raised as pure crop, boundary plantations, wind break, sand shelterbelts where it has been found to profitable after allowing for depression in crop yields. Agroforestry interventions should examine the management in terms of farmer livelihood strategies rather than trying to merely increase supply of forest products. There is a need to improve market information and reduce market constraints to tree growing. Public investment is needed in research to improve the genetic quality of the planting material, and to refine existing agroforestry practices/models. Enhanced investment is required for improving extension services and develop mechanisms for transferring agroforestry technology from the research institutes to the farmers and otheruseragencies.

Farming on sloping lands of Asia

Some success stories about sloping upland farming in Asia reveal that by taking advantage of specific niches, ecologically stable and economically productive production systems can be evolved. The three examples of success stories reflect a better understanding of the niche potentials of sloping marginal uplands and the opportunities for using them in sustainable ways. Fruit farming on sloping lands extends the availability of agricultural land resources. Cardamom farming demonstrates the use of local biodiversity as a good source of new crops for developing niche-based upland farming. Sea buck thorn example shows that soil and water conservation can be complementary to food security and poverty alleviation and those forests can also provide benefits of economically productive farming. The widening economic gaps between upstream and downstream communities in many densely populated regions could be socially and environmentally explosive

IFS models for the Andaman Nicobar Islands

(i) **Coconut-cum-fodder-cum-milch cattle:** Mixed farming by raising fodder grass coconut has been founded profitable; grass in coconut has been found profitable, grass like hybrid Napier or leguminous fodder like Stylo in coconut gardens can support 4- 5 dairy animals. Animal supply large qualities of cattle manure applied to coconut gardens improve the soil fertility.

(ii) **Coconut – cum fish culture in salt affected lands:** Coconut grows successfully even in salt affected paddy fields if the fields are within the approach of brackish water. The bunds of fields at water entrance gate should be raised to maintain the required level of water. Field bunds should also be raised as per the requirement. Water will be exchanged and thus fish raising can be taken up together with coconut in dry period. In rainy season, salinity tolerant rice varieties could be cultivated in the same field with coconut and fish culture.

(iii) **Fruits- fodder- milch cattle:** The space available between the fruit trees can be utilized for growing fodder crops such as cow peas, rice bean, green gram and black gram. The fodder could be used for feeding the cattle and cattle manure could be applied to the fruit trees and fodder crops.

(iv) **Cocunut-cum-fodder-cum-fish or prawn culture**: This system is suitable for tidal marshy area. The coconut palms are grown on platforms raised upto the highest tidal level and fish and prawn in channels between the coconut plantations. On platform with coconut trees, fodder crops like Guatemala grass, thin Napier and guinea grass could be raised successfully with an average forage yield of 290, 190 and 184 q/ha respectively under high salinity conditions.

5 ROLE OF INTEGRATED FARMING SYSTEM IN AGRICULTURE AND LIVELIHOOD SECURITY

Farming system represents an appropriate combination of farm enterprises viz. cropping systems, horticulture, livestock, fishery, forestry, poultry and utilization of other means available with the farmers to raise profitability. It interacts adequately with environment without dislocating the ecological and socio-economic balance on one hand and attempts to meet the national goal on the other. The farming system in its real sense will help in different ways to lift the economy of agriculture and standard of living of the farmers of the country as a whole. To sustain and satisfy as many of his needs, farmers include crop production, livestock, poultry, fisheries, beekeeping etc. in their farms. Earlier subsistence was the important objective of farming. Farmers took many activities such as planting of fruit trees in their farms or on the common lands just for the welfare of mankind without expecting anything in return. The kind of farming where subsistence and welfare of mankind were the main objectives is designated as "mixed farming". Farming today is also a mix of enterprises. However, higher profitability without altering the ecological balance is important in farming. A set of agricultural activities organized while preserving land productivity, environmental quality and maintaining desirable level of biological diversity and ecological stability is designated as "Farming system". Here the emphasis is mainly on a system rather than on gross output.

In other words "farming system" is a resource management strategy to achieve economic and sustain agricultural production to meet diverse requirement of the farm household while preserving the resource base and maintaining high environmental quality. The farming system in its real sense will help to lift the economy of agriculture and standard of living of the farmers.

Farming system specially refers to a group/combination of enterprises in which the products and or the byproducts of one enterprise serve as the inputs for production of other enterprise. The waste of dairying like dung, urine, refuse etc. is used for preparation of FYM, which is an input in cropping systems. The straw obtained from the crops is used as fodder for cattle.

Farming system takes into account the combination needs of the family–the economic factors like relative profitability of the technically feasible enterprises, availability of farm resources, infrastructure and institutions such as irrigation, marketing facilities including storage and transportation and credit besides the agro-biological consideration, namely, inter-dependence, if any among various technically feasible enterprises and the performance of individual farmers.

Farming is defined as the way in which the farm resources are allocated to the needs and priorities of the farmers in his local circumstances which include

(i) Agro-climatic condition such as the quantity, distribution and reliability of rainfall, temperature, etc.

(ii) Soil type and topography and

(iii) Economic and institutional circumstances like market opportunities, prices, institutional and infrastructure facilities and technology.

ADVANTAGES OF INTEGRATED FARMING SYSTEM

Integrated Farming System (IFS) a component of FSR (Farming System Research), introduces a change in the farming techniques for maximum production in the cropping pattern and takes care of optimal utilization of resources. The farm wastes are better recycled for productive purposes in the IFS. Unlike the SFS, IFS's activity is focussed round a few selected, interdependent, interrelated and often interlinking production systems based on a few crops, animals and related subsidiary professions. IFS envisages harnessing the complementarities and synergies among different agricultural subsystems/enterprises and augmenting the total productivity, sustainability and gainful employment.

Sustainability is the objective utilization of inputs without impairing the quality of the environment with which it interacts. Therefore, it is clear that farming system is a process in which sustainability of production is the objective.

The overall objective is to evolve technically feasible and economically viable farming system models by integrating cropping with allied enterprises for irrigated, rainy, hilly and coastal areas with a view to generate income and employment from the farm.

The advantages of IFS include pooling and sharing of resources/inputs, efficient use of family labour, conservation, preservation and utilization of farm biomass including non-conventional feed and fodder resources, effective use of manure/animal waste, regulation of soil fertility and health, income and employment generation for many people and increase in economic resources. It improves space utilization and provides diversified products.

IFS is part of the strategy to ensure sustainable use of the natural resources for the benefit of present and future generations.

(i) Productivity: IFS provides an opportunity to increase economic yield per unit area per unit time by virtue of intensification of crop and allied enterprises.

(ii) Profitability: Use waste material of one component at the least cost. Thus reduction of cost of production and form the linkage of utilization of waste material and elimination of middleman interference in most inputs used. Working out net profit/BC ratio is increased.

(iii) Potentiality or Sustainability: Organic supplementation through effective utilization of by-products of linked component is done, thus providing an opportunity to sustain the potentiality of production base for much longer periods.

(iv) **Balanced food:** Components of varied nature are linked to produce different sources of nutrition.

(v) **Environmental safety:** In integrated farming system waste materials are effectively recycled by linking appropriate components, thus minimizing environment pollution.

(vi) **Recycling:** Effective recycling of waste material (crop residues and livestock wastes) in integrated farming system. Therefore, there is less reliance to outside inputs – fertilizers, agrochemicals, feeds, energy, etc.

(vii) **Income rounds the year:** Due to interaction of enterprises with crops, eggs, milk, mushroom, honey, cocoons silkworm, it provides flow of money to the farmer round the year. There is higher net return to land and labour resources of the farming family.

(viii) **Adoption of new technology:** Resourceful farmers (big farmer) fully utilize technology. In integrated farming system there is linkage of farmers with dairy/mushroom / sericulture / vegetable etc. Money flow round the year gives an inducement to the small/ original farmers to go for the adoption of technologies.

(ix) **Saving energy:** To identify an alternative source to reduce our dependence on fossil energy source within a short time with effective recycling technique the organic wastes available in the system can be utilized to generate biogas. Energy crisis can be postponed to a later period.

(x) **Meeting fodder crisis:** Every piece of land area is effectively utilized. Plantation of perennial legume fodder trees on field borders and also fixing the atmospheric nitrogen. These practices will greatly relieve the problem of non-availability of quality fodder to the animal component linked.

(xi) **Solving fuel and timber crisis:** Linking agro-forestry appropriately the production level of fuel and industrial wood can be enhanced without determining effect on crop. This will also greatly reduce deforestation, preserving our natural ecosystem.

(xii) **Employment generation:** Combining crop with livestock enterprises would increase the labour requirement significantly and would help in reducing the problems of under employment to a great extent. IFS provides enough scope to employ family labour round the year.

(xiii) **Agro–industries:** When one of the produce linked in IFS is increased to commercial level there is surplus value adoption leading to the development of allied agro –industries.

(xiv) **Increasing input efficiency**: IFS provides good scope to use inputs in different components with greater efficiency and benefit : cost ratio.

FARMING SYSTEM COMPONENTS AND THEIR MAINTENANCE

Cropping systems: Cropping systems based on climate, soil and water availability have to be evolved for realizing the potential production levels through efficient use of available resources. The cropping system should provide enough food for the family, fodder to the cattle and generate sufficient cash income for domestic and cultivation expenses.

Interactions between component crops: In intensive cropping, crops are grown in association (intercropping) or in sequence (sequential cropping). In such situations there is possibility of interaction between the component crops. The interaction is mainly due to response of one species to the environment as modified by the presence of other species. Interaction may be competitive or non-competitive or complementary.

Factors such as light, water, nutrients, oxygen and CO_2 are required for plant growth. In mixed or intercropping situations, the component species compete for the growth factors. The close proximity of the species causes sub-optimal utilization of the growth factors and hence there is inequitable distribution of resources among the plants. Generally, competition will develop between two components or within the components.

Light: Intercropping can increase light inception by as much as 30-40%. When one component is taller than the other in an intercropping system, the taller component intercepts most of the solar radiation. In intercropping situation where the component crops have different growth durations, the peak demand for light would occur at different times. In such combinations, competition for light is less among the component crops and there is greater light use in intercropping than in pure stands.

In general the component crops under intercropping situations are grown in such a way that competition for light is minimized. Proper choice of crops and varieties, adjustment of planting density and pattern are the techniques to reduce competition and increase the light use efficiency.

Moisture and nutrients: Competition for water and nutrients results in two main types of effects on the less successful or suppressed component. First, the roots of the dominated crop may grow less on the sides of the aggressive component. The suppressed components adapt to such conditions by increased capacity for uptake. Also, if one part of the root system is on the depleted side, the remaining part shows compensatory activity and vigour. Secondly, plants affected by competition for soil factors are likely to have increased root/shoot ratio.

Allelopathy: Allelopathy is any direct or indirect harmful effect that one plant has on another through the release of chemical substances or toxins into the root environment. Some crops may be unsuitable to be grown as intercrops because they may produce and excrete toxins into the soil which are harmful to other components.

Annidation refers to complementary interaction which occurs both in space and time.

Annidation in space: The canopies of component crops may occupy different vertical layers with taller components tolerant to strong light and high evaporative demand and shorter components favouring shade and high relative humidity. Thus, one component crop helps the other. Multistoreyed cropping in coconut gardens and planting of shady trees in coffee, tea and cocoa plantations use this principle. Similarly, root systems of component crops exploit nutrients from different layers thus utilizing the resources efficiently. Generally, one component with shallow root system and another with deep root system are selected for intercropping, e.g. in Setaria (shallow) + red gram (deep) intercropping system.

Annidation in time: When two crops of widely varying duration are planted, their peak demands for light and nutrients are likely to occur at different periods, thus reducing competition. When the early maturing crop is harvested, conditions become favourable for the late maturing crop. This has been observed to occur in sorghum + red gram, groundnut + red gram and maize + green gram intercropping systems.

Other complementary effects: In an intercropping system, involving a legume and a non-legume, part of the nitrogen fixed in the root nodule of the legume may become available to the non-legume component. The presence of rhizosphere microflora and mycorrhiza on one species may lead to mobilization and greater availability of nutrients not only to the species concerned, but also to the associated species. Another example is the provision of physical support by one species to the other in intercropping system. Erect crop plants may improve the yield of a climber as in the case of coconut + pepper, maize + beans. The taller component acts as a wind barrier protecting the short crop as in maize + groundnut, onion + castor and turmeric + castor.

TYPES OF ENTERPRISES

- Independent enterprises
- Joint enterprises
- Competitive enterprises
- Supplementary enterprises and
- Complementary enterprises.
- **Independent enterprises:** Independent enterprises are those which have no direct bearing on each other, an increase in the level of one neither help not hinders the level of the other. In such cases each product should be treated separately e.g. production of maize and berseem for fodder independently.
- **Joint enterprises:** Joint products are those which are produced together, e.g. cotton and cottonseed, wheat and straw, etc. The quantity of one product decides the quantity of the other products. In case of joint products there is no economic decision to make with respect to the combination of products and two products can be treated as one.
- **Competitive enterprises:** Competitive enterprises are those which compete for use of the farmers' limited resources; use of resources to produce more of the one necessitates a sacrifice in the quantity of other product. When enterprises are competitive three things determine the exact combination of the product, which would be most profitable.

  (a) The rate at which one enterprises substitutes for the other.

(b) Prices of the products and

(c) The cost of producing the product.

The rate at which one product substitutes for another is known as the marginal rate of substitution.

The products can have:

(a) Constant rates of substitution.

(b) Decreasing rate of substitution

(c) Increasing rates of substitution e.g. paddy-sorghum, paddy-groundnut

Supplementary enterprises: Two products are said to be supplementary when an increase in the level of one does not adversely affect the production of the other but adds to the income of the farm i.e enterprise which does not complete with each other but adds to the total income. For example, on many small farms, dairy enterprise or a poultry enterprise may be supplementary to the main crop enterprises because they utilize surplus family labour and shelter available and perhaps even some feeds, which would otherwise go waste. Some times enterprises are supplementary for one resource but competitive for another. In such cases the relationship should be treated as one of competitive. Even though they are supplementary to one another in respect of other sources e.g. mixed crops.

Complementary enterprises: Complementary enterprises are those which add to the production of each other, e.g., Berseem and Brassica crops. Two products are complementary when the transfer of available input for the production of the one product to the production of the other results in increase in the production of both products. Then two crops are complementary enterprises, the use of resources for the two crops result in the increased production of both the crops.

Two enterprises do not remain complementary over all possible combinations. They become competitive at some stages. When both complementary and competitive relationship occur the complementary relationship occurs first and is then followed by competitive relationship.

Resource use efficiency in Indian agriculture

Resource use efficiency in agriculture is defined to include the concepts of technical efficiency, allocative efficiency and environmental efficiency. Public investment, subsidies and credit for agriculture are used in an efficient manner. There are large -scale inter-regional as well as inter-farm variations in factor productivity due to varying influence of different factors in different regions. A number of management factors such as timeliness and method of sowing, transplanting, irrigation and application of right doses of inputs and input-mix play an important role in influencing inter-farm variation in crop productivity.

Growing marginalization and fragmentation of land holdings coupled with rising incidence of informal tenancies and poor rural infrastructure such as road, electricity, markets and education affect factor productivity. The availability of good quality irrigation water coupled with flexibility of irrigation and drainage system and appropriate methods of application as well as pricing of irrigation water is crucial for sustainable use of land and water resources.

Three types of efficiency are identified in the literature viz. technical efficiency, allocative efficiency and overall or economic efficiency (Farrell, 1957; Olayide & Heady, 1982) are defied as;

(a) **Technical efficiency** is the ability of a firm to produce a given level of output with minimum quantity of inputs under a given technology.

(b) **Allocative efficiency** is a measure of the degree of success in achieving the best combination of different inputs in producing a specific level of output considering the relative prices of these inputs.

(c) **Economic efficiency** is a product of technical and allocative efficiency (Olayide & Heady, 1982).

In one sense, the efficiency of a firm is its success in producing as large an amount of output as possible from given sets of inputs. Maximum efficiency of a firm is attained when it becomes impossible to reshuffle a given resource combination without decreasing the total output.

Major categories of farming system

The delineation of the major farming systems provides a useful framework within which appropriate agricultural development strategies and interventions can be determined. The decision to adopt very broad farming systems inevitably results in a considerable degree of heterogeneity within any single system. However, the alternative of identifying numerous, discrete, micro-level farming systems in each developing country - which could result in hundreds or even thousands of systems worldwide would complicate the interpretation of appropriate regional and global strategic responses and detract from the overall impact of the analysis. Only the major farming systems have, therefore, been identified and then mapped in order to estimate the magnitudes of their populations and resource bases. Each of these broad systems is characterized by a typical farm type or household livelihood pattern, although significant sub-types are described where appropriate.

The classification of the farming systems of developing regions, may be based on the following criteria:

- available natural resource base, including water, land, grazing areas and forest; climate, of which altitude is one important determinant; landscape, including slope; farm size, tenure and organization; and
- dominant pattern of farm activities and household livelihoods, including field crops, livestock, trees, aquaculture, hunting and gathering, processing and off-farm activities; and taking into account the main technologies used, which determine the intensity of production and integration of crops, livestock and other activities.

Based on these criteria, the following eight broad categories of farming system have been distinguished:

- Irrigated farming systems, embracing a broad range of food and cash crop production;
- Wetland rice based farming systems, dependent upon monsoon rains supplemented by irrigation;
- Rainfed farming systems in humid areas of high resource potential, characterized by a crop activity (notably root crops, cereals, industrial tree crops - both small scale and plantation and commercial horticulture) or mixed crop-livestock systems;
- Rainfed farming systems in steep and highland areas, which are often mixed crop-livestock systems;
- Rainfed farming systems in dry or cold low potential areas, with mixed crop-livestock and pastoral systems merging into sparse and often dispersed systems with very low current productivity or potential because of extreme aridity or cold;
- Dualistic (mixed large commercial and small holder) farming systems, across a variety of ecologies and with diverse production patterns;
- Coastal artisanal fishing, often mixed farming systems; and
- Urban-based farming systems, typically focused on horticultural and livestock production.

VIEW OF SCIENTISTS REGARDING IFS

Ngambeki *et al.* (1992) demonstrated the profitability of the system by integrating livestock into a crop-based farming through increased financial benefits and a better use of intermediate farm resources such as manure, draft power, and crop residues. Singh *et al.* (1993) and Singh *et al.* (1997) observed that the integration of various enterprises on various sizes of land holdings tend to be more profitable than arable farming alone, and generate more employment. Jayanthi *et al.*, (1994) also reported that integrated farming of crop, poultry and fish culture generated 453 additional man-days over arable farming on 0.40 ha land whereas on 1 ha it was between 559 and 630 man days with almost uniform distribution throughout the year compared to 182 man days in arable farming. Rangasamy *et al.* (1996) concluded the integration of poultry, fish and mushroom with rice cultivation over a five-year period increases the net farm income and on-farm labour when compared with the conventional rice cropping system and also the comparative analysis suggested that diversification and integration of resource management can be productive, profitable and manageable, given access to labour and secure tenure. Itnal *et al.* (1999) stated that integration of two or more appropriate combination of enterprises like crop, dairy, piggery, fishery, poultry, bee keeping etc. for each farm according to the availability of resources helps to sustain and satisfy the necessities of the farmer. Ashby (2001) indicated that the reliance upon a few crops in combination with a high risk of crop failure due to a range of factors like disease, drought, etc. exposes farmers to a high degree of variability with respect to yields and income and therefore risk. Thamrongwarangkul (2001) reported that the diversification of farming activities should invariably improve the utilization of labour, reduce unemployment in areas where there is a surplus of underutilized labour and provide a source of living for those households that operate their farm as a full time occupation.

Radhamani *et al.* (2003) reviewed several studies on the financial viability of IFS and concluded that they positively influenced the economic viability of the IFS. Bosma *et al.* (2005) identified that the farmers who have transformed their rice mono-culture to rice-based farming systems including rice, upland crops, livestock and aquaculture on the same farm, allowing better use of farm resources, thereby improving farm income as well as safeguarding the environment. Tipraqsa *et al.* (2007) reported the advantages of IFS like increased productivity, capital saving, family labour employment and income generation. Prein (2002) and Nhan *et al.* (2007) concluded that the integration of 2 bullocks + 1 cow + 1 buffalo and 10 goats along with other subsidiaries like poultry and duck is the most beneficial system which can supplement the income of tribal people to improve their socio-economic status. Nageswaran *et al.* (2009) reported the average annual net revenue per acre of IFS was more than 2.5 times than that of CFS in Cuddalore district of Tamilnadu. And also in the event of failure of any crop due to delay or heavy rainfall, other enterprises in IFS would tend to compensate which is absent in conventional farming.

Channabasavanna *et al.* (2009) found that the integration of crop with fish, poultry and goat resulted in higher productivity than adoption of conventional rice-rice alone and also 26.3% higher productivity was reported in an IFS while compared to conventional rice-rice system. Biswas (2010) reported that the farming system revolves around better utilization of time, money, resources and family labour and also the farm family gets scope for gainful employment round the year thereby ensuring good income and higher standard of living even from the small holdings. Jagadeeshwara *et al.* (2011) reported that the productivity of IFS was 26.3% higher than the conventional system. Among the various components the productivity was maximum in crop yield (46.32%), closely followed by horticulture (16.77%), dairy (42.26%) and piggery (8.07%) in the southern Karnataka state. Poorani *et al.* (2011) reported that the IFS increased the productivity, profitability, employment generation by 48, 40 and 45 per cent respectively than the existing conventional farming system in Palladam district of Western Zone of Tamilnadu.

Jayanthi *et al.* (2003) reported the findings of net returns obtained from all the components was Rs. 22,887 with an increase of 32.3% higher returns than conventional rice-rice system. Ramrao *et al.* (2005) developed a crop livestock mixed farming model of 1.5 acre small scale holders with the employment generation of 571 man days, net income of Rs. 58,456 per year against crop farming alone with employment generation of 385 man days and net returns of Rs. 18,300 per year only. Ramrao *et al.* (2006) noticed that the mixed farming of 2 bullocks+ 1 cow+ 1buffalo + 10 goats+ 10 poultry and 10 ducks gave a net return of Rs 33,076 compared to Rs 7843 from arable farming. Ramasamy *et al.* (2008) reported that the income from integrated crop+ livestock + goat + poultry was Rs. 98,270 than Rs. 28,600 in traditional farming system. Similarly income of Rs. 99,209 in IFS with the crop + livestock + goat

+ poultry than conventional farming system (CFS). Nageswaran *et al.* (2009) found that the annual net revenue per acre is higher for IFS as compared to CFS: the average net annual revenues per acre of IFS and CFS are Rs. 11,662.57 and Rs.4, 553.31 respectively. Annual employment per acre is turned out to be 185.78 person days in IFS and that of CFS 89.3 persons respectively. Channabasavanna *et al.* (2009) seen the benefit cost ratio of 1.97 in IFS than conventional system which is of 1.64. Among the various components of Palladam district of goat recorded the highest benefit cost ratio (2.75) followed by fish (2.23), vegetables (2.00) whereas poultry showed the lowest benefit cost ratio (1.13) as a result of high cost of maintenance. Tripathi *et al.* (2010) reported that the integration of 7 different enterprises namely, crop+ fish+ goat+ vermicompost+ fruit production+ spice production+ agro-forestry obtained the net return to the tune of Rs. 2,30,329 annually with the Benefit Cost Ratio (BCR) of 1.07:1 and also reported that the maximum per cent contribution of the enterprise is the fish production (68.53%) followed by vermicomposting (9.90%), spices (8.46%) and animal production (7.40%). The BCR was found to be highest for the spice production (1.83:1) after fishery (2.25:1) followed by the vermicomposting (1.45:1).

LIVELIHOOD SECURITY STATUS AMONG FARMERS DUE TO INTEGRATED FARMING SYSTEMS

Conceptually 'livelihood' denotes the means, activities, entitlements and assets by which people make a living. Assets are defined as natural/biological (i.e. land, water, common-property resources, flora and fauna), social (i.e. community, family and social networks), political (i.e. participation, empowerment-sometimes included in the social category), human (i.e. education, labour, health and nutrition), physical (i.e. roads, clinics, markets, schools and bridges) and economic (i.e. jobs, saving and credit). The sustainability of livelihoods becomes a function of how men and women utilize asset portfolios on both a short and long-term basis. Sustainable livelihoods are those that are able to cope with and recover from shocks and stresses such as drought, civil war and policy failure through adaptive and coping strategies (Jirli *et al.*, 2008). Capability, equity and sustainability combine in the concept of sustainable livelihood. The concept Sustainable Rural Livelihood (SRL) is an attempt to go beyond the conventional definitions and approaches to poverty eradication. These had been found to be too narrow they focused only on certain aspects or manifestations of poverty, such as low income, or did not consider other vital aspects of poverty such as vulnerability and social inclusion. It is now recognized that more attention must be given to the various factors and processes which either constrain or enhance poor people's ability to make a living in an economically, ecologically and socially sustainable manner. The SRL concept offers a more coherent and integrated approach to poverty alleviation. To achieve sustainable rural livelihoods different livelihood capitals such as human capital, physical capital, natural capital, social capital and financial capital would play a greater role to cope with shocks and stresses and maintain or enhance the individuals capabilities and assets both in present and in the future without degrading the natural resource base. Veluswamy and Manoharan

(1998) perceived that beneficiaries were found to have increased in outside contact and knowledge about various development programmes after their participation in NGO programme. In the areas where development programme was implemented there is the chance for more savings and clearing debts, so as to have a follow-up programme "small savings" may be encouraged.

Behera and Mahapatra (1999) conducted research on a pond-based integrated farming system in Bhubaneshwar (Odisha). The integrated farming system consisted of pisciculture, field and horticultural crop, poultry, mushroom, apiculture and biogas enterprises. Apiculture produced the highest returns (7.94 per rupee invested), followed by pisciculture (5.46 per rupee invested). Among the crop enterprises, best returns were obtained with multi storey cropping involving pumpkin, ridge gourd, and poi as ground storey; pineapple, colocasia, ginger and turmeric as first storey; and coconut as second storey. Poultry and mushroom enterprises fetched low returns. The highest level of employment (180 mandays/year) was achieved in mushroom cultivation. The integrated farming system generated a net income of Rs. 58,360/year and an employment of 573 mandays/year on a small piece of land (1.25 ha), ensuring a high standard of living for small and marginal farmers. Radha *et al.* (2000) who conducted research on three agricultural and livestock-based farming systems viz., dairy, poultry and sheep rearing clearly revealed that all the farming system generated more than three times additional employment over arable farming. The net return was higher in agriculture+dairy (Rs. 35,293) followed by agriculture+poultry (Rs. 26,830) and agriculture+sheep rearing (Rs. 14,665). Among different farming systems, the agriculture+dairy was proved to be more promising than others. The main reason for high return that stover of maize/jowar for fodder and their grains for feed as well as sugarcane tops to feed cattle and buffalo were available at the farm. Reddy (2001) found that improvements in the household income and employment were statistically significant in all the sample villages with the total livelihoods assets (financial capital), while fuel wood and water availability was not found significant in all the villages except Mallapuram (natural capital); human capital indicators had improved significantly whereas social capital had not changed significantly due to watershed intervention. Sivamurugan (2001) stated that integration of cropping with dairy+biogas+mushroom generated the highest employment of 875 mandays/year.

Singh (2002) observed that in mixed farming on 1 ha land, employment generation was between 559 and 630 mandays with almost uniform distribution throughout the year compared to 182 mandays in arable farming with more labour employment during July, October-November and April-May periods, peak periods of agricultural operations. On small holdings of irrigated land, mixed farming generated 2 to 3 times more employment compared to arable farming. Balusamy *et al.* (2003) studied that net profit increased from Rs. 22,971/ha/annum in rice alone to Rs. 31,788/ha/annum in rice+fish+azolla. In Telangana, the major crops grown were rice, maize, jowar, groundnut, sugarcane and cotton and other components include buffalo, goat, sheep and poultry. Gill (2004) conducted a study on integrated farming systems on farmers fields in Punjab conditions showed that gross profit was found to increase from Rs. 81,200/ha/annum in cropping (Rice-Wheat) alone to Rs. 1,54,000/ha/annum in crop+dairy and Rs. 1,13,200 in fish+piggery system of farming.

Smitha (2005) indicated that majority of small farmers (58%) were placed at high level of livelihood security index while most of the marginal farmers (87%) belonged to medium level category. In labour class, they were more vulnerable as about 43% of the households in this category were low level of livelihood security and 57% belonged to medium level category. Rautaray *et al.* (2005) observed that a net profit from rice-rice-rice cropping sequence from 5000 m^2 area/annum was Rs. 9,468 vis-a-vis Rs.11,226 from integrated rice-fish farming. Further, net returns from integrated farming were expected to increase in the subsequent years from the produce of fruit plants (coconut, papaya and guava) and agro-forestry crop (teak). Dolli (2006) conducted a study on sustainability of Natural Resource Management (NRM) in watershed development project in Bijapur district revealed the impact of NRM was positive and significant on livelihood of participating families in the projects. Significant impact was also seen with respect to human, physical, natural, social, financial capitals and also food security of the participating farmers both in case of SHG and non-SHG members. Dwivedi (2007) concluded that economic returns from agri-horticultural system it was increased by 16.50 to 136.20% than sole cropping under different fruit crops. Availability of fuel wood, fodder, fruit, small timber and food grains from the same piece of the land increased. Standard of living increased in terms of better food and clothing, pucca house, pucca well and cemented irrigation channels, purchase of a motor bike and payment of loans taken from the Regional Rural Bank.

Gautam *et al.* (2007) concluded that the increase in farmers income upshot improvement of his livelihood security. Higher biological productivity translates into higher income for small farmers. Oilseed and pulses if included in cropping system, gave more income from the farm. It is also substantiated in Rajasthan where monocultures of pearl millet gave Rs. 2,480 of net profit per acre, whereas a bio-diverse farm of pearl millet-moth bean-sesame gave Rs. 12,045 a difference of nearly Rs. 10,000 per acre. The raised income resulted into higher purchasing capacity of the small and marginal farmers. Farmers can expand their earnings to purchase the essentials of their living and also the luxurious things. They can avail good education and health to their family members. This increase in pecuniary returns definitely increases their living standards. Use of modern crop production technologies is not only beneficial to the farmers engaged in farming but also to the small and marginal is farmers and landless workers who depend upon farm labour. Basavaraj (2008) showed that the overall livelihood status was high after undertaking income generating activities by most of the KAWAD project beneficiaries which was reflected by improvement in human capital, physical capital, social capital, financial capital and food security. The income earned by the members of the family was not sufficient for two square meals before undertaking the income generating activities. However, after undertaking income generating activities they could earn more and utilize the increased income for other purposes.

Jayanthi *et al.* (2009) concluded that integrated farming systems for different situations enhances productivity, profitability and nutrition security of the farmers and sustains soil productivity through recycling of crop and animal waste from the enterprises involved. The most notable advantage of utilizing low-cost material at the farm level for recycling was that it will certainly reduce the production cost and ultimately improve the farm income considerably. The mean maize grain equivalent yield was about 9,417 kg/acre/year under traditional cropping system whereas under IFS, the maize grain equivalent yield was about 22,754 kg/acre/year. As compared to traditional cropping system, IFS brought increased revenue, which might be due to resource recycling. The net return from inclusion of allied enterprises in IFS is about Rs. 60,141/acre/year and the increase in income over traditional cropping system was about 43.60%. IFS treatments generated more mandays of employment compared with the traditional system involving cropping and dairy. Cropping in traditional system generated 25 mandays/acre/year, while the various cropping systems under IFS generated 49 mandays of employment. A maximum of 183 mandays/acre/year was generated from animal components in IFS, whereas in traditional cropping system it was only about 80 mandays/acre/year. In the traditional cropping system, the residue generated was less as compared with IFS. The system of crop+milch cows+goat+guineafowl+bio-compost and vermicompost could provide better bioresource utilization and recycling. Based on the farmer participatory research, it was concluded that IFS approach was better than traditional system in its contribution to productivity, profitability, economics and employment generation for small and marginal farmers of Tamil Nadu.

Santhakumari *et al.* (2010) studied the impact of integrating interventions like nutrition gardening, livestock rearing, product diversification and allied income generation activities on food and nutritional security of small and marginal coconut homesteads. Regarding food security, the results revealed that about 67.00% of the farmers were in the completely secure category during the pre-project period, which was further enhanced to 96.00% with the project efforts. In case of nutritional security, the percentage of farmers under the completely secure category was only 8 before the project, which showed a remarkable increase to 72% after the project. After the project period, none of the farmer fell under the totally insecure category for both food and nutritional security. Ugwumba *et al.* (2010) in their study highlighted the impact of IFS on farm cash income. Majority of the farmers in the study area practiced partial integration. Results revealed that all types of IFS were on the average profitable. Net farm income realized by farmers who maintained crop-livestock-fish integration was the highest. Implying that farmers who want to achieve full integration and thus earn more income and escape from poverty will target the combination of more enterprises including crops, livestock, fisheries, processing and even biogas. Farm cash income was positively influenced by farmer's age, level of education, years of experience and type of integration. It was, however, negatively influenced by household size, cost of farm inputs and gender of farmer. Farm cash income can be improved by policy towards measures that will reduce cost of inputs and increase farmers knowledge and technical skills. Such measures may include subsidization of inputs and enlightenment campaigns in form of trainings, workshops and seminars. Singh *et al.* (2011) observed that gross and net returns under IFS averaged over the year were 3,29,400/ha/year and 1,35,820/ha/year, respectively, which more than crops alone (1,24,230 and 74,430 / ha/year). This probably was because of inclusion of more enterprising vegetables and flower crops, strengthening of livestock and fishery units, optimum recycling of farm waste and crop residues and better management of all the enterprises of the model maintaining higher level of production. As regard livelihood security, the IFS approach adopted in the model met almost all the domestic family needs of food, fodder, fuel, etc. and saved a sizable amount of money ranging from an average value of Rs. 46.66,000 in first four years to as much as Rs. 77.93,000 in fifth year, respectively, to meet the other liabilities of the family including education, health etc. This saving was exclusive of all the fixed and running cost of IFS model and money required for household food and fodder requirements of a family.

Gopinath *et al.* (2013) studied employment generation through different components of farming systems in Kothapalli and Seethagondhi clusters of Adilabad district in Andhra Pradesh. In Kothapalli, integrated farming system involving crop (2.0 ha)+dairy (2 cows)+small ruminants (100 sheep) generated the highest employment (725 mandays/year) compared to other farming systems. In Seetha gondhi, integrated farming system involving crop (2 ha)+draught animals (2 bullocks)+dairy (1 cow)+small ruminants (40 goats) generated the highest employment (672 mandays/ year) compared to other farming systems. Among the farming systems of marginal farmers, integrated farming system involving crop production (groundnut+pigeon pea) and rearing of small ruminants (90 sheep and 30 goats) was found better with a net return of Rs. 1,57,855/year compared to other farming systems. Similarly, among the three farming systems of small farmers, integrated farming system involving crop production (groundnut+pigeon pea) and livestock rearing (2 desi cows and 100 sheep) gave higher net return Rs. 1,09,650/year compared to other farming systems.

6 RESOURCE CYCLING AND FLOW OF ENERGY IN DIFFERENT FARMING SYSTEMS

The growth rate of agriculture in the recent past has been very slow in spite of the rapid economic growth in India. It is projected that in our country population will touch 1370 million by 2030 and to 1600 million by 2050. To meet the demand, we have to produce 289 and 349 mt of food grains during the respective periods. The current scenario in the country indicates that area under cultivation may further dwindle and more than 20% of current cultivable area will be converted for non-agricultural purposes by 2030.

The operational farm holding in India is declining and over 85 million out of 105 million are below the size of 1 ha. Due to ever increasing population and decline in per capita availability of land in the country, practically there is no scope for horizontal expansion of land for agriculture. Only vertical expansion is possible by integrating farming components requiring lesser space and time and ensuring reasonable returns to farm families. The Integrated Farming Systems (IFS) therefore assumes greater importance for sound management of farm resources to enhance the farm productivity and reduce the resource cycling and flow of energy in different farming systems environmental degradation, improve the quality of life of resource poor farmers and maintain sustainability. In order to sustain a positive growth rate in agriculture, a holistic approach is the need of the hour. Farming system is a mix of farm enterprises in which farm families allocate resources for efficient utilization of the existing enterprises for enhancing productivity and profitability of the farm. These farm enterprises are crop, livestock, aquaculture, agro-forestry, agri-horticulture and sericulture.

In such diversified farming, though crop and other enterprises coexist, the thrust is mainly to minimize the risk, while in IFS a judicious mix of one or more enterprises along with cropping, there exist a complementary effect through effective recycling of wastes and crop residues which encompasses additional source of income to farmers. IFS activity is focused around a few selected interdependent, interrelated and interlinking production system based on crops, animals and related subsidiary professions.

Integrated farming system approach is not only a reliable way of obtaining fairly high productivity with considerable scope for resource recycling, but also concept of ecological soundness leading to sustainable agriculture. With increasing energy crisis due to shrinking of non-renewable fossil-fuel based sources, the fertilizer nutrient cost have increased steeply and with gradual withdrawal of fertilizer subsidy. It is expected to have further hike in the cost of fertilizers. This will leave the farmers with no option but to fully explore the potential alternate sources of plant nutrients at least for the partial substitution of the fertilizer nutrients for individual crops and in the cropping systems.

Definition of Farming System

'Farming' is a process of harnessing solar energy in the form of economic plant and animal products. *'System'* implies a set of interrelated practices and processes organized into functional entity, i.e. an arrangement of components or parts that interact according to some process and transforms inputs into outputs.

PRESENT STATUS OF FARMING SYSTEM RESEARCH

The preliminary investigations clearly elucidated that integration of agricultural enterprises viz., crop, livestock, fishery, forestry etc. have great potential towards improvement in the agricultural economy. These enterprises not only supplement the income of the farmer by increasing the per unit productivity but also ensure the rational use of the resources and further create employment avenues. The following of suitable crop choice criteria having deep and shallow root system, inclusion of legume crop as catch, cover and fodder crops and adoption of bio-intensive complimentary cropping system along with other enterprise will certainly prove as a self sustained production system with least cost of production. The farming system is governed by various forces viz., physical environment, socio economic conditions, political forces under various institutional and operational constraints and above all government favorable policies, which may keep the food security intact and livelihood fully protected.

In traditional Chinese system, the animal houses were constructed over a pond so that animal waste fell directly into the water fueling the pond ecosystem, which the fish could then feast on for food. Not only were the fish harvested but the pond water, now with extra nutrients was used for irrigation in crops. The maximum return (Rs 79,064/ha) was earned from fisheries + piggery + poultry as compared to Rs 38,500/ha from the rice-wheat system and registered 48.6% gain. This also generated additional employment of about 500 man days/ha/annum.

For poor people, it starts small with ducks and chickens; then a few goats are kept for milk or fattening and to slaughter for the day of sacrifice; next a milch cow; then a bullock for ploughing in cooperation with another one buffalo family; then two bullocks. These can be used to plough the fields of others–a very lucrative business in the planting season. In India, one would add a milch buffalo at the apex of desirable animals on the farm. In the Vietnamese concept, the pigs will be the second step in the ladder. The concept means to start with small livestock and women and then the household will step by step get out of poverty. The poorest households kept only poultry and these households were those most dependent on common property resources for their living (e.g. use and sale of firewood from the forest). A similar stratification has been reported in several studies from Asia. Survey on farming systems in the country as a whole revealed that milch animals; cows and buffaloes irrespective of breed and productivity is the first choice of the farmers as an integral part of their farming system. However, from the economic point of view, vegetables and fruits (mango and banana in many parts of the country) followed by bee keeping, sericulture, mushroom and fish cultivation was the most enterprising components of any of the farming systems prevalent in the country. The average yield gaps between 27 pre-dominant and 37 diversified farming systems were examined across the agro-climatic zones through detailed survey on characterization of on-farm farming systems. Diversification of farming system by integration of enterprises in varied farming situations of the country enabled to enhance total production in terms of rice equivalent yield ranging from 9.2% in eastern Himalayan region to as high as 366% in Western-plain and Ghat region when compared to prevailing farming systems of the region. A number of success stories on IFS models including Sukhomajari Watershed of Chandigarh, Fakot Watershed in hilly areas of Uttarakhand. Jayanthi models for almost all the situations of Tamil Nadu, WTCER model for coastal and irrigated alluvial lands of Orissa, Darshan Singh Model for irrigated conditions of Punjab, PDCSR model, for western Uttar Pradesh and many more in different parts of the country suggest that farmers' income can be increased manifold by way of diversification of enterprises in a farming system mode for sustainability and economic viability of small and marginal category of farmers.

In view of serious limitations of horizontal expansion of land for agriculture, only alternative left is vertical expansion through various farm enterprises requiring less space and time but give high productivity and ensuring periodic income especially for the small and marginal farmers. The highlights about the research investigations carried out in India towards farming system outcome are discussed to conceptualize its significance towards farming community livelihood. In a study conducted at ICAR Research Complex, Goa, it was revealed that rice-brinjal crop rotation is the best in terms of productivity and profitability owing to higher yield of brinjal. The system yielded a total productivity of 11.22 t/ha rice grain equivalent yield with a net return of Rs.46, 440/ha. Further, with the integration of mushroom and poultry production (based on the resources availability within the system) the system productivity was increased to 21, 487 kg/ha especially with rice-brinjal rotation leading to an additional returns of Rs 30,865/ha with integration. In addition, the system approach was found to sustainable as reflected from the changes in soil organic carbon and indicated by sustainability yield index.

In Haryana, Singh *et al.* 1993 conducted studies of various farming systems on 1 ha of irrigated and 1.5 ha of unirrigated land and found that under irrigated conditions of mixed farming with crossbred cows yielded the highest net profit (Rs. 20,581/-) followed by mixed farming with buffaloes (Rs. 6,218/-) and lowest in arable farming (Rs. 4,615/-). In another study conducted with 240 farmers of Rohtak (wheat-sugarcane), Hisar (wheat-cotton) and Bhiwani (*gram*-bajra) districts in Haryana which represented zones of different crop rotations revealed that maximum returns (Rs/ha) of 12,593, 6,746 and 2,317 were obtained from 1 ha with buffaloes in Rohtak, Hisar and Bhiwani, respectively. The highest net returns from Rohtak was attributed to the existence of a better soil fertility type and of irrigation facilities coupled with better control measures compared to other zones. In terms of total mandays, Rohtak had the highest employment potential followed by Hisar and Bhiwani. The employment potential under mixed farming conditions was predominantly from livestock rather than crop production.

Another study involving cropping, poultry, pigeon, goat and fishery was conducted under wetland conditions of Tamil Nadu by three years results revealed that integration of crop with fish (400 reared in 3 ponds of 0.04 ha each), poultry (20 babkok layer bird), pigeon (40 pairs), and goat (Tellichery breed of 20 female and 1 male in 0.03 ha deep litter system) resulted in higher productivity, higher economic return of Rs. 1,31,118 (mean of 3 year). Integration of enterprises created employment opportunities where in comparison to 369 man days/year generated in cropping alone system, cropping with fish and goat created additional 207 man days/annum. The resources were recycled in such a way that fish were fed with poultry, pigeon and goat dropping. Similarly, extra poultry, pigeon and goat manure and composted crop residue of banana and sugarcane were applied to the crops. The four conventional cropping system tried were rice-rice-black gram, maize-rice-black gram, maize-rice sunhemp and rice-rice-sunhemp.

Balusamy *et al.* (2003) explained that rice + Azolla-cum-fish culture is one of the economical option in such type of area. Monoculture system relies mainly on external inputs while in integrated system, recycling of nutrients takes place that help in reducing the cost of production for economic yield. The fish in rice field utilized the untapped aquatic productivity of rice ecosystem as the rice bottom is highly fertilized on account of the production of zoo, and phytoplankton and these resources are fully utilized by the fish. The data clearly advocated the beneficial effect of Azolla on rice+fish. The gross income obtained in rice + Azolla + fish was 25.7 % more over the rice crop and 6.9 % more over the rice + fish. The net income followed the same trend. Thus rice + Azolla + fish on an average gave Rs 8,817/ha more over the rice monoculture and Rs.3,219/ha over the rice + fish. This model was proposed for extensive scale adoption in Tamil Nadu.

The intensive farming systems of developed countries, such as United Kingdom seek to maximize yield through what is usually described by agricultural economists as Best Management Practice (BMP), which involves the most efficient use of all inputs, including fertilizers, herbicides, seed varieties, and precision agricultural techniques (Goulding *et al.* 2008). Fertilizers have been central to this approach, which has resulted in a tremendous increase in productivity over the last 40 years. For example, the efficient use of improved fertilizers, combined with new varieties of wheat and the successful use of crop protection chemicals, has increased grain yields from 3 tons per hectare to approximately 10 to 11 tons per hectare today (Goulding *et al*, 2008).

DETERMINANTS OF FARMING SYSTEMS

Farming enterprises include crops/cropping systems (field, horticultural, plantation), cattle rearing, poultry, fish, piggery, sericulture, mushroom, bee keeping, agroforestry, biogas, vermicompost, etc. A combination of one or more enterprises with cropping when carefully chosen, planned and executed, give greater returns than single enterprise(s) specially for small and marginal farmers. Farm as a unit to be considered and planned for effective integration of the enterprises to be combined with crop production activities. In the IFS, it is always emphasized to combine cropping with other enterprises/activities.

The key categories of determinants influencing farming system are as follows:

- **Natural resources and climate:** The interaction of natural resources, climate and population determines the physical basis for farming systems. The increased variability of climate, and thus agricultural productivity, substantially increases the risk faced by farmers, with the concomitant reduction in investment and input use.
- **Science and technology:** Investment in agricultural science and technology has expanded rapidly during the last four decades. During this period, major technical and institutional reforms occurred, which shaped the pattern of technology development and dissemination.
- **Trade liberalization and market development:** Markets have a critical role to play in agricultural development as they form the linkages between farm, rural and urban economics upon which the development processes depend. As a result of the reduction of impediments to international trade and investment, the process of trade liberalization is already generating changes in the structure of production at all levels-including small holder-farming systems in many developing countries.
- **Policies, institutions and public goods:** The development of dynamic farming systems requires a conducive policy environment. Moreover, the establishment of the farm-rural-urban linkages requires effective demand. Policy makers have increasingly shifted their attention to the potential to increase the efficiency of service delivery through the restructuring of institutions.
- **Information and human capital:** The evolution of farming systems based upon increasing specialization (e.g. large-scale broiler units) or integrated intensification (e.g. rice-fish-ducks) has required extra knowledge on the part of farm operators. The need for better information and enhanced human capital has also increased, as production systems have become more integrated with regional, national and international market systems.
- **Indigenous technological knowledge:** Indigenous technical knowledge is the knowledge that people in a given community has developed over times and continues to develop. It is based on experience, often tested over long periods of use, adapted to local culture and environment, dynamic and changing, and lays emphasis on minimizing risks rather than maximizing profits.

Criteria for enterprise selection

The basic points that are to be considered while choosing appropriate enterprise in Integrated farming system are,

- Soil and climatic features of an area/locality.

- Social status of the family and social customs prevailing in the locality.
- Economical condition of the farmer (return/income from the existing farming system).
- Resource availability at farm and present level of utilization of resources.
- Economics of proposed integrated Farming System and credit facilities.
- Farmer's managerial skill.
- Household demand.
- Institutional infrastructure and technological knowhow.
- Market facilities.

Integrated Farming System deals with utilization of wastes and residues. It may be possible to reach the same level of yield with proportionately less input in the integrated farming and yield would be interestingly more sustainable because the waste of one enterprise becomes the input of another leaving almost nothing to pollute the environment or to degrade the resource base. To put this concept to practice efficiently it is necessary to study linkages and complementarities of different enterprises in various farming systems. (Fig. 6.1).

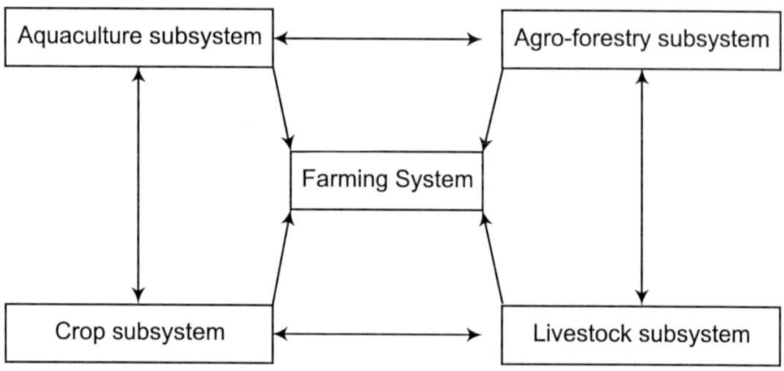

Fig. 6.1. Linkages of different enterprises in farming system

The key categories of determinants influencing farming system are as follows:

(i) Natural resources and climate: The interaction of natural resources, climate and population determines the physical basis for farming systems. The increased variability of climate, and thus agricultural productivity, substantially increases the risk faced by farmers, with the concomitant reduction in investment and input use. Certain soil and water regimes are suitable only for given type of crops. Similarly, some of the physical and geographical features e.g. drainage characteristics, elevations and slopes as well as climatic factors e.g. total rainfall and its distribution, minimum and maximum temperature, humidity and intensity of sunlight etc. are other factors which have to be taken in to considerations while making decision with respect to selection of enterprise for a farming systems.

(ii) Science and technology: Investment in agricultural science and technology has expanded rapidly during the last four decades. During this period, major technical and institutional reforms occurred, which shaped the pattern of technology development and dissemination. The research-driven growth in developing countries has been Green Revolution, where it achieved considerable achievement in the field of food grain production and for this the policy and other aspects supported the farming system for such achievement . Research has been focused principally upon intensifying crop and livestock production. There has been far less research on integrated technologies for diversifying the livelihoods of small farmers in developing countries and increasing the sustainability of land use.

Despite these weaknesses, the natural and global research agenda is gradually moving from a focus on individual crop performance to a growing acceptance of the importance of increased system productivity. There has been a lot of emphasis in recent agriculture of targeting technologies towards women farmers and poorer households.

(iii) **Trade liberalization and market development:** Markets have a critical role to play in agricultural development as they form the linkages between farm, rural and urban economics upon which the development processes depend. As a result of the reduction of impediments to international trade and investment, the process of trade liberalization is already generating changes in the structure of production at all levels including small-scale holding in many developing countries. Not only the market development is accelerating, but patterns of production and natural resources usage are also changing profoundly in response to market forces.

The availability of new post-harvest transport technologies will also change demand patterns due to delivery of new products or established products in new forms to markets, where they have been previously unattainable.

(iv) **Policies, institutions and public goods:** The development of dynamic farming systems requires a conducive policy environment. Moreover, the establishment of the farm-rural-urban linkages requires effective demand. Policy makers have increasingly shifted their attention to the potential to increase the efficiency of service delivery through the restructuring of institutions. The production incentives have dramatic effect on farming systems. Policies on land ownership, water management and taxation reform etc. have a great bearing on types of farming system in a region or area.

(v) **Information and human capital:** The evolution of farming systems based upon increasing specialization (e.g. large-scale broiler units) or integrated intensification (e.g. rice-fish-ducks) has required extra knowledge on the part of farm operators. The need for better information and enhanced human capital has also increased, as production systems have become more integrated with regional, national and international market systems. Lack of education, information and training is frequently a key limiting factor to smallholder development. Many observers anticipated an information revolution i.e. bridge the gap of knowledge between scientists and farmers will be very key factor for agricultural growth of these small farmers. Whereas in the past, many development efforts failed because planners had a poor understanding regarding the role women to be played in farming and household food security. Greater efforts are being made to take account of their actual situations. It is increasingly recognized and acknowledged by development workers that the empowerment of women is the key to raising levels of child and family nutrition, improving the production and distribution of food and agricultural products, and enhancing the living conditions of rural populations. It has been concluded that, if women in Africa received the same amount of education as men, farm yield would rise by between 7 and 22% (FAO, 1990).

Similarly, better access to credit, land and extension services would enable women to make an even greater contribution to eliminating rural hunger and poverty. As gender bias is progressively eliminated during coming years, productivity within many farming systems will be transformed.

(vi) Indigenous technological knowledge (ITK): Indigenous technical knowledge is the knowledge that people in a given community has developed over times, and continues to develop. It is based on experience, often tested over long period of use, adapted to local culture and environment, dynamic and changing, and lays emphasis on minimizing risks rather than maximizing profits. The ITK covers a wide spectrum – soil water and nutrient management; pasture and fodder management; crop cultivation; plant protection; farm equipment, farm power, post-harvest preservation and management; agro-forestry; biodiversity conservation and also exploitation; animal rearing and health care; animal products preservation and management; fisheries and fish preservation; and ethnic foods and homestead management. Thus, the ITK of a farmer has a great influence in managing the farm and farming system.

Agriculture has been the basic source of subsistence for mankind over thousands of years and it provides livelihood to half of the world population. Indian agriculture is known for its multi-functionalities of providing employment, livelihood, food, nutrient and ecological securities. Integrated farming system comprising of crop and other allied enterprises has been sustainable over centuries. The system could be able to meet the food needs of ever increasing population till recently. Gangwar (1993) considered farming systems research as a highly location specific research which represents an appropriate combination of farm enterprises and the means available to the farmers to raise their profitability. Further, the author described it as an interdisciplinary approach, which seeks to develop agricultural technology within the context of the whole farm system as distinct from reliance on interventions based on individual commodities or factors.

Farming system represents an appropriate combination of farm enterprises viz. cropping systems, horticulture, livestock, fishery, forestry, poultry and the means available with the farmers to raise them for profitability. It interacts adequately with environment without dislocating the ecological and socio-economic balance on one hand and attempt to meet the national goal on the other. The farming system in its real sense will help in different ways to lift the economy of agriculture and standard of living of the farmers of the country as a whole.

Integrated Farming System (IFS) is a complex related matrix of soil, plants, animals, implements, power, labour, capital and other on and off farm inputs influenced to varying degrees by political, economical, institutional and social factors that operate at many levels. The farming system therefore, refers to the farm as an entity to inter-dependent farming enterprises carried out on the farm as shown in Fig. 6.2.

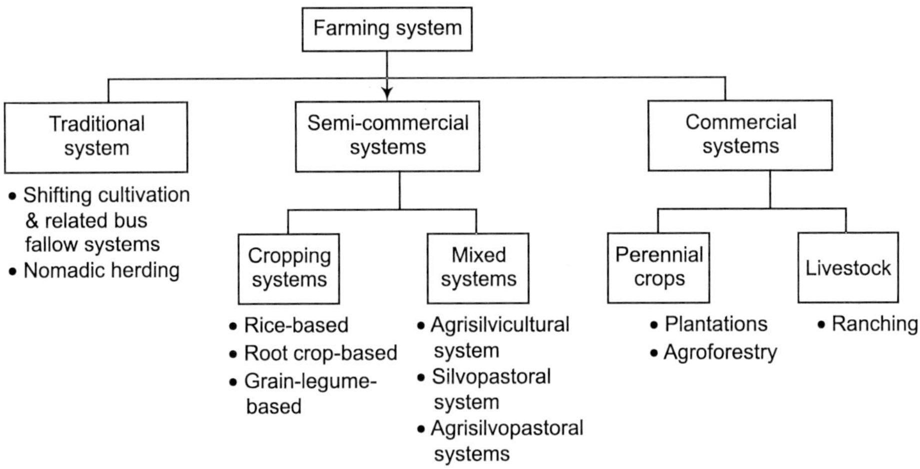

Fig. 6.2. Showing interdependent farming enterprises

Recycling of organic wastes such as crop residues, dung and urine from domesticated animals and wastage from slaughter house, human excreta and sewage, bio mass of weeds, organic wastes from fruit and vegetables production and household wastes, sugarcane trash, oilcakes, press mud and fly ash from thermal power plants. Material not suitable for direct application can better apply by composting and vermicompost.

The ultimate goal of sustainable agriculture is to develop a farming system that is productive and profitable, conserve the natural resources base, protect the environment and enhance health and safety, and to do so over the long term.

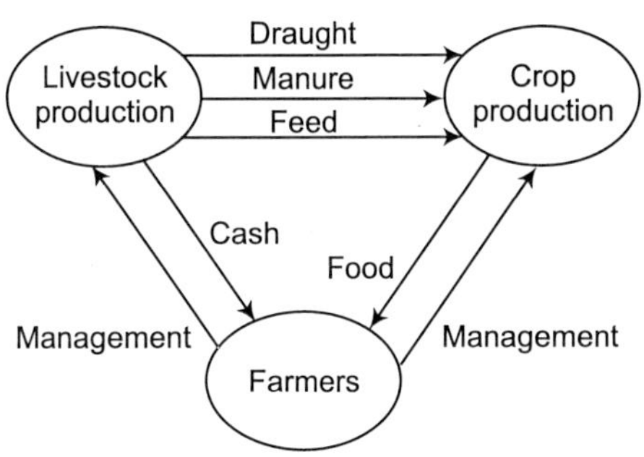

Fig. 6.3. Components of integrated farming system

FARMING SYSTEMS FOR ENSURING SUSTAINABILITY

Low Input Sustainable Agriculture (LISA)

In this system minimal use of external production inputs is made. The production costs are obviously lower. The overall risk of the farmer is considerably reduced. Besides, the above advantages, pollution of surface and ground water is avoided and healthy food with very little or no pesticide residues are ensured. These systems held promise for both short-and long-term profitability. However, the system suffers from one serious drawback – continuation of low external input agriculture will perpetuate to the vicious circle of low inputs and hence low yields with ever increasing population pressure can ill offered. High input system, on the other hand, will fail sooner or later, as they are not economically and environmentally sustainable. The optimal input farming system has the premise of low input per unit of output and lays emphasis on low or diminishing returns.

Four factors are most important for better growth of plant:

(a) Soil – 45%

(b) Organic matter – 5%

(c) Air – 25%

(d) Water -25%

Organic farming and waste recycling

Organic farming is the backbone of sustainable agriculture. Organic farming mainly depends on organic recycling: Industrial agriculture chemicals like fertilizer, pesticide, herbicide etc are not used or used to minimum extent. Organic farming may result in compatible performance as conventional agriculture and crops growth with high organic manure application could tolerate the pest and disease attack better. It is sound and sustainable way of growing more food.

Organic recyclable wastes include – crop residues, waste, farm industrial waste, multiple and sewage wastes. They are valuable sources of plant nutrient and humans in tropical and subtropical soils found in India, there is general deficiency of organic carbon and plant nutrients due to rapid loss of this component by biodegradation. To overcome these losses, extensive utilization of organic residues in agriculture is essential. In addition they also protect the soil from erosion.

In India, there is a general potential for utilization of crop residues/ straw of major crops. About 141.2 MT. straws available that contribute about 0.7, 0.84, and 2.1 MT .N.P.K respectively. If considering 50% crop residue utilized as animal feed, the rest can recycle. Crop residue has wide C: N ratio due to this, immobilization nutrients takes place. Care should be taken that before use, composting with efficient microbial inoculants may be done.

Important organic manures are

(i) **Sugarcane trash compost:** Fresh sugarcane trash contains 0.36% N with a wide C:N ratio of 122:1.The composted trash contains higher content of N (1.09) with reduced C:N ratio (20:1)

(ii) **Biogas slurry:** Organic manures from animal wastes are very important nutrient sources in building up soil fertility. In India, estimated production of dung and urine is about 1002 and 658 million tones respectively. They contribute about 5.7 million tones of N P K with proper utilization. Biogas plants have advantages of both fuel (gas) and fertilizer (slurry) Dry slurry contains abort 1.8% N, 1.10 % P_2O_5 and 1.50% K_2O.

(iii) **Vermicompost:** The average nutrient content of vermicompost is much higher than that of F.Y.M. Vermicompost contains 1.60 % N, 0.54% P_2O_5 and 0.80% K_2O. The C:N ratio of vermicompost is much lower (1:6) than the FYM (1:30).

(iv) **Industrial wastes:** Among the industrial by-products, spent wash from distilleries and molasses and press mud from sugar factories have good manurial value. It is important to use only well decomposed press mud at 10 tons / ha. Addition of press mud improves the soil fertility. It is a good reclamation agent in saline and sodic soils. Coir waste is the by product of the coil industry. Well decomposed coir waste of 12.5 t/ha with recommended fertilizer to groundnut and maize increase yield.

(v) **Municipal and sewage wastes:** This is one of the important components of organic wastes. In India, the total municipal refuse is about 12 million tons / annum containing 0.5 %N.0.3%P.& 0.3%K. Sewage (liquid portion) sludge (solid portion) is available to an extent of 4 million tons /annum containing 3.0 % N, 2.0 % P & 0.3% K . Such organic waste must be used carefully as it may contains metals that are hazardous to plants, animals and human beings.

(vi) **Crop residues:** Residues left after the harvest of the economic portion are called crop residues /straw. Cereal straw and residues contains about 0.5% N, 0.6% P and 1.5 % K. The crop residues can be recycled by way of incorporation compost making or mulch material.

(vii) **Rice husk:** It is a major by-product of the rice milling industry. It is a poor source of manure and nutrient 0.3 % N, 0.2 % P and 0.3 K. Rice husk should be incorporated into the wet soil and can be used in saline and alkaline soils to improve the physical condition. It can also be used as a bedding material for animals.

IFS IN VARIOUS ECOSYSTEMS

Low irrigated and uplands

Crop- Livestock- Fish Farming (Wetland)

Asia has been the cradle of integrated crop-livestock-fish farming systems, which have evolved since the inception of human civilization particularly when human settlements started moving inland leaving the river banks. Though there are several successful practices of integrated fish farming in Asian countries including India, the system of farming using synergizing scientific integration of agriculture, aquaculture and livestock farming are not yet widespread in the region. Further, large-scale integration of carp culture with irrigation and sewage utilization are to be viewed seriously both for economic and ecological reasons (Sinha, 1979).

The wet land of paddy field is congenial to many fish both for spawning and for pasture. Those breeding in paddy fields have adhesive eggs and are normally laid on green plants to facilitate more oxygen for the developing embryo whereas shallow water spawners and the nest builders get favourable conditions of breeding in paddy fields. The flooded field has considerable quantities of putrifying plants giving rise to enormous amount of plankton and in fact serves as a richly laid table for fry and fingerlings. The fish while controlling the excessive growth of plankton, which compete with the paddy, also control zootecton, insects, molluscs, the submerged and floating weeds harbouring the above and adversely affecting paddy. Fish fertilize through its fecal matter and also overturns the submerged soil normally under reduced stage thus making available more nutrient and oxygen to the root of paddy, acting like a biological plough (Sinha, 1985).

Evolved on the principles of productive recycling of farm wastes, fish- livestock farming systems are recognized as highly assured technologies for fish cultivation. In these technologies, predetermined quantum of livestock waste obtained by rearing the livestock in the pond area is applied in pond to raise the fish crop without any other additional supply of nutrients. Similarly, with the integrated poultry - fish farming system, the fish crop is integrated using only poultry droppings or dip litter by rearing the poultry either directly over the pond or on the pond embankment.

Dyke-Pond Systems

In many parts of Asia, the productive use of land and water resources has been integrated by transforming wetlands into ponds separated by cultivable ridges. Overall integrated farming systems that include semi-intensive aquaculture are less risky for the resource-poor farmer than intensive fish farms, because of their efficiency derived from synergism, their diversity of produce, and their environmental soundness. In many traditional systems aquaculture goes beyond fish production and cash income as pond water and pond biota perform many ecological, social, and cultural services on an integrated farm. Thus aquaculture and water management act as an engine driving the sustainability of the entire farming system (Lightfoot 1990).

An example of integrated agriculture–aquaculture system is the dyke-pond system, which has existed for centuries in South China. The history of the dyke-pond system may be traced back to the middle of the ninth century in the Pearl River delta region of China.

The dyke - pond system serves two major functions: (a) achievement of a general ecosystem balance through the harmonization of well-coordinated activities and functions embedded in the ecosystem, and (b) transformation and regeneration of organic substances based on a multi-layer trophic ecosystem structure, which helps contribute successfully to sustainable economic development. The system contains two interrelated systems of dyke and pond; the dyke is the land ecosystem for the growth of crop whereas the pond is the water ecosystem, consisting of fish and aquatic plants. The dyke-pond system can be of various kinds depending on the crops planted on the dyke such as mulberry dyke-fishpond, sugarcane dyke-fishpond, banana dyke-fishpond, and so on. The input and output of material and energy in the dyke-pond system are basically balanced. (Ruddle & Chung, 1988, Korn, 1996).

Experience showed that the economic return from the integrated mulberry dyke-fish pond system were greater than those obtained from cultivating fruit trees or rice on the dyke. It was found that such integrated management is beneficial to mulberry and fish, as well as for development of sericulture.

Moreover, pond mud enriched with silkworm excrement and other wastes can be used to fertilize the pond and feed the fish. Mulberry leaves are fed to the silkworms, whose excreta are used as fish food, and the fertile pond mud, consisting of fish excreta, organic matter, and chemical elements, is brought up from the bottom and used as manure for the mulberry trees. In this system, the mulberry tree represents the first trophic level. Photosynthesis takes place in its leaves, which are fed to the silkworms for yarn production (the primary consumers) whose excreta and chrysalises are in turn fed to fish (the secondary consumers). The aquatic organisms in the pond are the reduction agents that decompose fish excreta and algae, break down the organic matter in the pond, and produce nitrogen, phosphorus, and potassium. The pond supplies the dyke with the fertile mud, which after decomposition is used as a primary source of fertilizer for the mulberry trees. Moreover, the leaf fodder of mulberry is reported to be rich in crude protein, ether extract, calcium, ascorbic acid, potassium, iron and thus can be profitably utilized as a supplement to poor quality roughages. Mulberry leaves can also be used in poultry ration. Incorporation of dried mulberry leaves in layers mash to the extent of 6 % showed an increase in egg production with desirable yolk colour without any adverse effect on body weight and egg quality (Narayana & Setty, 1977). Mulberry leaves, owing to their high carotene content, can form a valuable source of vitamin A for the health of poultry birds and increased egg production.

In development of ecologically stable and environmentally sound integrated farming system models to replace or to improve traditional farming systems would maximize the productivity of the farming system. The goals of integrated farming system are to sustain agricultural production, maintain farm income, safeguard the environment and respond to consumer concerns about food quality issues. Location specific systems are adopted by farmers based on the suitability and situations of the region.

In India such integrated systems have not been systematically evolved so far, however in many places, the pond mud is used for terrestrial crop and pond embankment for papaya, coconut, banana plantation and at times for growing vegetables. Mulberry trees are very economical and can be incorporated in IFS to improve the profit of farmers (Fig 6.4).

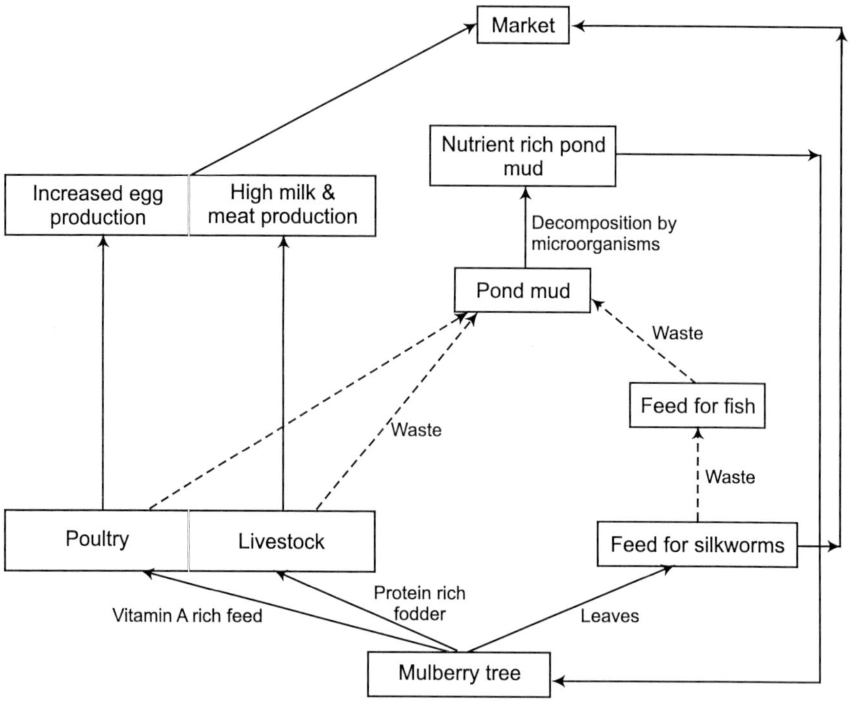

Fig. 6.4. Mulberry Dike-Pond System

For regular flow of income and effective recycling of wastes, besides cropping, dairying, mushroom production and tree cultivation would be more ideal for garden land farmers of North Western zone of Tamil Nadu (Rangasamy *et al*, (1994). An integrated farming system model for dryland conditions of Coimbatore for one hectare was suggested by Sivasankaran *et al.* (1995) which included sorghum + cowpea for grain/ seed or fodder, *Leucaena leucocephala* + *Cenchrus ciliaris* and *Acacia senegal* + *Cenchrus ciliaris* for fodder and an animal component of 5 female + 1 male goat.

Tiwari *et al.* (1999) suggested a farming system model having 2 crossbred cows + 15 goats + 10 poultry birds + 10 ducks on 0.25 hectare area including area for the farmers family living + fish (0.15 hectare pond) along with crop cultivation (1.1 hectare) was the best system. This system gave a net income of 44, 911 year-1 with a cost return of 1:2 and employment generation of 380 man-days. Whereas, arable farming alone recorded a net return of 24,093. Ganeche *et al* (2000) suggested the integration of Tellicherry goats (5+1) and dairy unit (1 cow) with fruit tree based system for higher productivity.

Theodore *et al.* (2001 and 2002) in a more-or-less similar study reported that dairy made the difference between diversification (rice + dairy + banana) and non-diversification (rice + banana) in the study area, which influenced significantly the economy and ecology of the wetland farmers. Similar report was made earlier by Singh *et al.* (1996).

Cropping with integration of fish and goat recorded higher productivity of rice grain equivalent yield of 39,610 kg ha^{-1} and higher net return of Rs. 1,26,564 per ha per year (Jayanthi *et al.* 2002).

Mixed crop-livestock systems are the dominant form of agricultural production in India. Integrating crops and livestock on the same farms help small farmers to diversify the sources of income and employment. Livestock act as a storehouse of capital and an insurance against crop production risks, a coping mechanism against livelihood shocks as well as a vital source of dietary protein (Parthasarathy Rao & Hall, 2003).

Southern arid and semi-arid region comprises about 20% of geographical area and more prone to disasters like flood and drought which influences scarcity of feed, fodder, water and outbreaks of diseases. Diversification through goat component in their farming system is the better option for self-sustenance (Gokhale *et al.* 2002).

Mkhabela (2005) reported that, vegetable production constitutes an important sub-sector of the agricultural economy of KwaZulu-Natal. Most vegetables are cultivated in mixed-cropping types of farming systems. Small-scale commercial vegetable cultivation is prominent because of its favourable climate and the long-established irrigation scheme.

Integration of crop and livestock enterprises offers highest level of synergy to the farmers of three different agro-climatic zones of Australia (Fleming *et al.* 2007). Most rural households in the semi-arid regions of Sub-Saharan Africa practise mixed crop-livestock farming. With population growth and the subsequent scarcity of land available for extensive farming, the only option available for these households is to intensify production (Miyuki Iiyama *et al.* 2007).

The water resources of Bangladesh are vast, covering 4.3 million hectares of inland water and 480 km of coast line with a sub-tropical climate and suitable soil and water conditions and offer very high potential for aquaculture development. Pond aquaculture is an integral component of the farming system in Bangladesh (Anon. 2008).

The majority of rural households in Gansu Province of China are engaged in semi subsistence farming activities and agricultural development based on livestock intensification programs has the potential to influence a large portion of the population. Livestock production in the Huan County of Qingyang of China has increased by 103% from 2000-2007 (Brown *et al.* 2009).

The largest land area of the Greater Horn of Africa is under rangelands (65%). They support substantial population in their livelihoods and contribute considerable amounts to the national budgets through livestock production (Notenbaert *et al.* 2009).

Small ruminants are playing an important role in the economy of small holding farmers in the mixed farming systems of Southern Ethiopia. However, their relative contribution compared with crop production varied between sites, crops being the major contributor in fertile areas with reasonably higher landholding (Getahun Legesse *et al.* 2010).

Kenya depends on agriculture diversification due to diminishing farm size. Small hold-led agriculture is an important aspect in Kenya and therefore diversification needs to be promoted particularly for the rural population. The inclusion of high valued agricultural products augment diversification portfolio and therefore increased income (Wanyama *et al.* 2010).

Lindsay Bell and Andrew Moore (2012) found that in Australia, mixed crop-livestock farming has also been a major and long standing feature of agricultural land use. In 2010, about 0.35-0.40 M km^2 of land, a third of the agricultural zone, was occupied by farms operating both cropping and livestock enterprises.

From the available literature, it can be concluded that crop-animal systems have positive and economic benefits which promote sustainable agriculture and environmental protection. Adoption of animal components along with crops varies according to the climate, soil, water, economic, social, diet and other factors of the locality.

MODEL FOR RESOURCE ENERGY FLOW IN IFS

Integrated farming system at Siruguppa, Karnataka, under canal irrigation of Tungabhadra project during 2003-04 to 2005-06, involved cropping (rice, maize, sunflower, vegetables), fishery, poultry and goat as the integrated system, and cropping (rice-rice) alone as the control (Channabasavanna et al 2009). In one hectare area of integrated farming system, an area of 0.73 ha was allotted for crop component (rice-rice, maize - sunflower sequence, vegetables), 0.06 ha for fish pond and 0.21 for goats (including fodder area). Poly culture fingerlings @ 10,000/ha (rohu 20%, catla 30% and mrigal 40%) were released into the pond (600 m^2). Thirty poultry birds (giriraj) were maintained in the poultry shed constructed on the fish pond. Goats (10 females + 2 males) were maintained in a shed constructed separately. This was compared with the conventional rice-rice system. To sustain the productivity the residues obtained in the system was recycled. Poultry droppings was allowed to drop into the pond directly which served as the source of food for fish. Fishes were harvested after completing one year using drag net. The nutrient rich pond water was used as source of irrigation for crops.

The integration of crop with fish, poultry and goat resulted in higher productivity than adoption of conventional rice-rice alone. Integrated farming system showed 26.3 % higher productivity over conventional rice-rice system.

Dairy forms an important component in farming system especially under garden land conditions. Crop + livestock farming system is a broad system that is suitable for different ecosystems. Integrating fish with crop + dairy farming system is ideal for wetland system. Crop + poultry is also ideal in wetland system. Therefore, adding poultry component would help recycling resources better and generating additional income. About 0.04 ha has allotted to the fish pond. Fish pond has to be excavated to a depth of 1.5 m. Fifteen days old fingerlings of different fish species have been stocked in varied proportions (silver carp/rohu/mrigal/grass carp at 4:2:3:1 ratio with a total of 7500 fingerlings/ha). No artificial feeding has been done. Harvest of fish commences from the tenth month. Washings of the cowshed have been diverted to the the fish pond. A poultry shed with a plinth area of 2.2 m^2 has been erected at one corner of the fish pond. Bottom of the poultry shed has been provided with wire mesh (3m × 3m) to facilitate free fall of poultry droppings into the fish pond. Biogas can be used for cooking, lighting and as a substitute for diesel. Poultry feed has been managed with recycling of farm wastes and the produce obtained from the crop components. Alfalfa + grass mixture cultivated around the fish pond was also used for feed of poultry as well as fish. The birds will start laying eggs around 22nd week upto 72 weeks, when they were culled out. Each bird was fed with feed of 100-200 g/day. The feed ingredients were maize, rice bran, groundnut cake, fish meal, shell grit, dicalcium phosphate and vitamins.

Resource flow model of integrated farming system - irrigated upland (0.4 ha) for Western Zone of Tamil Nadu was prepared by Jayanthi *et al.* 2007 and presented in Fig 6.5.

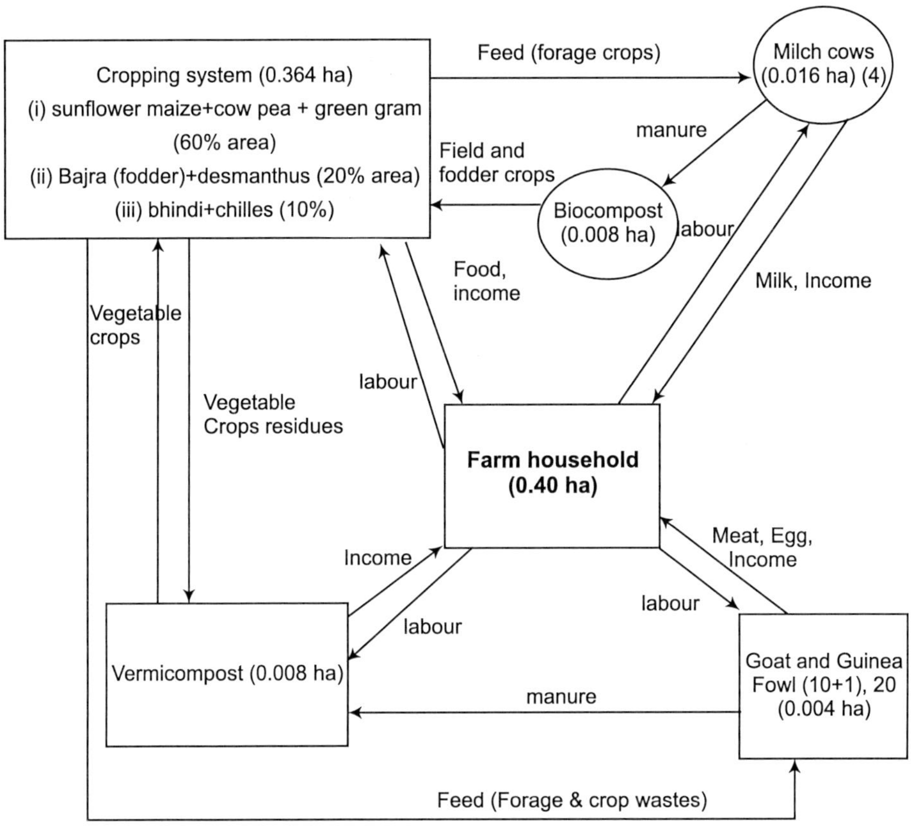

Fig. 6.5. Resource flow model of Integrated Farming System of irrigated upland (0.4 ha) for western zone of Tamil Nadu

Resource and energy flow in crops + goat + agroforestry IFS model under drylands is presented below:

IFS model has been proposed for 1 ha area. Crop and goat association is common in drylands. Goats are essential browsers and eat plants which any other animal won't touch. They eat more of tree fodder and hence 40- 50% of green fodder should contain tree leaf fodder, the rest with other grass species. Therefore, integrating agroforestry with crops + goats farming system is a better proposition. About 0.3 ha has been alloted to agroforestry which can support 8+1 units of goats nicely. Composting or vermicomposting activity has also been proposed for effective recycling of farm waste and weeds.

Conclusion

IFS enables the agricultural production system sustainable, profitable and productive. About 95 % of nutritional requirement of the system is self sustained through resource recycling. As the number of enterprises are increased, the profit margin increases but simultaneously coupled with increase in cost of production and employment generation though the profit increase was marginal. Further, it is evident that profit margin varied with the ecosystem (rainfed/irrigated), management skill, and socio-economic conditions. On an average profit margin on account of IFS varied from Rs 15,000 to Rs 1,50,000/ha/annum. Simultaneously it takes care of the food and nutritional security of the farming family. The study further revealed improvement in the net profit margin varying from 30-50 %. The resource characterization study reveled that/ha improvement in profitability varied from Rs 20,000 to 25,000 under irrigated condition, resource recycling improve led to 5 to 10 q/ha crop yield increase, generate 50-75 man days/ family/ year and reduce the cost of production by Rs.500-1,000/ha. Therefore, there is an urgent need to promote the IFS concept under all agro-climatic conditions of the country.

7 CONSERVATION AGRICULTURE STRATEGIES IN CROPPING SYSTEM

Conservation agriculture (CA) is generally referred to a way of practising agriculture that primarily includes no tillage, incorporation or surface application of crop residues and follow-up of better crop rotations. However, there are many other management practices that qualify the definition of CA. The Food and Agriculture Organization (FAO), Rome, Italy defined the concept as 'resource saving agricultural crop production that strives to achieve acceptable profits together with high and sustained production levels while concurrently conserving the environment'. Tillage interventions are reduced to the bare minimum besides low external inputs which will lead to better biological processes in the soil. The widely practised CA in terms of minimum/zero tillage, viz. zero/low tillage is predominant in South America (47%), the United States and Canada (39%), Australia (9%) and the rest elsewhere. In the Indian context, CA is popular in the rice–wheat system in the Indo-Gangetic Plains (IGP) and of late is getting popular in the Deccan Plateau in the rice–rice system with the second crop of rice getting replaced by maize. Under Indian context, CA has its special role as it can revive the halted system of monotonous conventional rice wheat system, besides providing resilience under climate change and burgeoning population scenario

Some Distinguished Features of Conventional and Conservation Agriculture

Conventional agriculture	Conservation agriculture
Excessive mechanical tillage and soil erosion	No till or minimum tillage with more dependence on biological tillage
High wind and soil erosion	Low wind and soil erosion
Residue burning is frequent	Surface retention of residues is predominant
Water infiltration is low	Infiltration of water is high
Green manuring (incorporated)	Brown manuring/cover crop (surface retention)
More weed seed bank development	Weed problem during early stage but decrease with time
Free-wheeling of farm machinery, increased compaction	Controlled traffic, compaction in tramline and no compaction in planted or crop area

PRINCIPLES OF CONSERVATION AGRICULTURE (CA)

Adoption of CA for enhancing Resource Use Efficiency (RUE) and crop productivity is the need of the hour as a powerful tool for management of natural resources and to achieve sustainability in agriculture. Conservation agriculture basically relies on 3 principles, which are interdependent and strongly linked and must be considered together for appropriate design, planning and implementation processes. These are:

Minimal mechanical soil disturbance: The soil biological activity produces very stable soil aggregates as well as various sizes of pores, allowing air and water infiltration. This process can be called "biological tillage" and it is not compatible with mechanical tillage. With mechanical soil disturbance, the biological soil structuring processes will disappear. Minimum soil disturbance provides/maintains optimum proportions of respiration gases in the rooting-zone, moderate organic matter oxidation, porosity for water movement, retention and release and limits the re-exposure of weed seeds and their germination.

Permanent organic soil cover: A permanent soil cover is important to protect the soil against the deleterious effects of exposure to rain and sun; to provide the micro and macro organisms in the soil with a constant supply of "food"; and alter the microclimate in the soil for optimal growth and development of soil organisms, including plant roots. In turn it improves soil aggregation, soil biological activity and soil biodiversity and carbon sequestration.

Diversified crop rotations: The rotation of crops is not only necessary to offer a diverse "diet" to the soil micro-organisms, but also for exploring different soil layers for nutrients that have been leached to deeper layers that can be "recycled" by the crops in rotation. Furthermore, a diversity of crops in rotation leads to a diverse soil flora and fauna. Cropping sequence and rotations involving legumes helps in minimal rates of build-up of population of pest species, through life cycle disruption, biological nitrogen fixation, control of off-site pollution and enhancing biodiversity.

The area under conservation agriculture is increasing globally (Fig 7.1) with highest area of about 26 mha in USA followed by Brazil and Argentina (25 mha each)

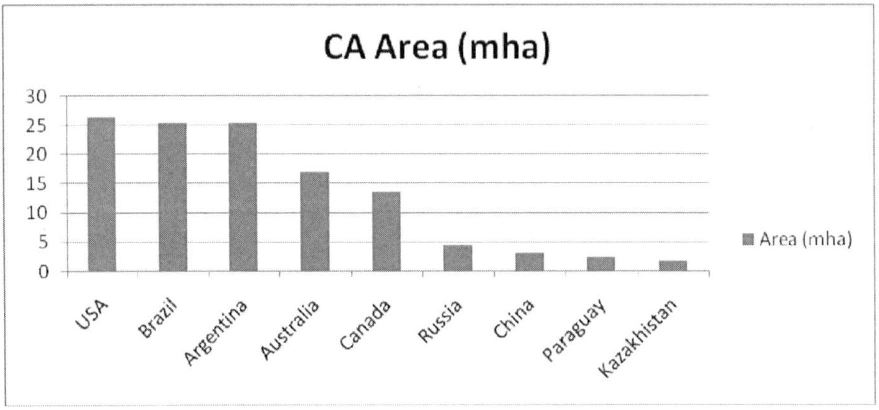

Fig. 7.1. Global adoption of conservation agriculture (FAO, 2014)

CA FOR ENHANCING AGRICULTURE PRODUCTIVITY

CA and Rainfed Agriculture

Rainfed or dryland agriculture is prominent in India, accounting for about 56% of the total cropped area and contributes 87.5% coarse cereals, 87.5% pulses, 77% oilseeds and 65.7% cotton of the country's total production. The rainfed lands are also the areas that are prone to the ill-effects of climate change and experience hardships like drought and frequent crop failures. Further, the Indian climate, which is predominantly tropical and subtropical, encourages degradation of the land mass. In fact, the rainfed regions, especially arid and semi-arid areas particularly represented by Alfisols, Aridisols and related soil groups are characterized by light texture, low organic carbon and low depth soils and low moisture, erratic rainfall, soil erosion and low use of fertilizer due to poor income levels and low productivity. Other soil groups such as Vertisols and Inceptisols also experience diversity of soil-related constraints. CA practices are likely to address the above constraints, if adopted on a long-term basis. It is therefore important to evaluate the feasibility and potential of CA in rainfed regions. While for developed economies like USA, saving the surface soils might be the prime drive for adoption of CA, for the developing economies, beside protecting the lands, other additional emerging agrarian concerns like shortage of water, labour, energy, deteriorating soil health and climate change are also important. Given the large area under rainfed agriculture which is associated with poor soils, low cropping intensity and low productivity, it would be interesting to analyze the techno-economic feasibility of different CA practices in these areas.

Extent of Land Degradation and Role of CA in its Management

Degradation of land, both arable and non-arable is a phenomenon influenced by natural weathering and human interventions. Of late, the latter is adding to the degradation at a faster pace than the former. In the context of conserving the land for agriculture and livelihood, it is necessary to comprehend the extent of degradation across the geo-political regions. Among the States (provinces) in India, Jammu & Kashmir and Nagaland have maximum share of their land (94%) under degradation (Table 7.1). This is primarily due to large areas under mountains, cold deserts and other such degraded lands. In the agriculturally prominent States like Uttar Pradesh (63%), Madhya Pradesh (50%) and Karnataka (46%), the extent of land degradation is of great concern. Ecological agriculture either through CA or any other form will help check land degradation. Rainfed areas which have greater need of CA are spread across different states of India. Among the states, the extent of rainfed area to the total cultivated area is the highest in Jharkhand (91%) followed by Maharashtra (81%) and Chhattisgarh (74%). On the other hand, agriculturally important states in the Deccan Plateau like Andhra Pradesh and Karnataka have 53% and 71% area under rainfed agriculture respectively. Land degradation in the country is posing a big threat to the natural resources, resulting in almost 5 billion tonnes soil and 6 million tonnes of nutrient loss every year.

Table 7.1: Area under degraded land across the various States of India

State/province	Degraded land (ha)	Geographical area (ha)	Degraded land (%)
Andhra Pradesh	10,306,989	27,506,544	37
Arunachal Pradesh	3,391,291	8,374,215	40
Assam	4,642,071	7,843,860	59
Bihar	1,659,322	9,416,395	18
Chhattisgarh	4,947,419	13,518,883	37
Gujarat	3,350,636	19,612,517	17
Haryana	582,307	4,421,858	13
Himachal Pradesh	2,946,569	5,567,137	53
Jammu & Kashmir	20,856,943	22,223,695	94
Jharkhand	4,290,431	7,971,624	54
Karnataka	8,740,358	19,179,155	46
Kerala	2.599.450	3.886,749	67
Madhya Pradesh	15,338,850	30,824,985	50
Maharashtra	10,837,506	30,758,377	35
Orissa	5,024,846	15,570,900	32
Punjab	552,733	5,036,513	11
Rajasthan	21,013,097	34,224,212	61
Tamil Nadu	3,850,948	13,005,861	30
Uttar Pradesh	15,270,634	24,092,803	63
Uttarakhand	3,215,069	5,348,204	60
West Bengal	2,237,615	8,875,874	25
India	154,553,121	327,869,491	47

National remote sensing center

The loss of precious top soil is comparatively much more in dryland areas because of low moisture and poor soil covering vegetation and residues. It is a common phenomenon in the Indian subcontinent where the water bodies get clogged and choked with silt, which is mainly the sediment coming along with the run-off. Economically, the topsoil loss not only deprives the production, but also adds to the cost of desilting of the water bodies. Hence, protection of the topsoil, especially in rainfed regions is most crucial for sustainable agricultural production. The rainfed cropping systems are mostly single-cropped in the red soil areas, whereas in the black soil regions a second crop is taken on the residual moisture. In black soils, farmers (agriculture producers) keep lands fallow also during *Kharif* and grow *Rabi* crops on conserved moisture. The rainfall ranges from < 500 mm in arid to 1000 mm in dry sub-humid areas. Alfisols, Vertisols, Inceptisols, Entisols and Oxisols are the major soil orders. Soils are sloppy and highly degraded due to continuous erosion by water and wind. Sealing, crusting, sub-surface hard pans and cracking are the key factors which cause high erosion and impede infiltration of rainfall. The choice and type of tillage largely depend on the soil type and rainfall.

Leaving crop residue on the surface is another important component of CA, but in rainfed areas due to its competing uses as fodder, little or no residues are available for surface application. The key principles of rainfed agriculture rely on soil and water conservation which are integral components of CA. Though conserving both soil and water are equally important in low to medium rainfall regions, more priority is given for conservation of rainwater by facilitating better infiltration and reduced run-off. That is why deep tillage once in three years is suggested to promote greater infiltration of rainwater and to control weeds. However, practices like chiselling can meet the objective of breaking the hard pan without soil inversion associated with deep tillage. Leaving surface residue is the key to control runoff, soil erosion and hard setting in rainfed areas, which are the key problems. In view of the shortage of residues in rainfed areas in arid and semi-arid regions, several alternative strategies have emerged for generation of residues either through in situ cultivation and incorporation as a cover crop or harvesting from perennial plants grown on bunds and adding the green leaves as manure cum mulch. Agroforestry and alley cropping systems are other options where biomass generation can be integrated along with crop production. Thus, the concept of CA has to be understood in a broader perspective in arid and semi-arid areas, which includes an array of practices like reduced tillage, land treatments for in situ and ex situ water conservation, on-farm and off-farm biomass generation and alternate land-use systems and also the energy conservation. Here, reduced or minimum tillage with residue retention on the surface, which is also called conservation tillage, is more appropriate than zero tillage which is emphasized in irrigated agriculture. Besides conservation tillage, a number of practices, viz. intercropping systems with pulse, soil and water conservation techniques, water harvesting and recycling, selective mechanization and alternate land-use systems have similar resource conservation effects like CA. As most of the dry lands are single-cropped regions, the length of the growing period is around 4–6 months, setting aside 6–8 months as the off-season, allowing land to be barren and exposed to weathering and soil loss. Therefore, it is imperative that some vegetation or stubbles are maintained on the soil to prevent such weathering. The potential of such measures is high in sub-humid and humid regions where the stubbles have greater chance of remaining in the same soil.

Residue Retention under Conservation Agriculture and its Effects

Residue retention is a key component of modern conservation agriculture systems. However, conflicting evidence exists on the role of residue and its architecture on the capture of rainfall and subsequent water storage in the soil in rainfed environments. Swella *et al.* (2015) conducted experiments in 2010 and 2011, on a sandy soil, to investigate the effect of crop residue height and architecture on the capture of rainfall and spatial variability of soil water across the standing-residue rows. They hypothesized that: (1) high residue rates, particularly horizontal residue, prevent water reaching the ground during small rainfall events, but maximise soil water retention during larger rainfall events; and (2) tall standing-residue increases the capture of rainfall and reduces evaporation from the soil surface, but increases spatial variability of soil water. Further, they also investigated the effect of combinations of vertical and horizontal residues on water capture (using simulated rainfall, sprinkler irrigation and rainfall events), and the quantity and spatial distribution of water stored in the soil. The results revealed that the quantity of water intercepted by residue in the inter-row position increased with rate of horizontal residue and standing-residue height. However, the quantity of stored water concurrently increased under and adjacent to the standing-residue row, suggested that some of this 'lost' water was intercepted by the high-cut residue, much of which infiltrated deeper into the soil, where it was less prone to evaporation. The quantity of stored soil water increased with increasing residue cut-height and horizontal residue, with bare soil having the least water. Likewise it was found that most soil water was stored with straw cut high (0.25–0.3 m) in combination with at least 2 t ha^{-1} horizontal residue, with soil directly under the standing straw row having the most water. Finally they concluded that retaining of tall standing-residue in combination with horizontal residues can increase soil water capture in the no-till system.

CA and soil enzymatic activity: The effect of long-term conservation agriculture practices on biological soil quality indices in rainfed Alfisols, using linear scoring technique and GMeanBSQI approach was conducted by Sharma *et al.* 2016. They have noticed with minimum tillage, surface application of GL @ 2 t ha^{-1} and N @ 90 kg ha^{-1} superior in positively influencing the activity of arylsulphatase, urease, dehydrogenase enzymes. Addition of organic amendments and adoption of management practices that increase soil organic matter led to increased enzyme activity. They further indicated that conservation agriculture practices have significantly contributed in influencing soil enzyme activities and labile pools of carbon, which in turn have played an important part in influencing the biological soil quality, expressed as GMeanBSQI.

Effect of CA on greenhouse gas (GHG) emissions under dry climates:_Tellez-Rio *et al.* (2017) carried out studies on conservation tillage and crop rotations: the impact of these practices on greenhouse gas (GHG) emissions and crop, particularly under dry climates. They conducted 2 year field experiment under rainfed conditions to evaluate the effect of three long-term (17–18 years) tillage systems (Conventional Tillage (CT), Minimum Tillage (MT) and No Tillage (NT)) and two cropping systems (rotational wheat (*Triticum aestivum* L.) preceded by fallow and wheat in monoculture), on nitrous oxide (N_2O) and methane (CH_4) emissions, during two field campaigns. They have measured soil mineral N, water-filled pore space, dissolved organic carbon (C) and grain yield and yield-scaled N_2O emissions, N surplus and Global Warming Potentials (GWP). The results revealed that, No tillage only decreased cumulative N_2O losses (compared to MT/CT) during campaign 1 (the driest campaign with least fertilizer N input), while tillage did not affect CH_4 oxidation. The GWP demonstrated that the enhancement of C stocks under NT caused this tillage management to decrease overall CO_2 equivalent emissions. Monoculture increased N_2O fluxes during campaign 2 (normal year and conventional N input) and decreased CH_4 uptake, as opposed to rotational wheat. They found that NT and CT treatments yielded most during the dry and the normal campaign, and the yield-scaled N_2O emissions followed the same tendency and minimum tillage was not an adequate tillage management considering the GWP and the yield-scaled N_2O emissions(which were 39% lower in NT with respect to MT). Further, regarding the crop effect, wheat in rotation resulted in a 32% increase in grain yield and 31% mitigation of yield-scaled N_2O emissions and low cumulative N_2O fluxes (<250 g N_2O-N ha^{-1} campaign^{-1}) highlighted the relevance of soil organic C and CO_2 emissions from inputs and operations in rainfed semi-arid cropping systems. They suggested that NT and crop rotation can be recommended as good agricultural practices in order to establish an optimal balance between GHGs fluxes, GWP, yield-scaled N_2O emissions and N surpluses.

Long term studies on benefits of conservation agriculture: In a study by Farooq *et al.* (2011) on analysis of 25 experiments found that a slight increase in CA crop yields over time relative to CVT, indicating that CA can compete with CVT on a purely crop production basis and also has well-established environmental benefits. Similarly, Kumar *et al.* (2011) studied the effect of the CA practices under different types of rainfall, soil and production system were identified based on the analysis of studies undertaken in different agro-climatic regions of India. They estimated net benefits from different suggested CA practices estimated at current prices for the year 2010 based on the results of long-term experiments conducted at the Central Research Institute for Dryland Agriculture, Hyderabad, and centres of its All-India Coordinated Research Project on Dryland Agriculture, which ranged from Rs. 325 to Rs. 18,000 ha^{-1}. Further, the capital requirement of various CA practices was also estimated. Their analysis showed that CA in its broader context not only improves soil health but also gives higher net returns per unit of land to the farmers. However they have suggested that CA practices need to be adopted selectively in different rainfall, soils and agro-ecological situations. The advantage of CA practices can be availed in rainfed agro-ecology, if practised appropriately over long-term basis. They opined that higher carbon sequestration through CA practices may also give additional benefits in terms of carbon emission reduction, if made easily tradable in future. Finally, they suggested for promotion of CA practices across diverse agro ecologies requires appropriate policy, and institutional and technology support. Kurothe *et al.* (2014) investigated four practices including conventional tillage (CT), ridge farming tillage (RFT), no tillage (NT) and stubble mulch farming tillage (SMFT) for 11 years (1990–1991 to 2001–2002) under semiarid rainfed conditions in western India on a very deep, sandy loam soil. Green gram (*Vigna radiata*) – mustard (*Brassica juncea*) sequential cropping and pearl millet (*Pennisetum glaucum*) + pigeon pea (*Cajanus cajan*) intercropping systems were tested for the first four years (first phase of the experiment). Cowpea (*Vigna unguiculata*)–mustard sequential cropping and cowpea + castor (*Ricinus communis*) intercropping systems used for the following seven years (second phase of the experiment). They observed runoff, soil losses, sediment concentrations, crop yields, soil organic carbon, bulk density and water stable aggregates for all the treatment combinations. The results revealed that, RFT and SMFT were both effective in reducing runoff and soil loss and RFT, NT and SMFT reduced runoff by 69.4, 16.2 and 59.6% respectively compared with CT. Further, they noticed average soil loss in NT was 37.2% less than CT and average sediment concentration of runoff during June–July greater than in August–October for all treatments and average sediment concentrations were greater under CT and RFT. They recorded highest average yield of all crops except green gram under SMFT. They observed surface soil organic carbon (SOC), bulk density and water stable macro-aggregates all significantly greater under NT at the end of the experiment, but reverting to uniform tillage negated this effect and micro-aggregates built-up under SMFT were relatively more stable than those all under NT. They finally conclude that in the semi-arid sub-tropical agro-ecosystem of Gujarat (western India) adoption of SMFT can significantly improve and stabilize the crop yields and reverse land degradation process.

In another study, Pittelkow *et al.* (2015) have conducted a global meta-analysis using 5,463 paired yield observations from 610 studies to compare no-till, the original and central concept of conservation agriculture, with conventional tillage practices across 48 crops and 63 countries. Their results showed that no-till reduces yields, yet this response is variable and under certain conditions no-till can produce equivalent or greater yields than conventional tillage. They suggested that importantly, when no-till is combined with the other two conservation agriculture principles of residue retention and crop rotation, its negative impacts are minimized and no-till in combination with the other two principles significantly increases rainfed crop productivity in dry climates, suggesting that it may become an important climate-change adaptation strategy for ever-drier regions of the world. But, they have suggested that any expansion of conservation agriculture should be done with caution in these areas, as implementation of the other two principles is often challenging in resource-poor and vulnerable smallholder farming systems, thereby increasing the likelihood of yield losses rather than gains. Melero *et al.* (2008) analyses have shown that success of conservation agriculture in Southern Africa depends on the promotion of other good agronomic practices such as targeted fertilizer based on a meta-analysis of long-term effects of conservation agriculture on maize grain yield under rain-fed conditions.

PERFORMANCE OF DIFFERENT CROPPING SYSTEMS WITH CA

Maize

Kutu *et al.* (2012) studied role of Conservation Agriculture under irrigated and dryland conditions during 2007/2008-summer cropping season to determine a suitable soil-crop management practice for increase maize yield. They studied consisted of tillage practices (conventional, minimum and zero), cropping systems (sole and intercrop plots) and fertilizer regimes (unfertilized control, low, adjusted low and optimum) as treatments. Minimum and zero tillage practices constituted the conservation agriculture tillage practices while supplementation of low fertilizer rate with seed inoculation using growth enhancing microbial inoculant constituted the adjusted low fertilizer rate. They have found that fertilizer application gave a significant ($P < 0.05$) maize grain yield increase at both trial sites and maize grain yield ranged from 1254 to 3683 kg ha^{-1} and from 1738 to 3199 kg ha^{-1} under supplementary irrigation and dryland non-irrigated condition, respectively. They have noticed highest maize grain yield obtained at optimum fertilizer rate did not differ significantly from that of adjusted low rate under dryland and inherent low soil nutrients condition. Further, they also observed although the adjusted low fertilizer rate gave no maize grain yield advantage over the low rate in soil with fairly high native nutrients, it however resulted in better crop responses and conservation of some of the soil properties studied under low-nutrients soil status. Finally, they concluded that Zero tillage as a conservation agriculture practice gives the highest maize grain yield of 2805 and 2776 kg ha-1 under supplementary irrigation and dryland conditions, respectively.

In another study, Pradhan *et al.* (2016) has assessed the effects over 3 years (2011–2014) of reduced tillage, intercropping, and cover cropping practices customized for maize-based production systems in upland areas of Odisha, India. They focused on crop yield, system productivity and profitability through maize equivalent yield and dominance analysis. Their results showed that maize grain yield did not differ significantly over time or among conservation agriculture production systems (CAPSs) treatments while cowpea yield was considered as an additional yield in intercropping systems. Further, they noticed mustard and horsegram grown in plots after maize cowpea intercropping recorded higher grain yields of 25 and 37%, respectively, as compared to those without intercropping. They opined that, in overall, the full CAPS implementation, i.e., minimum tillage, maize–cowpea intercropping and mustard residue retention had significantly higher system productivity and net benefits than traditional farmer practices, i.e., conventional tillage, sole maize cropping, and no mustard residue retention.

Wheat

Colecchia *et al.* (2015) conducted conservation agriculture experiment in Southern Italy on durum wheat continuous cropping with the aim to assess the durum wheat productivity and grain quality in reduced soil tillage systems. They noticed interactions among experimental treatments and climate revealed a close relationship among grain yield, grain quality and wheat growth conditions. The results revealed that conventional tillage (CT, plowing and 2 disc harrowing) showed in the 2-year period higher grain production than reduced tillage treatments, minimum (MT, 1 disc harrowing) and No tillage (NT), especially for good crop water availability (3.29 t ha^{-1} of grain yield in CT, 2.67 in MT and 2.54 in NT). Further, the amount of rainfall (above the average in both years) and its distribution in the growing seasons (more regular in the first year) strongly influenced wheat-grain quality indices (11.97% of protein content in the first year and 9.82% in the second one). Also, the wheat quality was more sensitive to the *"Year × Tillage"* interaction, with differences among tillage's more evident in the second year and favourable to NT and MT. They opined spectral vegetation indexes (NDVI and TVI) measurements at flowering, have been useful to support farmers in N-late application for improving grain wheat quality.

Potential benefits recognized with conservation agriculture are summarized as below

Reduction in cost of production by Rs 2,000 to 3,000 ha^{-1} ($ 33 to 50); zero tillage technology farmers were able to save on land preparation costs by about Rs. 2,500 ($41.7) per ha and reduce diesel consumption by 50 – 60 litres per ha	(Malik *et al.* 2005; Sharma *et al.* 2005).
Enhancement of soil quality, i.e. soil physical, chemical and biological conditions	(Jat *et al.* 2009a; Gathala *et al.* 2011b)
Enhancement, in the long term Carbon sequestration and build-up in soil organic matter constitute a practical strategy to mitigate Greenhouse Gas emissions and impart greater resilience to production systems to climate change related aberrations	(Saharawat *et al.* 2012)
Reduction of the incidence of weeds, such as *Phalaris minor* in wheat	(Malik *et al.* 2005)

Enhancement of water and nutrient use efficiency	(Jat *et al.* 2012
Enhancement of production and productivity (4% – 10%)	(Gathala *et al.* 2011a)
Advanced sowing date	(Malik *et al.* 2005)
Reduction in greenhouse gas emission and improved environmental sustainability	(Pathak *et al.* 2011)
Opportunities for crop diversification and intensification-for example in sugarcane based systems, mustard, chickpea, pigeonpea etc.	(Jat *et al.* 2005)
Avoiding crop residue burning reduces loss of nutrients, and environmental pollution, which reduces a serious health hazard	(Sidhu *et al.* 2007)

CONSTRAINTS OF CONSERVATION AGRICULTURE

Conservation agriculture systems have not been tried or promoted in other major agro-eco regions like rainfed semi-arid tropics and the arid regions of the mountain agro-ecosystems.

- The widespread use of crop residues for livestock feed and fuel: Specially under rainfed situations, farmers face a scarcity of crop residues due to less biomass production of different crops.
- There is competition between CA practice and livestock feeding for crop residue. This is a major constraint for promotion of CA.
- Lack of appropriate seeders especially for small and medium scale farmers: successful adoption will call for accelerated effort in developing, standardizing and promoting quality machinery aimed at a range of crop and cropping sequences. These would include the development of permanent bed and furrow planting systems and harvest operations to manage crop residues.
- Burning of crop residues: For timely sowing of the next crop and without machinery for sowing under CA systems, farmers prefer to sow the crop in time by burning the residue. This has become a common feature in the rice-wheat system in north India. This creates environmental problems for the region.
- Lack of knowledge about the potential of CA to agriculture leaders, extension agents and farmers: This implies that the whole range of practices in conservation agriculture, including planting and harvesting,
- Water and nutrient management, diseases and pest control etc. need to be evolved, evaluated and matched in the context of new systems.
- Skilled and scientific manpower: Managing conservation agriculture systems, will call for enhanced capacity of scientists to address problems from a systems perspective and to be able to work in close partnerships with farmers and other stakeholders. Strengthened knowledge and information sharing mechanisms are needed.
- In semiarid regions (300–500 mm annual rainfall), success of conservation agriculture depends on the ability of farmers to retain crop residues and to ensure adequate weed control

- In theory, reduced tillage and surface cover increase soil water available for crop growth by increasing infiltration and by limiting run-off and evaporation losses. However, mulching is not positive in all circumstances; under continuous rainfall, mulches have little effect on soil water status
- Prolonged dry periods may also cause the benefits of mulching to diminish due to continued evaporation. Intensive rainfall in mulched fields can cause waterlogging because of reduced evaporation leading to reduced soil aeration.

HIGH EXTERNAL INPUT AGRICULTURE (HEIA)

The farming pattern heavily depends upon external and chemical inputs.The focus of agricultural development and research has mainly been on maximizing yields coupled with increasing specialization of production. Primarily, one or two commodity, driven development, lack of diversity in the farming practices, as a result, there is greater risk of failure and price fluctuation. The number of products and commodities are minimal.

Distinguishing characteristics of HEIA

Advantages	Disadvantages
Agricultural production could be rapidly increased to meet the demand for food for the increasing population.	Collapse of environmental balance due to lack of biodiversity by planting a few cash crops.
As a result of availability of adequate food stuffs many problems related to diseases caused by malnutrition and deficiency were prevented or reduced.	Although the need for high capital investment, the large scale farmers benefited while small scale farmers who were short of capital ran into debt.
Mechanization solves the problem of labour shortage.	The farmers in developing countries had to encounter a series of environmental, social economic and political problems as a result of the use of High External Input Agricultural practices.
Income and profit margins of the products increased	Dependence on imported machinery, chemical fertilizers, pesticides, hybrid seeds and other inputs.
Productivity of land increased with Increased market facilities for production.	Increase in soil erosion due to constant furrowing by machinery; Extensive use of pesticides disturb the natural mechanism of controlling pests and diseases as the artificial pesticides kill both pests and their natural enemies. Use of artificial agro-chemicals adversely affect the soil pH, cation exchange capacity, soil structure, soil texture and soil organisms. Consequently the microbial activities of the soil tend to reduce forming dead soil.
New improved varieties gave yields within a short period of time.	Neglecting environmental friendly traditional varieties of seeds and their genetic resources face extinction due to introduction of hybrid varieties. Conventional agricultural knowledge and techniques are neglected.

HEIA-based Green Revolution Shortcomings

- Failure to foresee the fall of prices of surplus agricultural commodities due to the increase in production caused by the use of chemical fertilizer, fuel, pesticides, hybrid seeds etc., which were costly.
- Failure to realize long-term dependency of the farmers on pesticides and chemical fertilizers and the consequent effect on ecology and human life. As a result of mechanization, use of new improved seed varieties and provision of irrigation water, the agricultural production in Punjab, Haryana and Uttar Pradesh of India increased tremendously. Simultaneously many problems cropped up. Frequent irrigation facilities had to be provided to the new improved varieties of crops. Due to their inability to withstand shortage of water, the farmers pumped water from agricultural wells. Excess pumping of water caused lowering of ground water table. Consequent upon frequent puddling of soil an impenetrable layers adjacent to the surfaces was formed and prevented sucking of water and nutrients. Due to this restriction the farmers had to use more chemical fertilizers on this soil surface. Subsequently imbalance of nutrients and deficiencies of micronutrients in the soil became a problem, excessive use of pesticides for plant protection caused contamination of food and plant residues threatening the health of human and animal lives. Although the production had increased rapidly the demand for excess agricultural production in the market had not commensurately increased. The growers encountered with the problems in selling of their products and at the same time prices of food items were reduced.
- Quick response for growth and production and relatively lesser quantities required for one application promoted farmers to apply artificial manure. But the farmers were not aware of long-term consequences of chemical fertilizer.
- The efficiency of chemical fertilizer was below the expected level. It is found that 40% - 50% of nitrogen in the chemical fertilizer applied in the tropical countries is wasted. This wastage further aggravates due to factors such as high rainfall, severe droughts, soil erosion and low content of organic matter in the soil.
- The chemical fertilizer directly affects on the nutrient balance and the biological activities of the soil. It disturbs the soil structure and expedites decomposition of soil organic matter. Application of acidic nitrogen fertilizer like ammonium sulphate decreases the soil PH value and reduces the ability to use phosphorus by plants.
- Continuous application of nitrogen, phosphorus and potassium fertilizer creates micronutrients deficiencies in the soil. The plant display deficiencies of zinc, iron, copper, magnesium, manganese, molybdenum and boron. The reasons for this is the barrier imposed by chemical fertilizer for micronutrients to enter into the soil. This situation results in the reduction of crop production and susceptibility to pests and diseases.

- *Greenhouse gas emission*. In addition, as a result of using chemical fertilizers the developed and developing countries face global problems. The nitrous oxide gas emitted enters into the atmosphere. Nitrous oxide gas so entered into the atmosphere adversely affects the ozone layer and causes greenhouse effect.

LOW EXTERNAL INPUT AND SUSTAINABLE AGRICULTURE (LEISA)

Agricultural practices with lesser inputs have been developed by integrating selected traditional basic principles with new technological knowledge. This system which is abbreviated as LEISA is frequently used for low external input and Sustainable Agriculture. The term which will be referred to as LEISA in the books follow an integrated agricultural system consisting of bio-dynamics, environmental, natural, organic and regeneration sub systems.

Ecological Principles of LEISA

- Creating a favourable condition for growth and sustenance of plant by stimulating of soil microorganisms as far as possible and adding organic matter sufficiently.
- Maintaining nutrient content at optimum level thus assuring the balance of nutrients in the soil by nitrogen fixation, utilization of nutrients available in the deep soil layers, promotion of recycling process and addition of external fertilizer as and when necessary to complement deficient nutrients.
- Controlling the micro climatic conditions to minimize loss of resources due to sunlight, air and water. Use of biological and mechanical methods to prevent soil erosion.
- Minimizing loss of resources caused by pests and diseases. Integration of pest control methods giving priority to natural biological control of pests by natural enemies on the principle that prevention is better than eradication.
- Promoting biodiversity and complexity, Stimulating synergetic and symbiotic conditions between plants/ plants and plants/animals.

The objective of LEISA system is to maintain the agricultural production at an optimum level using less external inputs in a eco-friendly environment. To achieve this objective LEISA practices concentrate heavily on the following factors:

- Maintaining a living soil
- Creating biodiversity
- Recycling of resources.
- Natural pest management

Some Promising LEISA Techniques and Practices

1. **Nutrient management:** Nutrient management is managing the amount, source, placement, form, and timing of the application of nutrients and soil amendments to ensure adequate soil fertility for plant production and

to minimize the potential for environmental degradation, particularly water quality impairment. Nutrient management has taken on new connotations in recent times. Soil fertility traditionally dealt with supplying and managing nutrients to meet crop production requirements, focusing on optimization of agronomic production and economic returns to crop production (LEISA, 2006). Contemporary nutrient management deals with these same production concerns, but recognizes that ways of farming must now balance the limits of soil and crop nutrient use with the demands of intensive animal production. Farmers in the hill agriculture follow a system of close integration of crop, livestock and forestry/grassland management. Farmers maintain traditional practices such as terracing, manure management, legumes inter cropping, and mulching where appropriate. Farm yard manure quality can be increased by better decomposition and the N-content can be increased by at least 2 to 3 times from about 0.5% N to 1.5% N through proper management of urine and manure. Liquid manure can be prepared from urine and various plants extracts rich in minerals or secondary plant compounds. These "manure teas" were shown to be effective liquid fertilizers on crops such as vegetables and also for organic pest and disease management. Local marketing systems for such "manure teas" are emerging in some areas. The use of urea fertilizer declined in several areas due to liquid manure use. In Nepal, some of the LEISA practices are: Bio-pesticide (cow urine, neem products, tobacco, ash etc), composting (pit), tree plantation, green manuring, vermicomposting, liquid compost, NADEP compost, seed treatment, seed production.

Management practices for integrated pest management are listed below:

- use of resistant varieties
- crop rotation
- optimal use of biological control organisms
- certified seeds
- protective seed treatments
- disease-free transplants or rootstock
- timeliness of crop cultivation
- improved timing of pesticide applications
- removal or 'plow down' of infested plant material.

2. **Integrated pest management (IPM):** IPM is an ecologically-based approach to pest(animal and weed) control that utilizes a multi-disciplinary knowledge of crop/pest relationships, establishment of acceptable economic thresholds for pest populations and constant field monitoring for potential problems. Management may include such practices (LEISA, 2006).
3. **Crop residue management and conservation tillage:** Conservation tillage is a term that covers a broad range of soil management systems that leave residue cover on the soil surface, substantially reducing the effects of soil erosion from wind and water. These practices minimize nutrient loss, decreased

water storage capacity, crop damage, and improve soil quality. The soil is left undisturbed from harvest to planting except for nutrient amendment. Weed control is accomplished primarily with herbicides, limited cultivation, and, in more sustainable systems, with cover crops. The National Crop Residue Management Survey (Conservation Technology Information Center (CTIC)) specifies that 30% or more of crop residue must be left after planting to qualify as a conservation tillage system. Some specific types of conservation tillage are Minimum tillage, Zone tillage, No-till, Ridge-till, Mulch-till, Reduced-till, Strip-till, Rotational tillage and Crop residue management.

4. **Converting farm wastes into useful organic manure under LEISA system:** The ingredients required are green leaves, dry leaves, weeds from adjacent farms, cowdung and cow urine, fodder wastes from cattle sheds, gobar gas slurry, coir pith, tank silt, quarry dust and rock phosphate. These items are put in alternate layers (cowdung in between layers) in a heap and left for 45 days. The heap is turned once or twice. In 45 days, the items get-semi decomposed. The partially decomposed material can be used in two ways- for vermicompost preparation, which gets converted to vermicompost in 45 days, and the other way is to add coir pith, tank silt and quarry dust to semi decomposed material, which turns into good quality manure in 45 days.
5. **Green manuring and cover crops:** Green manure and cover crop species should fit the agroecological condition. In general, these crops should have the characteristics as: easy establishment, vigorous growth under local conditions, ability to cover weeds quickly, ability to either fix atmospheric nitrogen or concentrate plenty of phosphorus, should also have multiple uses. The main problem in adoption of green manuring is that the land on which green manure crop is raised could have been profitably utilized for growing a crop of economic importance. Another problem is adequate soil moisture either through rainfall or irrigation is essential for in-situ decomposition of green manure crop. Under rainfed condition, if sufficient rainfall is not received, proper decomposition may not take place, and the germination of the succeeding crop is hampered. There is a possibility of incidence of diseases and pests, even nematodes.
6. **Practices for land degradation:** A good plant cover is very important for preventing soil degradation and achieving soil rehabilitation. A vegetative cover has a number of beneficial effects on soil aeration, soil moisture and organic matter content, physical characteristics and biological activity in the soil. In addition, a plant cover protects against soil erosion. Cover crops can be quite aggressive creepers and may compete with the main crop. On the other hand if leguminous plants are used as cover crops they add nutrients because they fix atmospheric nitrogen and make it available for the crop. Selection of crops is very important while dealing with degraded soils. Sodic soils in South Indian region were corrected for their soil pH by including cucumber in the cropping pattern. Similarly, raising Eucalyptus plantation

along the canal bunds helped in solving the drainage problem in low-lying delta areas. Farmers follow their own traditional practices in maintaining soil physical structure and health. Practices like application of sand, groundnut shell, sal leaves, retention of sunflower stalks are a few among those followed by the farmers for mulching in South Indian condition.

7. **Intercropping**: Intercropping is defined as growing of two or more crops simultaneously on the same piece of land; crop diversification is in both temporal and spatial dimension; there is intercrop competition during all or part of the crop growth. There are many types of intercropping viz., mixed intercropping, row intercropping, strip inter cropping and relay intercropping.
8. **Organic manuring:** Organic manures are organic materials derived from animal, human and plant residues which contain nutrients in complex organic forms. They are the sources of plant nutrients. They release nutrients after their decomposition. They provide organic acids that help to dissolve soil nutrients and make them available for the plants. Organic manures can be grouped into bulky organic manures and concentrated organic manures based on the concentration of the nutrients.

LEIA vis-a-vis HEIA

Characteristics of HEIA	**Characteristics of LEIA**
The farming pattern depends heavily on external and chemical inputs.	LEIA relies on the optimal use of natural processes. Although yields have increased substantially, contributing to raising total production, farmers and the environment have had to pay the price for keeping up with this development.
The focus of agricultural development and research has mainly been on maximizing yields coupled with increasing specialization of production	The focus is on the sustainability of farming system
There is a great damage to the environment	Environmentally sound and has the potential to contribute to the long-term sustainability of agriculture.
The continuing drop in prices of farm produce and the rising costs of agricultural inputs have made farming increasingly unprofitable	Greater emphasis is on the long-term sustenance and balance between the profit and livelihood.
HEIA depends on the higher production and profit, without consideration of the local needs and local market	Sustainable ecological practices depend largely on local agro-ecological conditions and on local socio-economic circumstances, as well as on farmers' individual needs and aspirations.

Primarily one or two commodity driven development, lack of diversity in the farming practices, as a result, there is greater risk of failure and price fluctuation. The number of products and commodities are very minimum.	One way of LEIA is to diversification of farms; with a range of crops and/or animals, farmers will suffer less from price fluctuations or drops in yield of single crops. Maintaining diversity will also provide a farm family with a range of products to eat or sell throughout a large part of the year.
Under HEIA system, soil quality deteriorates, and there is resurgence of pests, lack of resilience in the soil-plant system	LEIA maintains a healthy soil, recycling nutrients on the farm, and utilizing approaches such as integrated pest management (IPM).
In HEIA, there is lack of use of indigenous technologies.	Best bet technologies, for example, are soil and water conservation (terraces, ditches, and vegetation strips on sloping land), better timing of operations, improved crop spacing and densities, manure or compost and water application based on local conditions.

8

SUSTAINABLE AGRICULTURE

The phrase 'sustainable agriculture' was reportedly coined by the Australian agricultural scientist Gordon McClymont. Wes Jackson is credited with the first publication of the expression in his 1980 book *New Roots for Agriculture*. The term became popular in the late 1980s

It has been defined as "an integrated system of plant and animal production practices having a site-specific application that will last over the long term", for example to satisfy human food and fiber needs, to enhance environmental quality and the natural resource base upon which the agricultural economy depends, to make the most efficient use of non-renewable and on-farm resources and integrate natural biological cycles and controls, to sustain the economic viability of farm operations, and to enhance the quality of life for farmers and society as a whole

In simplest terms, sustainable agriculture is the production of food, fiber, or other plant or animal products using farming techniques that protect the environment, public health, human communities, and animal welfare. This form of agriculture enables us to produce healthful food without compromising future generations' ability to do the same.

Sustainable agriculture is a type of agriculture that focuses on producing long-term crops and livestock while having minimal effects on the environment. This type of agriculture tries to find a good balance between the need for food production and the preservation of the ecological system within the environment. In addition to producing food, there are several overall goals associated with sustainable agriculture, including conserving water, reducing the use of fertilizers and pesticides, and promoting biodiversity in crops grown and the ecosystem. Sustainable agriculture also focuses on maintaining economic stability of farms and helping farmers improve their techniques and quality of life.

There are many farming strategies that are used that help make agriculture more sustainable. Some of the most common techniques include growing plants that can create their own nutrients to reduce the use of fertilizers and rotating crops in fields, which minimizes pesticide use because the crops are changing frequently. Another common technique is mixing crops, which reduces the risk of a disease destroying a whole crop and decreases the need for pesticides and herbicides. Sustainable farmers also utilize water management systems, such as drip irrigation, that waste less water.

In short, Sustainable Agriculture is:

- **Economically Viable:** If it is not profitable, it is not sustainable.
- **Socially Supportive:** The quality of life of farmers, farm families and farm communities is important.
- **Ecologically Sound:** We must preserve the resource base that sustains us all.

The word "sustainable" has become very popular in recent years and it is now used to describe a lot of things. Simply put, sustainable agriculture is the production of plant and animal products, including food, in a way which uses farming techniques that protect the environment, public health, communities, and the welfare of animals. Sustainable agriculture allows us to produce and enjoy healthy foods without compromising the ability of future generations to do the same. The key to sustainable agriculture is finding the right balance between the need for food production and the preservation of environmental ecosystems. Sustainable agriculture also promotes economic stability for farms and helps farmers to better their quality of life. Agriculture continues to be the biggest employer in the world with 40% of the world's population working in it.

"*Sustainable agriculture is the act of farming using principles of ecology, the study of relationships between organisms and their environment. It has been defined as "an integrated system of plant and animal production practices having a site-specific application that will last over the long term.*"

All human population prefer to consume natural chemical free food instead of food that is sprayed with pesticides and chemical fertilizers. Sustainable agriculture differs greatly from industrial agriculture where large volumes of crops as well as livestock are produced for sale using industrial techniques. Industrial agriculture relies heavily on pesticides and chemical fertilizers and other chemical enhancers. In the past decade the majority of food we ate has been grown in this manner. In 1996 only 20% of the corn in the United States was genetically modified, that number had reached 88% by 2006. However in the last couple of years, due to the negative aspects of the technique, there has been a slight shift towards the use of sustainable agricultural methods.

Key Principles

There are several key principles associated with sustainability in agriculture:

- The incorporation of biological and ecological processes into agricultural and food production practices. For example, these processes could include nutrient cycling, soil regeneration, and nitrogen fixation.
- Using decreased amounts of non-renewable and unsustainable inputs, particularly the ones that are environmentally harmful.
- Using the expertise of farmers to both productively work the land as well as to promote the self-reliance and self-sufficiency of farmers.
- Solving agricultural and natural resource problems through the cooperation and collaboration of people with different skills. The problems tackled include pest management and irrigation.

BENEFITS OF SUSTAINABLE AGRICULTURE

1. **Contributes to environmental conservation:** The environment plays a huge role in fulfilling our basic needs to sustain life. In turn, it is our duty to look after the environment so that future generations are not deprived of their needs. Sustainable agriculture helps to replenish the land as well as other natural resources such as water and air. This replenishment ensures that these natural resources will be able for future generations to sustain life.
2. **Public health safety:** Sustainable agriculture avoids hazardous pesticides and fertilizers. As a result, farmers are able to produce fruits, vegetables and other crops that are safer for consumers, workers, and surrounding communities. Through careful and proper management of livestock waste, sustainable farmers are able to protect humans from exposure to pathogens, toxins, and other hazardous pollutants.
3. **Prevents pollution:** Sustainable agriculture means that any waste a farm produces remains inside the farms ecosystem. In this way the waste cannot cause pollution due to non-burning of crop residues.
4. **Reduction in cost:** The use of sustainable agriculture reduces the need for fossil fuels, resulting in significant cost savings in terms of purchasing as well as transporting them. This in turn lessens the overall costs involved in farming.
5. **Biodiversity:** Sustainable farms produces a wide variety of plants and animals resulting in biodiversity. During crop rotation, plants are seasonally rotated and this results in soil enrichment, prevention of diseases, and pest outbreaks.
6. **Beneficial to animals:** Sustainable agriculture results in animals being better cared for, as well as treated humanely and with respect. The natural behaviors of all living animals, including grazing or pecking, are catered for. As a result they develop in a natural way. Sustainable farmers and ranchers implement livestock husbandry practices that protect animals' health.
7. **Economically beneficial for farmers:** In exchange for engaging with sustainable farming methods, farmers receive a fair wage for their produce. This greatly reduces their reliance on government subsidies and strengthens rural communities. Organic farms typically require 2 ½ times less labour than factory farms yet yield 10 times the profit.
8. **Social equality:** Practicing sustainable agriculture techniques also benefits workers as they are offered a more competitive salary as well as benefits. They also work in humane and fair working conditions, which include a safe work environment, food, and adequate living conditions.
9. **Beneficial for environment:** Sustainable agriculture reduces the need for use of non-renewable energy resources and as a result benefits the environment.

Due to population increase, it is estimated that by 2050 we will need approximately 70% more food than is currently being produced in order to provide the estimated 9.6 billion world population with their recommended daily calorie intake. This is by no means a small challenge, but unlike many other sustainability challenges, everyone can play a part. We all need to eat, but by simply reducing food loss and waste, as well as eating diets that are lower impact, and investing in sustainable produce, we can make a difference. From countries, to companies, right down to consumers, we all have a role to play. The challenge is simply making people care in a world where we are surrounded by such abundance.

Sustainable agriculture practices

Over decades of science and practice, several key sustainable farming practices have emerged—for example:

- **Rotating crops and embracing diversity:** Planting a variety of crops can have many benefits, including healthier soil and improved pest control. Crop diversity practices include intercropping (growing a mix of crops in the same area) and complex multi-year crop rotations
- **Planting cover crops:** Cover crops, like clover or hairy vetch, are planted during off-season times when the soil might otherwise be left bare. These crops protect and build soil health by preventing erosion, replenishing soil nutrients, and keeping weeds in check and reducing the need for herbicides.
- **Reducing or eliminating tillage:** Traditional plowing (tillage) prepares fields for planting and prevents weed problems, but can cause a lot of soil loss. No-till or reduced till methods, which involve inserting seeds directly into undisturbed soil, can reduce erosion and improve soil health.
- **Applying integrated pest management (IPM):** A range of methods, including mechanical and biological controls, can be applied systematically to keep pest population under control while minimizing use of chemical pesticides.
- **Integrating livestock and crops:** Industrial agriculture tends to keep plant and animal production separate, with animals living far from the areas where their feed is produced, and crops growing far away from abundant manure fertilizers. A growing body of evidence shows that a smart integration of crop and animal production can be a recipe for more efficient, profitable farms.
- **Adopting agroforestry practices:** By mixing trees or shrubs into their operations, farmers can provide shade and shelter to protect plants, animals, and water resources, while also potentially offering additional income.
- **Managing whole systems and landscapes:** Sustainable farms treat uncultivated or less intensively cultivated areas, such as riparian buffers or prairie strips, as integral to the farm—valued for their role in controlling erosion, reducing nutrient runoff, and supporting pollinators and other biodiversity.

ELEMENTS FOR SUSTAINABLE AGRICULTURE AND RURAL DEVELOPMENT

Integrated activities

- **Government level:** Policies, instruments, development plans, agrarian reform, nutrition surveys, food quality and food security, data, monitoring, early warning systems
- **Rural community level:** Development of local organizations and capacity building for people's participation, training, extension
- **Area level:** For example, coastal zones, watersheds, river basins, agro-ecological zones
- **Production unit level:** Farming systems, diversification to increase incomes, creation of rural industries, credit and marketing
- **Consumer level:** Improving nutrition and food quality, adjusting dietary patterns, product marketing

Key natural resources

- **Land:** Land use planning, land management, soil conservation, land rehabilitation
- **Water:** Water conservation, irrigation improvements, water database development, water-users, associations
- **Plant and animal biological resources:** Conservation of genetic resources, development of varieties and breeds
- **Trees and forests:** Reduction of deforestation rates, sustainable forest management and wood harvesting, promotion of non-wood forest uses and industries, conservation of habitats, integrating trees in farming systems
- **Fisheries:** Reduction of fishing effort to maximize production, increasing aquaculture production, exploitation of new species

Key external inputs

- **Pest management:** Programmes and projects on integrated pest control. Control of pesticide use
- **Plant nutrition:** Programmes and projects for integrated plant nutrition
- **Rural energy:** National strategies and technology transfer for integrated rural energy development

Farming and natural resources

Sustainable agriculture can be understood as an ecosystem approach to agriculture. Practices that can cause long-term damage to soil include excessive tilling of the soil (leading to erosion) and irrigation without adequate drainage (leading to salinization). Long-term experiments have provided some of the best data on how various practices affect soil properties essential to sustainability. In the United States a federal agency, USDA-Natural Resources Conservation Service, specializes in providing technical and financial assistance for those interested in pursuing natural resource conservation and production agriculture as compatible goals.

The most important factors for an individual site are sun, air, soil, nutrients, and water. Of the five, water and soil quality and quantity are most amenable to human intervention through time and labor.

Although air and sunlight are available everywhere on earth, crops also depend on soil nutrients and the availability of water. When farmers grow and harvest crops, they remove some of these nutrients from the soil. Without replenishment, land suffers from nutrient depletion and becomes either unusable or suffers from reduced yields. Sustainable agriculture depends on replenishing the soil while minimizing the use or need of non-renewable resources, such as natural gas (used in converting atmospheric nitrogen into synthetic fertilizer), or mineral ores (e.g., phosphate). Possible sources of nitrogen that would, in principle, be available indefinitely, include:

(i) recycling crop waste and livestock or treated human manure

(ii) growing legume crops and forages such as peanuts or alfalfa that form symbioses with nitrogen-fixing bacteria called rhizobia

(iii) industrial production of nitrogen by the Haber process uses hydrogen, which is currently derived from natural gas (but this hydrogen could instead be made by electrolysis of water using electricity (perhaps from solar cells or windmills) or

(iv) genetically engineering (non-legume) crops to form nitrogen-fixing symbioses or fix nitrogen without microbial symbionts.

The last option was proposed in the 1970s, but is only gradually becoming feasible. Sustainable options for replacing other nutrient inputs such as phosphorus and potassium are more limited.

More realistic, and often overlooked, options include long-term crop rotations, returning to natural cycles that annually flood cultivated lands (returning lost nutrients indefinitely) such as the flooding of the Nile, the long-term use of biochar, and use of crop and livestock landraces that are adapted to less than ideal conditions such as pests, drought, or lack of nutrients. Crops that require high levels of soil nutrients can be cultivated in a more sustainable manner with appropriate fertilizer management practices.

Water

In some areas sufficient rainfall is available for crop growth, but many other areas require irrigation. For irrigation systems to be sustainable, they require proper management (to avoid salinization) and must not use more water from their source than is naturally replenishable. Otherwise, the water source effectively becomes a non-renewable resource. Improvements in water well drilling technology and submersible pumps, combined with the development of drip irrigation and low-pressure pivots, have made it possible to regularly achieve high crop yields in areas where reliance on rainfall alone had previously made successful agriculture unpredictable. However, this progress has come at a price. In many areas, such as the Ogallala Aquifer, the water is being used faster than it can be replenished.

Several steps must be taken to develop drought-resistant farming systems even in "normal" years with average rainfall. These measures include both policy and management actions:

- improving water conservation and storage measures,
- providing incentives for selection of drought-tolerant crop species,
- using reduced-volume irrigation systems,
- managing crops to reduce water loss, and
- not planting crops at all.

Indicators for sustainable water resource development are:

- **Internal renewable water resources:** This is the average annual flow of rivers and groundwater generated from endogenous precipitation, after ensuring that there is no double counting. It represents the maximum amount of water resource produced within the boundaries of a country. This value, which is expressed as an average on a yearly basis, is invariant in time (except in the case of proved climate change). The indicator can be expressed in three different units: in absolute terms (km^3/yr), in mm/yr (it is a measure of the humidity of the country), and as a function of population (m^3/person per year).
- **Global renewable water resources:** This is the sum of internal renewable water resources and incoming flow originating outside the country. Unlike internal resources, this value can vary with time if upstream development reduces water availability at the border. Treaties ensuring a specific flow to be reserved from upstream to downstream countries may be taken into account in the computation of global water resources in both countries.
- **Dependency ratio:** This is the proportion of the global renewable water resources originating outside the country, expressed in percentage. It is an expression of the level to which the water resources of a country depend on neighbouring countries.
- **Water withdrawal:** In view of the limitations described above, only gross water withdrawal can be computed systematically on a country basis as a measure of water use. Absolute or per-person value of yearly water withdrawal gives a measure of the importance of water in the country's economy. When expressed in percentage of water resources, it shows the degree of pressure on water resources. A rough estimate shows that if water withdrawal exceeds a quarter of global renewable water resources of a country, water can be considered a limiting factor to development and, reciprocally, the pressure on water resources can affect all sectors, from agriculture to environment and fisheries.

Soil

Soil erosion is fast becoming one of the world's severest problem. It is estimated that more than a thousand million tonnes of southern Africa's soil are eroded every year. Experts predict that crop yields will be halved within thirty to fifty years if erosion continues at present rates. Soil erosion is occurring worldwide. The phenomenon is being called *peak soil* as present large-scale factory farming techniques are jeopardizing humanity's ability to grow food in the present and in the future. Without efforts to improve soil management practices, the availability of arable soil will become increasingly problematic. Intensive agriculture reduces the carbon level in soil, impairing soil structure, crop growth and ecosystem functioning, and accelerating climate change. Soil management techniques include no-till farming, keyline design, windbreaks to reduce wind erosion, incorporating carbon-containing organic matter back into fields, reducing chemical fertilizers, and protecting soil from water run-off.

Land

As the global population increases and demand for food increases, there is pressure on land resources. In land use planning and management, considering the impacts of land use changes on factors such as soil erosion can support long-term agricultural sustainability, as shown by a study of Wadi Ziqlab, a dry area in the Middle East where farmers graze livestock and grow olives, vegetables, and grains.

Looking back over the 20th century shows that for people in poverty, following environmentally sound land practices has not always been a viable option due to many complex and challenging life circumstances. Currently, increased land degradation in developing countries may be connected with rural poverty among smallholder farmers when forced into unsustainable agricultural practices out of necessity.

Land is a finite resource on Earth. And although expansion of agricultural land can decrease biodiversity and contribute to deforestation, the picture is complex; for instance, a study examining the introduction of sheep by Norse settlers (Vikings) to the Faroe Islands of the North Atlantic concluded that, over time, the fine partitioning of land plots contributed more to soil erosion and degradation than grazing itself.

The Food and Agriculture Organisation of the United Nations estimates that in the coming decades, crop land will continue to be lost to industrial and urban development, along with reclamation of wet lands, and conversion of forest to cultivation, resulting in the loss of biodiversity and increased soil erosion. Many tools will be called upon to offset these projections. In Europe, one such tool is a geo-spatial data system called Soil Cons Web which is being developed to inform soil conservation minded decision making within agricultural sectors and other areas of land management.

Energy

Energy is used all the way down the food chain from farm to fork. In industrial agriculture, energy is used in on-farm mechanisation, food processing, storage, and transportation processes. It has therefore been found that energy prices are closely linked to food prices. Oil is also used as an input in agricultural chemicals. the International Energy Agency projects higher prices of non-renewable energy resources as a result of fossil fuel resources being depleted. It may therefore decrease global food security unless action is taken to 'decouple' fossil fuel energy from food production, with a move towards 'energy-smart' agricultural systems including renewable energy. The use of solar powered irrigation in Pakistan has come to be recognized as a leading example of energy use in creating a closed system for water irrigation in agricultural activity.

Economics

Socioeconomic aspects of sustainability are also partly understood. Regarding less concentrated farming, the best known analysis is Netting's study on smallholder systems through history.

Given the finite supply of natural resources at any specific cost and location, agriculture that is inefficient or damaging to needed resources may eventually exhaust the available resources or the ability to afford and acquire them. It may also generate negative externality, such as pollution as well as financial and production costs. There are several studies incorporating these negative externalities in an economic analysis concerning ecosystem services, biodiversity, land degradation and sustainable land management. These include The Economics of Ecosystems and Biodiversity study led by Pavan Sukhdev and the Economics of Land Degradation Initiative which seeks to establish an economic cost benefit analysis on the practice of sustainable land management and sustainable agriculture.

The way that crops are sold must be accounted for in the sustainability equation. Food sold locally does not require additional energy for transportation (including consumers). Food sold at a remote location, whether at a farmers' market or the supermarket, incurs a different set of energy cost for materials, labour, and transport.

Pursuing sustainable agriculture results in many localized benefits. Having the opportunities to sell products directly to consumers, rather than at wholesale or commodity prices, allows farmers to bring in optimal profit.

METHODS OF SUSTAINABLE AGRICULTURE

1. **Crop rotation:** Crop rotation is one of the most powerful techniques of sustainable agriculture. Its purpose is to avoid the consequences that come with planting the same crops in the same soil for years in a row. It helps tackle pest problems, as many pests prefer specific crops. If the pests have a steady food supply they can greatly increase their population size. Rotation breaks the reproduction cycles of pests. During rotation, farmers can plant certain crops, which replenish plant nutrients. These crops reduce the need for chemical fertilizers.
2. **Cover crops:** Many farmers choose to have crops planted in a field at all times and never leave it barren, this can cause unintended consequences. By planting cover crops, such as clover or oats, the farmer can achieve his goals of preventing soil erosion, suppressing the growth of weeds, and enhancing the quality of the soil. The use of cover crops also reduces the need for chemicals such as fertilizers.
3. **Soil enrichment:** Soil is a central component of agricultural ecosystems. Healthy soil is full of life, which can often be killed by the overuse of pesticides. Good soils can increase yields as well as creating more robust

crops. It is possible to maintain and enhance the quality of soil in many ways. Some examples include leaving crop residue in the field after a harvest, and the use of composted plant material or animal manure.

4. **Natural pest predators:** In order to maintain effective control over pests, it is important to view the farm as an ecosystem as opposed to a factory. For example, many birds and other animals are in fact natural predators of agricultural pests. Managing your farm so that it can harbor populations of these pest predators is an effective as well as a sophisticated technique. The use of chemical pesticides can result in the indiscriminate killing of pest predators.
5. **Bio intensive integrated pest management:** Integrated Pest Management (IPM). This is an approach, which really relies on biological as opposed to chemical methods. IMP also emphasizes the importance of crop rotation to combat pest management. Once a pest problem is identified, IPM will mean that chemical solutions will only be used as a last resort. Instead the appropriate responses would be the use of sterile males, and biocontrol agents such as ladybirds.

What grows where and how it is grown are a matter of choice. Two of the many possible practices of sustainable agriculture are crop rotation and soil amendment, both designed to ensure that crops being cultivated can obtain the necessary nutrients for healthy growth. Soil amendments would include using locally available compost from community recycling centers. These community recycling centers help produce the compost needed by the local organic farms.

Using community recycling from yard and kitchen waste utilizes a local area's commonly available resources. These resources in the past were thrown away into large waste disposal sites, are now used to produce low cost organic compost for organic farming. Other practices includes growing a diverse number of perennial crops in a single field, each of which would grow in separate season so as not to compete with each other for natural resources. This system would result in increased resistance to diseases and decreased effects of erosion and loss of nutrients in soil. Nitrogen fixation from legumes, for example, used in conjunction with plants that rely on nitrate from soil for growth, helps to allow the land to be reused annually. Legumes will grow for a season and replenish the soil with ammonium and nitrate, and the next season other plants can be seeded and grown in the field in preparation for harvest.

Monoculture, a method of growing only one crop at a time in a given field, is a very widespread practice, but there are questions about its sustainability, especially if the same crop is grown every year. Today it is realized to get around this problem local cities and farms can work together to produce the needed compost for the farmers around them. This combined with growing a mixture of crops (polyculture) sometimes reduces disease or pest problems but polyculture has rarely, if ever, been compared to the more widespread practice of growing different crops in successive years (crop rotation) with the same overall crop diversity. Such methods may also support sustainable weed management in that the development of herbicide-resistant weeds is reduced. Cropping systems that include a variety of crops (polyculture and/or rotation) may also replenish nitrogen (if legumes are included) and may also use resources such as sunlight, water, or nutrients more efficiently.

Replacing a natural ecosystem with a few specifically chosen plant varieties reduces the genetic diversity found in wildlife and makes the organisms susceptible to widespread disease. The Great Irish Famine (1845–1849) is a well-known example of the dangers of monoculture. In practice, there is no single approach to sustainable agriculture, as the precise goals and methods must be adapted to each individual case. There may be some techniques of farming that are inherently in conflict with the concept of sustainability, but there is widespread misunderstanding on effects of some practices. Today the growth of local farmers' markets offer small farms the ability to sell the products that they have grown back to the cities that they got the recycled compost from. This will help move people away from the slash-and-burn or slash-and-char techniques that are the characteristic feature of shifting cultivation. These are often cited as inherently destructive, yet slash-and-burn cultivation has been practiced in the Amazon for at least 6000 years.

There are also many ways to practice sustainable animal husbandry. Some of the key tools to grazing management include fencing off the grazing area into smaller areas called paddocks, lowering stock density, and moving the stock between paddocks frequently.

Sustainable Intensification

In the light of concerns about food security, human population growth and dwindling land suitable for agriculture, sustainable intensive farming practises are needed to maintain high crop yields, while maintaining soil health and ecosystem services. The capacity for ecosystem services to be strong enough to allow a reduction in use of synthetic, non renewable inputs whilst maintaining or even boosting yields has been the subject of much debate. Recent work in the globally important irrigated rice production system of east Asia has suggested that - in relation to pest management at least - promoting the ecosystem service of biological control using nectar plants can reduce the need for insecticides by 70% whilst delivering a 5% yield advantage compared with standard practice.

Soil Treatment

Soil steaming can be used as an ecological alternative to chemicals for soil sterilization. Different methods are available to induce steam into the soil in order to kill pests and increase soil health.

Solarizing is based on the same principle, used to increase the temperature of the soil to kill pathogens and pests.

Certain crops act as natural bio fumigants, releasing pest suppressing compounds. Mustard, radishes, and other plants in the brassica family are best known for this effect. There exist varieties of mustard shown to be almost as effective as synthetic fumigants at a similar or lesser cost.

Off-farm Impacts

A farm that is able to "produce perpetually", yet has negative effects on environmental quality elsewhere is not sustainable agriculture. An example of a case in which a global view may be warranted is over-application of synthetic fertilizer or animal manures, which can improve productivity of a farm but can pollute nearby rivers and coastal waters (eutrophication). The other extreme can also be undesirable, as the problem of low crop yields due to exhaustion of nutrients in the soil has been related to rainforest destruction, as in the case of slash and burn farming for livestock feed.In Asia, specific land for sustainable farming is about 12.5 acres which includes land for animal fodder, cereals productions lands for some cash crops and even recycling of related food crops.In some cases even a small unit of aquaculture is also included in this number (AARI-1996).

Sustainability affects overall production, which must increase to meet the increasing food and fiber requirements as the world's human population expands to a projected 9.3 billion people by 2050. Increased production may come from creating new farmland, which may ameliorate carbon dioxide emissions if done through reclamation of desert as in Israel and Palestine, or may worsen emissions if done through slash and burn farming, as in Brazil.

Anthropogenic Changes

As the earth is entering the Anthropocene, an epoch characterized by human impacts such as climate change, agriculture and agricultural development are at risk. Agriculture has an enormous environmental footprint, and is simultaneously leading to huge amounts of environmental changes globally and being hugely impacted by these global changes. Additionally, the human population is continuing to grow rapidly at a rate which will require an increase in food production globally. This is complicated by the fact that the earth is undergoing rising amounts of environmental risks. Sustainable agriculture provides a potential solution to enable agricultural systems to feed a growing population while successfully operating within the changing environmental conditions.

Social Development

In 2007, the United Nations reported on "Organic Agriculture and Food Security", stating that using organic and sustainable agriculture could be used as a tool to reach global food security without expanding land usage and reducing environmental impacts. Another way to define sustainable agriculture is to give attention to the "human and environmental aspects," because of the turn to a more unsustainable way of farming in U.S. agriculture. During the Great Depression in the United States many farming families were living in subhuman and hungry conditions and treated "sustainability as a resource-input and food-output equation." Though conditions have improved, the farming has not as much done so. There has been evidence provided by developing nations from the early 2000s stating that when people in their communities are not factored into the agricultural process that serious harm is done. Although global food security would most likely not drastically fall, these practices would impact, first hand, local, rural farming communities, making them unable to feed themselves and their families. The social scientist Charles Kellogg has stated that, "In a final effort, exploited people pass their suffering to the land." This turn to more unsustainable farming has seen suffering for many people. For if something is sustainable, it should be that way in all aspects of it, not just the crop yield or soil health. It has been seen in the developing country, the starving of rural farming communities are mainly due to their unsustainable farming methods. Sustainable agriculture mean the ability to permanently and continuously "feed its constituent populations."

There are a lot of opportunities that can increase farmers' profits, better communities, and continue sustainable practices. For example, in Uganda Genetically Modified Organisms (GMOs) were originally illegal, however, under stressful circumstances where Banana Bacterial Wilt (BBW) has the potential to wipe out 90% of yield they decided to explore GMOs as a possible solution. Therefore, as a result of the banana crisis in Uganda caused by the BBW, the government issued the National Biotechnology and Biosafety bill which will allow scientists that are part of the National Banana Research Program to start experimenting with genetically modified organisms. This effort has the potential to help local communities because a significant portion live off the food they grow themselves and it will keep their economy in check because their main sources of produce will remain stable.

Women

In the past 30 years (1978-2007) in the United States the number of women farm operators has tripled. Today, women operate 14% of farms, compared to 5.0 % in 1978. Much of the growth is due to women farming outside the "male dominated field of conventional agriculture". In community supported agriculture women represent 40% of farm operators, and 21% of organic farmers. With the change of laws in land ownership over the past century, women are now allowed the same freedom of land ownership that the men have.

International Policy

Sustainable agriculture has become a topic of interest in the international policy arena, especially with regards to its potential to reduce the risks associated with a changing climate and growing human population.

The Commission on Sustainable Agriculture and Climate Change, as part of its recommendations for policy makers on achieving food security in the face of climate change, urged that sustainable agriculture must be integrated into national and international policy. The Commission stressed that increasing weather variability and climate shocks will negatively affect agricultural yields, necessitating early action to drive change in agricultural production systems towards increasing resilience. It also called for dramatically increased investments in sustainable agriculture in the next decade, including in national research and development budgets, land rehabilitation, economic incentives, and infrastructure improvement.

Policy Ethics

Most agricultural professionals agree that there is a "moral obligation to pursue the goal of sustainability." The major debate comes from what system will provide a path to that goal. Because if an unsustainable method is used on a large scale it will have a massive negative effect on the environment and human population. The best way to create policy for agriculture is to be free of any bias. A good review would be done with "practical wisdom," a virtue identified by Aristotle, distinguishing practical wisdom from scientific knowledge, this coming from Nichomachean Ethics. The science of agriculture is called "agronomy", the root of this word relating to scientific law. Although agriculture may not fit well under scientific law, and may not be designed to be treated as an Aristotelian scientific knowledge, but more practical wisdom. Practical wisdom requires recognition of past failures in agriculture to better attain a more sustainable agricultural system.

Urban Planning

The use of available city space (e.g., rooftop gardens, community gardens, garden sharing, and other forms of urban agriculture) for cooperative food production may be able to contribute to sustainability. A recent idea (2014) is to create large, urban, technical facilities for vertical farming. Potential advantages include year-round production, isolation from pests and diseases, controllable resource recycling, and reduced transportation costs.

Increasing threats of climate change have influenced cities and public officials are thinking more proactively about the ways they can deliver services and food more efficiently.The environmental cost of transportation could be avoided if people take back their connection to fresh food. This raises questions; however, about the excess environmental costs associated with local farming vs more large scale operations which offer food security around the world.

Key Debates

There are several key debates involving sustainable agriculture:

Ecocentric vs Technocentric

The main debate on how sustainable agriculture might be achieved centers around two different approaches: an ecocentric approach and a technocentric approach. The ecocentric approach emphasizes no- or low-growth levels of human development, and focuses on organic and biodynamic farming techniques with the goal of changing consumption patterns, and resource allocation and usage. The technocentric approach argues that sustainability can be attained through a variety of strategies, from the view that state-led modification of the industrial system like conservation-oriented farming systems should be implemented, to the argument that biotechnology is the best way to meet the increasing demand for food.

Multifunctional Agriculture vs Ecosystem Services

There are different scientific communities that are looking at the topic of sustainable agriculture through two separate lenses: multifunctional agriculture (MFA) and ecosystem services (ES). While both of these frameworks are similar, they look at the function of agriculture in different lights. Those that employ the multifunctional agriculture philosophy focus on farm-centered approaches, and define function as being the outputs of agricultural activity. The central argument of MFA is that agriculture has other functions aside from the production of food and fiber, and therefore agriculture is a multifunctional enterprise. These additional functions include renewable natural resource management and conservation of landscape and biodiversity. On the other hand, ES focuses on service-centered approaches, and defines function as the provision of services to human beings. Specifically, ES posits that individuals and society as a whole receive benefits from ecosystems, which are called ecosystem services. Within the field of sustainable agriculture, the services that ecosystems provide include pollination, soil formation, and nutrient cycling, all of which are necessary functions for the production of food.

Barriers

Since World War II, dominant models of agriculture in the United States and the entire national food system have been characterized by a focus on monetary profitability at the expense of social and environmental integrity.

In sustainable agriculture, changes in lower rates of soil and nutrient loss, improved soil structure, and higher levels of beneficial microorganisms are not quick. The changes are not immediately evident to the operate when using sustainable agriculture. In conventional agriculture the benefits are easily visible with no weeds, pests, etc. and the "process of externalization" hides the costs to soil and ecosystems around it. A major barrier to sustainable agriculture is the lack of knowledge of its benefits. Many benefits are not visible, so they are often unknown.

Sustainable Agriculture can be Achieved by Adapting

- Mixed farming
- Mixed cropping
- Crop rotation
- Crop selection
- Varietal improvement.

SUSTAINABLE AGRICULTURE IN INDIA

A discussion on Indian agriculture is fraught with the possibility of it being redirected to issues not necessarily connected to agriculture. Yet, considering the dependence of our population on it, agriculture needs to be brought centre-stage.

The prime challenge is sustainability and related issues are as follows:

According to ICAR's reports, a staggering 37% of India's total geographical area of 328.73 million ha is affected due to land degradation. A staggering amount of soil gets degraded annually. This is an irony considering the fact that a big amount of our fiscal budget is spent on defending our soil even as we lose the very same soil to rain and wind!

Similarly, our water resources are also getting degraded. Only about 47.6 per cent of the net sown area is irrigated. Groundwater, which accounts for almost 60 per cent of the irrigated area in the country, is under severe strain. Subsidies for electricity consumed in agriculture have led to wasteful use of both energy and groundwater. This has led to depletion of the water table and deterioration of water quality.

Further, in the face of competing demands for groundwater due to population growth and industrialisation, the share of agriculture in available water resources is expected to fall from the current 83% to 68% by 2050.

Adverse climate change is also a risk to sustainability. It leads to drop in yield and lower quality produce, and increases incidence of attacks by pests and insects.

The figures emerging from Network Programme on Climate Change, ICAR, paint a disturbing scenario. A reduction of 4.5 to 9% in agriculture yields is expected in the medium term (2010-2039), and over 25% in the long term (2040 and beyond) if no corrective measures are taken. What this means for India's GDP growth in the medium term is a hit of up to 2% per annum.

Sustainable agriculture must involve measures for soil conservation, water conservation and irrigation, and mitigation of the adverse effects of climate change. Large-scale afforestation, encompassing commercial forestry, farm forestry, social forestry, captive plantation as also adoption of community-based forest management practices are needed.

Other measures for water conservation include rainwater harvesting, construction of check dams, farm ponds, and irrigation-efficient technology such as drip and sprinkler systems. In the long run, we have to revisit unsustainable crop patterns from the water-usage point of view.

We must swiftly adopt climate smart agricultural practices as spelt out by the FAO. This calls for using renewable sources such as bio fuels and solar; nitrogen-smart nutrient management, organic farming; carbon smart practices (agroforestry, horticulture, livestock management); weather smart extension services (ICT-based agro advisories) and so on.

Schemes Galore

The good news is that we have a well-defined array of schemes for almost all the threats crippling agriculture. The Pradhan Mantri Krishi Sichayee Yojana aims to extend the coverage of irrigation and improve water use efficiency. The Micro Irrigation Fund focuses on "more crop per drop". The Long Term Irrigation Fund has the potential to bring an additional 76 lakh ha area under irrigation.

The Rashtriya Krishi Vikas Yojana incentivises States to draw up plans for their agriculture sector more comprehensively, taking agro-climatic conditions, natural resource issues and technology into account. The Pradhan Mantri Fasal Bima Yojana is a comprehensive insurance scheme that covers the damage caused by natural calamities. Schemes like Dairy Processing and Infrastructure Development Fund reduce the dependence of farmers on agriculture.

Promoting Sustainable Agriculture in India

Agriculture sector, world over, has experienced a phenomenal growth since the mid-20th century. The growth, driven by Green Revolution technology, has made a significant dent on aggregate supply of food grains, ensuring food security to the growing population. The next stage of agricultural growth however, faces a serious challenge in terms of sustainability. Whereas the main problem faced by the developing countries in the south pertains to sustainability of resource use, the main challenge facing the developed economies in the north is overuse of chemical inputs. These problems have led to increasing awareness and a felt need for moving away from the input intensive agriculture perused during the Green revolution phase, to sustainable farming in different parts of the world.

While the need for a paradigmatic shift in the growth strategy is well recognized, the transition from input intensive to sustainable farming however, has certain inherent difficulties. Not with standing these limitations, policies in both the north as well as the south have led increased emphasis on promoting sustainable agriculture.

India can safely be characterized as an agricultural country despite the recent spurt in manufacturing and services and the declining share of agriculture in the national income, since majority of its workforce (~ 65%) are still engaged in agriculture and allied activities. It has been the noblest profession in India since the time immemorial and has been carried out on sustainable basis. It is only relatively recent phenomenon that large-scale forest areas, grazing lands and waste lands have been converted into croplands to support the rising population, which has caused ecological imbalance and atmospheric pollution. With no further scope for expansion of agricultural land efforts have been made to enhance the production of food grains using high-yielding variety of seeds, fertilizers and irrigation along with advanced farm equipments. However, so-called green revolution is confined to a few crops, viz, wheat, rice and maize and has been possible only in restricted areas, i.e., Punjab, Haryana and Western Uttar Pradesh and certain selected districts of Andhra Pradesh, Maharashtra and Tamilnadu.

Naturally, much work is needed to lift the agriculture to a level where it is least affected by vagaries of monsoon and needs little from outside the farm, i.e., lesser dependence on chemical fertilizers and water. The limited success of green revolution has been a mixed bag in that it has given rise to new set of problems: overuse of water and fertilizers. Excessive use of water results in water logging and salinization whereas excess of fertilizers and pesticide cause pollution of water bodies contamination of ground water. India has the largest area of irrigated land (55 million hectares) of which about one-third land is already degraded and 7 million hectare have been abandoned. In such a situation a renewable and lasting alternative, sustainable agriculture, has to emerge for successful agricultural revolution.

TYPES OF FARMING

We can compare three broad types of farming: traditional production systems, conventional modern agriculture (such as Green Revolution technologies), and sustainable agriculture. We can compare them across three dimensions: ecological, economic and social.

1. Ecological sustainability

Many traditional and most conventional farm practices are not ecologically sustainable: they overuse natural resources, reducing soil fertility, causing soil erosion, and contributing to global climatic change. Sustainable agriculture has several major advantages over both traditional and conventional practices:

- **Soil fertility:** A continuous fall in soil fertility is a major problem in many parts of India. Sustainable agriculture improves fertility and soil structure and prevents erosion, so would be an answer to this problem.
- **Water:** Irrigation is the biggest consumer of freshwater, and fertilizer and pesticides contaminate both surface- and groundwater. Sustainable agriculture increases the organic matter content of the topsoil, so raising its ability to retain and store water that falls as rain.
- **Biodiversity:** Sustainable agricultural practices frequently involve mixed cropping, so increasing the diversity of crops produced and raising the diversity of insects and other animals and plants in and around fields.
- **Pollution:** Pesticides are hazardous to human health as well as to the local ecology. Incorrect handling, storage and use of pesticides lead to health and pollution problems. Sustainable agriculture reduces or eliminates the use of hazardous chemicals; instead it controls pests with a variety of biological and agronomic measures and the use of natural substances.
- **Landscape:** Agriculture and forestry decides the rural landscape. Inappropriate use causes erosion, landslides and flooding, clogs irrigation channels, and reduces the ability of the land to support the local population. Impoverished rural people flock into the cities in search of jobs, forming unsightly, insanitary slums that further destroy the landscape. Rehabilitating ecologically damaged areas needs huge investments that few countries can afford. Sustainable agriculture avoids these problems by improving productivity, conserving the soil, avoiding the expansion of farming into unsuitable areas, and preserving rural jobs.
- **Climate:** The way agriculture is practiced contributes significantly to global climatic changes. Conventional agriculture contributes to the production of greenhouse gases in various ways: by reducing the amount of carbon stored in the soil and in vegetation, through the production of methane in irrigated fields, and through energy-intensive activities such as the production of artificial fertilizers. Adopting sustainable agriculture would reduce these impacts significantly.

2. Economic Sustainability

Agriculture cannot be sustainable unless it is economically viable over the long term. Conventional agriculture poses greater long-term economic risks than "sustainable" alternatives.

- **Export vs. local orientation:** Governments tend to view export-oriented production systems as more important than those that supply domestic demands. This is misguided. Focusing on exports alone involves hidden costs: in transport, in assuring local food security, etc. Policies should treat domestic demand and in particular food security (either by farmers producing food for themselves, or by selling produce for cash they can use to buy food) as equally important to the visible trade balance.
- **Debt:** The Green Revolution raised India's grain output significantly, but a vast number of small-scale farmers ran into a debt trap: they took out loans to raise their production, and then found they could not pay the money back. About 40,000 were so desperate that they committed suicide.
- **Risk:** Concentrating on specific commodities seems to promise high economic returns. But market production implies certain risks: markets change quickly, and international agricultural prices are dropping. Cheap foreign food may sweep into the national market, leaving Indian farmers without a market. As a World Trade Organization signatory, the Indian government is under pressure to deregulate and open its economy to the world market so cannot protect its farmers behind tariff walls.
- **Niche markets:** Organic agriculture is one of the strongest ways to farm in an environmentally sustainable way. The demand for certified organic products is increasing quickly, opening opportunities to expand sales of such products and to explore niche markets.
- **Employment:** Farming is the main source of employment for rural people. Trends towards specialization and mechanization may increase narrowly measured "efficiency", but they reduce employment on the land. The welfare costs of unemployment must be taken into account when designing national agricultural support programmes. Sustainable agriculture, with its emphasis on small-scale, labour-intensive activities, helps overcome these problems.

3. Social Sustainability

The social sustainability of farming techniques is related to the ideas of social acceptability and justice.

- **Inclusiveness:** Development cannot be sustainable unless it reduces poverty for the broad masses of people in India. The government must find ways to enable the rural poor to benefit from agricultural development.
- **Political unrest:** Gaps between the "haves" and "have-nots" feed a feeling of social injustice among those who feel neglected and excluded from development opportunities, as well as from better-off sympathizers. The result is a climate favorable to political opposition and even violence.

- **Local acceptance:** Many new technologies fail because they are based on practices or assumptions from outside. Sustainable agricultural practices usually are based on local social customs, traditions, norms and taboos, so local people are more likely to accept them and adapt them to their own needs.
- **Indigenous knowledge:** Sustainable agricultural practices often rely on traditional knowhow and local innovation. Local people have a wealth of knowledge about their environment, crops and livestock. They keep locally adapted breeds and crop varieties. They have social structures that manage and conserve common resources, help people in need, and maintain the social fabric. Rather than ignoring or replacing this knowledge, sustainable agricultural development seeks to build on it and enrich it with appropriate information from outside.
- **Gender:** In traditional agriculture, women traditionally bear the heaviest burdens in terms of labour. In modern conventional farming, too, men often benefit the most: they control what is grown and how the resulting income is spent. Sustainable agriculture attempts to ensure that the burdens and benefits are shared more equitably between men and women.
- **Food security:** Traditional farming techniques often fail to produce enough food, or enough variety of food for a balanced diet. Conventional modern farming focuses on a few commodities, so people still do not have a balanced diet. Sustainable agriculture improves food security by improving the quality and nutritional valuc of the food, and by producing a bigger range of produce throughout the year.
- **Participation:** Traditional society in India is driven by wealth and caste distinctions. Introducing conventional farming innovations tends to exacerbate these: the rich and higher-caste tend to benefit, while the poor and lower-caste are left out. Sustainable agricultural interventions consciously target the less well-off, and empower them so they can organize and speak with their own "voice", so promoting dialogue and democracy.

Policies for Sustainable Agriculture and Organic Farming

The Indian government's policies have always emphasized food grain self-sufficiency, which has not necessarily coincided with agricultural sustainability. The growth of agricultural production and productivity, which had risen significantly during 1970s and 1980s, declined during 1990s. These slowdowns have worsened since 2000, both overall agricultural production and food grains production have shown negative growth rates in 2000-01 to 2002-03 period. Decline in the growth rates of agricultural production and productivity is a serious issue considering the questions of food security, livelihood, and environment. As such, a critical examination of the approaches for sustainable agricultural development is necessary. This examination must be framed not only by India's ongoing need to ensure food self-sufficiency but also by the consequences of access to international markets.

Sustainable Agriculture – Possible Actions in India

- Improvement of existing production systems (e.g. altered crop rotations, introduction of green manuring, use of plant species adapted to specific locations)
- Improved protection of natural resources (e.g. erosion protection)
- Increase in efficiency of existing resources (e.g. irrigation, use of technology, basic and advanced training)
- Introduction of regenerative branches of business (e.g. horticulture or aquaculture)
- Introduction of a new production element in existing enterprises (such as fruit trees to stabilize terraced fields, fish-farming in rice fields)
- Optimization of post-harvest systems (e.g. storage)
- Increase the value of agricultural products through further processing (e.g. production of yoghurt from milk)
- Improvement of channels of distribution (e.g. market access, transport)
- Access to loans and other financial services
- Covering risk (e.g. through land law, support of producer groups)

MITIGATION TECHNOLOGIES

The second Law of Ecology – everything must go somewhere – relates directly to the awareness that emitting greenhouse gases into the atmosphere in increasing quantities *ad infinitum* will have a major visible effect in the not too distant future on climate, ecosystems and societies. Adverse signs are appearing around the globe that strongly suggest that rates of change are occurring faster than scientists have been anticipating: most glaciers around the globe are melting and sea level is rising, warm temperature ecosystems (Table 8.1)

Table 8.1: Key mitigation technologies and practices in agriculture and forestry, policies and measures, constraints and opportunities (IPCC, 2007b)

Key mitigation technologies and practices	Policies, measures and instruments shown to be environmentally effective	Key constraints	Key opportunities
Improved crop and grazing land management to increase soil carbon storage; restoration of cultivated peaty soils and degraded lands; improved rice cultivation; techniques and livestock and manure management to reduce CH4 emissions; improved nitrogen fertilizer application techniques to reduce N_2O emissions; dedicated energy crops to replace fossil fuel use; improved energy efficiency; improvements of crop yields	Financial incentives and regulations for improved land management; maintaining soil carbon content; efficient use of fertilizers and irrigation		May encourage synergy with sustainable development and with reducing vulnerability to climate change, thereby overcoming barriers to implementation

Afforestation; reforestation; forest management; reduced deforestation; harvested wood product management; use of forestry products for bioenergy to replace fossil fuel use; tree species improvement to increase biomass productivity and carbon sequestration; improved remote sensing technologies for analysis of vegetation/soil carbon sequestration potential and mapping land-use change; Landfill management and monitoring	Financial incentives (national and international) to increase forest area, to reduce deforestation and to maintain and manage forests; land-use regulation and enforcement	Constraints include lack of investment capital and land tenure issues	Can foster poverty alleviation

Moving upslope into higher altitudes and the area coverage of Arctic sea ice is rapidly decreasing. As scientists have learned in recent decades where those greenhouse gases are going and what they are doing to the earth's atmosphere and oceans, governments have started to seek ways both to reduce their sources and to increase the sinks for those gases and to prepare for foreseeable adverse impacts. Mitigation refers to technological change and substitution that reduce energy resource inputs and emissions per unit of output. Although several social, economic and technological policies would also lead to emission reduction, climate change mitigation encompasses implementing policies to reduce greenhouse gas emissions and to enhance sinks. Table 8.1 provides selected examples of mitigation technologies, policies and measures as well as constraints and opportunities for agriculture and forests as outlined in the IPCC (2007) Synthesis Report. Box 2 provides details of GHG emission and mitigation potential in food and agriculture sector. A Danish government action program defined mitigation in the same words as those of the IPCC: "[mitigation] is an intervention to reduce human-caused net emissions of greenhouse gases." Its report suggested some obvious measures that governments could pursue for mitigation:

- Reduction (at the source) in the use of fossil fuels (clean coal technology, renewable energies)

Capture of Methane from Landfills and Rice Paddies

Creation of sinks for storing carbon through natural resource management (carbon sequestration) (e.g. reducing tropical deforestation and increasing tree planting) Mitigation policies, which require identifying effective ways to reduce the amount of greenhouse gases produced and released into the atmosphere, are the first and foremost line of defense for reducing emissions before the worst consequences of global warming are allowed to occur. Although mitigation is the preferred path, it is also perhaps the most difficult to achieve in a way that would have positive global results in a short time. One reason is that implementation of the many suggested mitigation techniques (e.g. transfer of clean technologies, switch to alternative sources of energy (including nuclear), capture and sequestration of carbon and other greenhouse gases such as methane, reduction of fertilizer use, more efficient

Conclusion

Under the changing agricultural scenario, the agricultural technologies need a shift from production-oriented to profit-oriented sustainable farming. In this direction, the pace of adoption of resource conserving technologies (RCTs) by the Indian farmers is satisfactory to a larger extent but, under the present scenario, we are in the half way of conservation agriculture. The CA systems will leads to sustainable farming and will be the thrust of future farming.

The conditions for development of sustainable agriculture are becoming more and more favorable. New opportunities are opening the eyes of farmers, development workers, researchers and policy makers. They now see the potential and importance of these practices not only for their direct economic interest but also as the basis of further intensification and ecological sustainability. This does not mean that agrochemicals can be abandoned. Also, research has an important role to play. Bankers and funders should think of how best to make and credits accessible to poor farmers and women and to make investment in dry land farming possible. As conditions for farming will continue to change, the key to sustainable agriculture is the capacity of farmers and all other actors in agricultural development, as well as the wider society, to learn, experiment, adapt and cooperate in an effective way. To conclude, a small farm management to improve productivity, profitability and sustainability of the farming system will go a long way to ensure all round sustainability.

9 SCOPE OF HYDROPONICS TECHNIQUES IN AGRICULTURE

Hydroponics, by definition, is a method of growing plants in a water-based, nutrient-rich solution. Hydroponics does not usc soil, instead the root system is supported using an inert medium such as perlite, rockwool, clay pellets, peat moss, or vermiculite. The basic premise behind hydroponics is to allow the plants roots to come in direct contact with the nutrient solution, while also having access to oxygen, which is essential for proper growth.

The term hydroponics originates from the ancient Greek "hydros", meaning water, and "ponos", meaning work. It can sometimes be mistakenly referred to as aquaculture, but these terms are really more appropriately used for other branches of science that have nothing to do with gardening.

HISTORY OF HYDROPONICS AND SOIL LESS GARDENING

While it's easy to imagine this kind of process being labeled as a bunch of new age science fiction, hydroponics has actually been in use for thousands of years. The famous Hanging Gardens of Babylon, one of the seven wonders of the ancient world, are largely believed to have functioned according to hydroponic principles. Built around 600 B.C. in Babylonia, or Mesopotamia, the gardens were situated along the Euphrates River. The area suffered from a dry, arid climate that rarely saw rain, and it is believed that the lush gardens were watered using a chain pully system, which carried water up from the river and allowed it to trickle down to each step or landing of the garden structure.

During the 10th and 11th centuries, the Aztecs developed a system of floating gardens based on hydroponics. Driven out of their land, they settled at Lake Tenochtitlan. Unable to grow crops on the lake's marshy shore, they built rafts out of reeds and roots. These rafts were topped with a bit of soil from the bottom of the lake, and then floated out to the center of the water. Crops would grow on top of the rafts, their roots reaching through the rafts and down into the water. Marco Polo's writings indicate he witnessed similar floating gardens while visiting China in the late 13th century.

Formal research and publications on hydroponics didn't begin until the 17th century. Sir Francis Bacon, a British scientist, philosopher and politician did research, on soil less gardening in the 1620s. His work on the subject was published posthumously in 1627 and sparked an incredible wave of research into hydroponics.

In 1699, another English scientist, John Woodward, performed tests involving spearmint growth in various water solutions. He attempted to grow spearmint plants in rain water, river water and water that had been mixed with soil and then drained. He found that the mint grew faster and produced healthier plants in the water solution that had been mixed with soil. His conclusion was that plants would grow better in less pure water than they would in distilled water. We know today that his results were due to minerals that remained in the water after it had been mixed with the soil.

A Berkeley scientist, William Gericke, promoted the use of hydroponics in commercial agriculture. Using a process he called "aquaculture," he touted the benefits of soil less gardening by growing massive tomatoes in his home via water and nutrient solutions. After finding that the term "aquaculture" was already being used to describe the study of aquatic organisms, he coined the term "hydroponics", which we still use today.

Two other Berkeley scientists, Dennis Hoagland and Daniel Arnon, later expanded upon Gericke's research. In 1938, they published "*The Water Culture Method for Growing Plants without Soil*", which is widely considered to be one of the most important texts ever published about hydroponics. Several of the nutrient solutions they developed are still used today.

According to a 1938 Time magazine article, one of the first commercial uses of hydroponics occurred during this period based on the research taking place at Berkeley. Tanks of mineralized water were used to grow beans, tomatoes, and vegetables on tiny Wake Island, a small piece of land in the Pacific Ocean. This island was used as a refueling stop for Pan-Am Airways, and the food grown there was used successfully to feed the airline's staff and crew. Similar situations occurred during World War II, as hydroponics was used to grow crops for troops on barren Pacific Islands.

Why use a Hydroponic System?

After all, people have been growing food just fine for thousands, if not millions of years using good old-fashioned dirt. Hydroponics offers some significant benefits over traditional farming, and as word about these benefits spreads, more people will turn to hydroponics for their agricultural needs.

First, hydroponics offers people the ability to grow food in places where traditional agriculture simply isn't possible. In areas with arid climates, like Arizona and Israel, hydroponics has been in use for decades. This science allows the people of these areas to enjoy locally grown produce and to expand their food production. Similarly, hydroponics is useful in dense urban areas, where land is at a premium. In Tokyo, hydroponics is used in lieu of traditional soil-based plant growth. Hydroponics is also useful in remotes locales, such as Bermuda. With so little space available for planting, Bermudians have turned to hydroponic systems, which take around 20% of the land usually required for crop growth. This allows the citizens of the island to enjoy year-round local produce without the expense and delay of importation. Finally, areas that don't receive consistent sunlight or warm weather can benefit from hydroponics. Places like Alaska and Russia, where growing seasons are shorter, use hydroponic greenhouses, where light and temperature can be controlled to produce higher crop yields.

We also must consider the significant environmental benefits to hydroponics use. Hydroponics systems require only around 10% of the water that soil-based agriculture requires. This is due to the fact that hydroponic systems allow recycling and reuse of water and nutrient solutions, and the fact that no water is wasted. This can have quite an impact on areas where water is scarce, such as in the Middle East and parts of Africa. Similarly, hydroponics requires little or no pesticides and only around 25% of the nutrients and fertilizers required of soil-based plants. This represents not only a cost savings but also benefits the environment in that no chemicals are being released into the air. Finally, we must consider the environmental impacts of transportation. As hydroponics allows produce to be grown locally and requires fewer areas to import their crops, there is a reduction in both price and greenhouse gas emissions due to reduced transportation requirements.

Next, hydroponics offers us the benefit of a shorter harvest time. Plants grown in this manner have direct access to water and nutrients and therefore, are not forced to develop extensive root systems to allow them to find the nutrients they need. This saves time and produces healthier, lusher plants in about half the time as traditional agriculture.

So why isn't hydroponics taking over? This is due to several distinct disadvantages associated with these systems. The first is the high capital investment when compared with soil farming. Though hydroponics is typically much cheaper over time, it does require a substantial upfront cost to establish any sort of larger system. Next, there's the threat of power failure, which can cause pumps to stop working and ruin crops. Finally, many people fear that hydroponics requires substantial know-how and research, when in fact, it's very similar to traditional gardening. After all, plants rely on certain nutrients in order to grow, and these nutrients don't change, no matter which system you're using.

The Science behind Hydroponic Nutrients

Before we can take a look at how hydroponics works, we must first understand how plants themselves work. Generally speaking, plants need very little to grow. They can subsist on a simple blend of water, sunlight, carbon dioxide and mineral nutrients from the soil. Plants are able to transform light energy into chemical energy to form sugars that allow them to grow and sustain themselves. Thus, plants convert carbon dioxide, water and light into sugars and oxygen through a process called photosynthesis. The photosynthesis process requires that the plant has access to certain minerals, especially nitrogen, phosphorus and potassium. These nutrients can be naturally occurring in soil and are found in most commercial fertilizers. Notice that the soil itself is not required for plant growth: the plant simply needs the minerals from the soil. This is the basic premise behind hydroponics – all the elements required for plant growth are the same as with traditional soil-based gardening. Hydroponics simply takes away the soil requirements.

There are several different types of hydroponic systems, though each is based on the same initial concepts. Here we'll examine each type, discover how and why it's used and see which kinds of plants respond best to each method.

Ebb and Flow Systems require a medium, such as perlite, which serves no purpose other than to provide stability for the plant's roots. The plant derives no nutrients from the medium itself. Ebb and flow systems include a tray in which the plant is placed in a medium; below the tray in a separate container is a reservoir containing water and mineral solutions. The water from the reservoir is periodically pumped up into the tray. This floods the tray and allows the plants to absorb water and nutrients. Gradually, the water drains back into the reservoir due to gravity. Ebb and Flow systems work best with small plants like herbs and are typically used in smaller hydroponic setups, such as those in the home.

Nutrient Film Technique (NFT) is a water-based system that requires no soil or mediums. They're built using wooden channels, which support polyethylene film liners. Plants such as tomatoes and cucumbers are placed on the channels, and the nutrient enriched water is pumped to the high end of each channel. The channels slope down, and water is collected at the end to be pumped back through the system and reused. Only plants with large established root systems will work with this technique.

Drip Systems are set up almost identically to an ebb and flow system, although instead of water being pumped through one large tube, it's pumped through many small tubes and drains onto the top of the plants. This system is ideal for plants that don't yet have a developed root system, and like an ebb and flow system, works best with smaller plants.

Aeroponics is another water based system, which, like NFT, requires no medium. Plants are suspended on a tray, with their roots freely dangling below. The entire tray is placed into a box that has a small amount of water and nutrient solution in the bottom. A pump system is used to draw the water up, where it's sprayed in a fine mist onto the entire plant and root in a continuous manner. This system is the most difficult to set up and manage, but it has great potential for large commercial uses.

Wick Systems are similar to Ebb and Flow systems in that they're medium-based. Plants are placed into a tray filled with a medium such as perlite or rockwool. At the base of each root, a nylon rope is placed, which is allowed to dangle freely, extending beyond the bottom of the tray. The entire tray is then placed on top of a reservoir. The nylon ropes absorb the water and nutrients, wicking them up to the plant's roots. This system is desirable because it requires no pumps or other equipment to be purchased.

TECHNIQUES

There are two main variations for each medium, sub-irrigation and top irrigation. For all techniques, most hydroponic reservoirs are now built of plastic, but other materials have been used including concrete, glass, metal, vegetable solids, and wood. The containers should exclude light to prevent algae and fungal growth in the nutrient solution.

1. **Static solution culture:** In static solution culture, plants are grown in containers of nutrient solution, such as glass Mason jars (typically, in-home applications), plastic buckets, tubs, or tanks. The solution is usually gently aerated but may be un-aerated. If un-aerated, the solution level is kept low

enough that enough roots are above the solution so they get adequate oxygen. A hole is cut in the lid of the reservoir for each plant. A single reservoir can be dedicated to a single plant, or to various plants. Reservoir size can be increased as plant size increases. A home made system can be constructed from plastic food containers or glass canning jars with aeration provided by an aquarium pump, aquarium airline tubing and aquarium valves. Clear containers are covered with aluminium foil, butcher paper, black plastic, or other material to exclude light, thus helping to eliminate the formation of algae. The nutrient solution is changed either on a schedule, such as once per week, or when the concentration drops below a certain level as determined with an electrical conductivity meter. Whenever the solution is depleted below a certain level, either water or fresh nutrient solution is added. A Mariotte's bottle, or a float valve, can be used to automatically maintain the solution level. In raft solution culture, plants are placed in a sheet of buoyant plastic that is floated on the surface of the nutrient solution. That way, the solution level never drops below the roots.

2. **Continuous-flow solution culture:** In continuous-flow solution culture, the nutrient solution constantly flows past the roots. It is much easier to automate than the static solution culture because sampling and adjustments to the temperature and nutrient concentrations can be made in a large storage tank that has potential to serve thousands of plants. A popular variation is the nutrient film technique or NFT, where a very shallow stream of water containing all the dissolved nutrients required for plant growth is recirculated past the bare roots of plants in a watertight thick root mat, which develops in the bottom of the channel and has an upper surface that, although moist, is in the air. Subsequent to this, an abundant supply of oxygen is provided to the roots of the plants. A properly designed NFT system is based on using the right channel slope, the right flow rate, and the right channel length. The main advantage of the NFT system over other forms of hydroponics is that the plant roots are exposed to adequate supplies of water, oxygen, and nutrients. In all other forms of production, there is a conflict between the supply of these requirements, since excessive or deficient amounts of one results in an imbalance of one or both of the others. NFT, because of its design, provides a system where all three requirements for healthy plant growth can be met at the same time, provided that the simple concept of NFT is always remembered and practised. The result of these advantages is that higher yields of high-quality produce are obtained over an extended period of cropping. A downside of NFT is that it has very little buffering against interruptions in the flow (e.g., power outages). But, overall, it is probably one of the more productive techniques

The same design characteristics apply to all conventional NFT systems. While slopes along channels of 1:100 have been recommended, in practice it is difficult to build a base for channels that is sufficiently true to enable nutrient films to flow without ponding in locally depressed areas. As a consequence, it is recommended that slopes of 1:30 to 1:40 are used This allows for minor irregularities in the surface, but, even with these slopes, ponding and water logging may occur. The slope may be provided by the floor, benches or racks may hold the channels and provide the required slope. Both methods are used and depend on local requirements, often determined by the site and crop requirements.

As a general guide, flow rates for each gully should be one liter per minute. At planting, rates may be half this and the upper limit of 2 L/min appears about the maximum. Flow rates beyond these extremes are often associated with nutritional problems. Depressed growth rates of many crops have been observed when channels exceed 12 meters in length. On rapidly growing crops, tests have indicated that, while oxygen levels remain adequate, nitrogen may be depleted over the length of the gully. As a consequence, channel length should not exceed 10–15 meters. In situations where this is not possible, the reductions in growth can be eliminated by placing another nutrient feed halfway along the gully and halving the flow rates through each outlet

3. **Aeroponics:** Aeroponics is a system wherein roots are continuously or discontinuously kept in an environment saturated with fine drops (a mist or aerosol) of nutrient solution. The method requires no substrate and entails growing plants with their roots suspended in a deep air or growth chamber with the roots periodically wetted with a fine mist of atomized nutrients. Excellent aeration is the main advantage of aeroponics.

   Aeroponic techniques have proven to be commercially successful for propagation, seed germination, seed potato production, tomato production, leaf crops, and micro-greens. Since its inventor Richard Stoner commercialized aeroponic technology in 1983, aeroponics has been implemented as an alternative to water intensive hydroponic systems worldwide. The limitation of hydroponics is the fact that 1 kilogram (2.2 lb) of water can only hold 8 milligrams (0.12 gr) of air, no matter whether aerators are utilized or not.

   Another distinct advantage of aeroponics over hydroponics is that any species of plants can be grown in a true aeroponic system because the microenvironment of an aeroponic can be finely controlled. The limitation of hydroponics is that certain species of plants can only survive for so long in water before they become waterlogged. The advantage of aeroponics is that suspended aeroponic plants receive 100% of the available oxygen and carbon dioxide to the roots zone, stems, and leaves, thus accelerating biomass growth and reducing rooting times. NASA research has shown that aeroponically grown plants have an 80% increase in dry weight biomass (essential minerals) compared to hydroponically grown plants. Aeroponics used 65% less water than hydroponics. NASA also concluded that aeroponically grown plants require ¼ the nutrient input compared to hydroponics Unlike hydroponically grown plants, aeroponically grown plants will not suffer transplant shock when transplanted to soil, and offers growers the ability to reduce the spread of disease and pathogens. Aeroponics is also widely used in laboratory studies of plant physiology and plant pathology. Aeroponic techniques have been given special attention from NASA since a mist is easier to handle than a liquid in a zero-gravity environment

4. **Fogponics:** Fogponics is a derivation of aeroponics wherein the nutrient solution is aerosolized by a diaphragm vibrating at ultrasonic frequencies. Solution droplets produced by this method tend to be 5–10 μm in diameter, smaller than those produced by forcing a nutrient solution through pressurized nozzles, as in aeroponics. The smaller size of the droplets allows them to diffuse through the air more easily, and deliver nutrients to the roots without limiting their access to oxygen.

5. **Passive sub-irrigation:** Passive sub-irrigation, also known as passive hydroponics, semi-hydroponics, or *hydroculture*, is a method wherein plants are grown in an inert porous medium that transports water and fertilizer to the roots by capillary action from a separate reservoir as necessary, reducing labor and providing a constant supply of water to the roots. In the simplest method, the pot sits in a shallow solution of fertilizer and water or on a capillary mat saturated with nutrient solution. The various hydroponic media available, such as expanded clay and coconut husk, contain more air space than more traditional potting mixes, delivering increased oxygen to the roots, which is important in epiphytic plants such as orchids and bromeliads, whose roots are exposed to the air in nature. Additional advantages of passive hydroponics are the reduction of root rot and the additional ambient humidity provided through evaporations.

   Hydroculture compared to traditional farming in terms of crops yield per area in a controlled environment was roughly 10 times more efficient than traditional farming, uses 13 times less water in one crop cycle then traditional farming, but on average uses 100 times more kilojoules per kilogram of energy than traditional farming.

6. **Ebb and flow or flood and drain sub-irrigation:** In its simplest form, there is a tray above a reservoir of nutrient solution. Either the tray is filled with growing medium (clay granules being the most common) and then plant directly or place the pot over medium, stand in the tray. At regular intervals, a simple timer causes a pump to fill the upper tray with nutrient solution, after which the solution drains back down into the reservoir. This keeps the medium regularly flushed with nutrients and air. Once the upper tray fills past the drain stop, it begins recirculating the water until the timer turns the pump off, and the water in the upper tray drains back into the reservoirs.

7. **Run-to-waste:** In a run-to-waste system, nutrient and water solution is periodically applied to the medium surface. The method was invented in Bengal in 1946; for this reason it is sometimes referred to as "The Bengal System".

   This method can be set up in various configurations. In its simplest form, a nutrient-and-water solution is manually applied one or more times per day to a container of inert growing media, such as rockwool, perlite, vermiculite, coco fibre, or sand. In a slightly more complex system, it is automated with a delivery pump, a timer and irrigation tubing to deliver nutrient solution with a delivery frequency that is governed by the key parameters of plant size, plant growing stage, climate, substrate, and substrate conductivity, pH, and water content.

In a commercial setting, watering frequency is multifactorial and governed by computers or PLCs. Commercial hydroponics production of large plants like tomatoes, cucumber, and peppers uses one form or another of run-to-waste hydroponics.

In environmentally responsible uses, the nutrient-rich waste is collected and processed through an on-site filtration system to be used many times, making the system very productive.

Some bonsai are also grown in soil-free substrates (typically consisting of akadama, grit, diatomaceous earth and other inorganic components) and have their water and nutrients provided in a run-to-waste form.

8. **Deep water culture:** The hydroponic method of plant production by means of suspending the plant roots in a solution of nutrient-rich, oxygenated water. Traditional methods favor the use of plastic buckets and large containers with the plant contained in a net pot suspended from the centre of the lid and the roots suspended in the nutrient solution. The solution is oxygen saturated by an air pump combined with porous stones. With this method, the plants grow much faster because of the high amount of oxygen that the roots receive.

9. **Top-fed deep water culture:** *Top-fed* deep water culture is a technique involving delivering highly oxygenated nutrient solution direct to the root zone of plants. While deep water culture involves the plant roots hanging down into a reservoir of nutrient solution, in top-fed deep water culture the solution is pumped from the reservoir up to the roots (top feeding). The water is released over the plant's roots and then runs back into the reservoir below in a constantly recirculating system. As with deep water culture, there is an airstone in the reservoir that pumps air into the water via a hose from outside the reservoir. The airstone helps add oxygen to the water. Both the airstone and the water pump run 24 hours a day.

   The biggest advantage of top-fed deep water culture over standard deep water culture is increased growth during the first few weeks With deep water culture, there is a time when the roots have not reached the water yet. With top-fed deep water culture, the roots get easy access to water from the beginning and will grow to the reservoir below much more quickly than with a deep water culture system. Once the roots have reached the reservoir below, there is not a huge advantage with top-fed deep water culture over standard deep water culture. However, due to the quicker growth in the beginning, grow time can be reduced by a few weeks

10. **Rotary:** A rotary hydroponic garden is a style of commercial hydroponics created within a circular frame which rotates continuously during the entire growth cycle of whatever plant is being grown.

    While system specifics vary, systems typically rotate once per hour, giving a plant 24 full turns within the circle each 24-hour period. Within the center of each rotary hydroponic garden can be a high intensity grow light, designed to simulate sunlight, often with the assistance of a mechanized timer.

    Each day, as the plants rotate, they are periodically watered with a hydroponic growth solution to provide all nutrients necessary for robust growth. Due to the plants continuous fight against gravity, plants typically mature much more quickly than when grown in soil or other traditional hydroponic growing systems. Due to the small foot print a rotary hydroponic system has, it allows for more plant material to be grown per square foot of floor space than other traditional hydroponic systems.

SUBSTRATES

One of the most obvious decisions hydroponic farmers have to make is which medium they should use. Different media are appropriate for different growing techniques.

1. **Expanded clay aggregate:** Baked clay pellets are suitable for hydroponic systems in which all nutrients are carefully controlled in a water solution. The clay pellets are inert, pH neutral and do not contain any nutrient value.

   The clay is formed into round pellets and fired in rotary kilns at 1,200 °C (2,190 °F). This causes the clay to expand, like popcorn, and become porous. It is light in weight, and does not compact over time. The shape of an individual pellet can be irregular or uniform depending on brand and manufacturing process. The manufacturers consider expanded clay to be an ecologically sustainable and re-usable growing medium because of its ability to be cleaned and sterilized, typically by washing in solutions of white vinegar, chlorine bleach, or hydrogen peroxide (H_2O_2), and rinsing completely.

   Another view is that clay pebbles are best not reused even when they are cleaned, due to root growth that may enter the medium. Breaking open a clay pebble after a crop has been shown reveal this growth.

2. **Growstones:** Growstones, made from glass waste, have both more air and water retention space than perlite and peat. This aggregate holds more water than parboiled rice hulls. Growstones by volume consist of 0.5 to 5% calcium carbonate for a standard 5.1 kg bag of grow stones that corresponds to 25.8 to 258 grams of calcium carbonate. The remainder is soda-lime glass.

3. **Coir peat:** Coco peat, also known as coir or coco, is the leftover material after the fibres have been removed from the outermost shell (bolster) of the coconut. Coir is a 100% natural grow and flowering medium. Coconut coir is colonized with trichoderma fungi, which protects roots and stimulates root growth. It is extremely difficult to over-water coir due to its perfect air-to-water ratio; plant

roots thrive in this environment. Coir has a high cation exchange, meaning it can store unused minerals to be released to the plant as and when it requires it. Coir is available in many forms; most common is coco peat, which has the appearance and texture of soil but contains no mineral content.

4. **Rice husks:** Parboiled rice husks (PBH) are an agricultural byproduct that would otherwise have little use. They decay over time, and allow drainage, and even retain less water than growstones. A study showed that rice husks did not affect the effects of plant growth regulators.
5. **Perlite:** Perlite is a volcanic rock that has been superheated into very lightweight expanded glass pebbles. It is used loose or in plastic sleeves immersed in the water. It is also used in potting soil mixes to decrease soil density. Perlite has similar properties and uses to vermiculite but, in general, holds more air and less water and is buoyant.
6. **Vermiculite:** Like perlite, vermiculite is a mineral that has been superheated until it has expanded into light pebbles. Vermiculite holds more water than perlite and has a natural "wicking" property that can draw water and nutrients in a passive hydroponic system. If too much water and not enough air surrounds the plants roots, it is possible to gradually lower the medium's water-retention capability by mixing in increasing quantities of perlite.
7. **Pumice:** Like perlite, pumice is a lightweight, mined volcanic rock that finds application in hydroponics.
8. **Sand:** Sand is cheap and easily available. However, it is heavy, does not hold water very well, and it must be sterilized between uses
9. **Gravel:** The same type that is used in aquariums, though any small gravel can be used, provided it is washed first. Indeed, plants growing in a typical traditional gravel filter bed, with water circulated using electric powerhead pumps, are in effect being grown using gravel hydroponics. Gravel is inexpensive, easy to keep clean, drains well and will not become waterlogged. However, it is also heavy, and, if the system does not provide continuous water, the plant roots may dry out.
10. **Wood fibre:** Wood fibre, produced from steam friction of wood, is a very efficient organic substrate for hydroponics. It has the advantage that it keeps its structure for a very long time. Wood wool (i.e. wood slivers) have been used since the earliest days of the hydroponics research. However, more recent research suggests that wood fibre may have detrimental effects on "plant growth regulators".
11. **Sheep wool:** Wool from shearing sheep is a little-used yet promising renewable growing medium. In a study comparing wool with peat slabs, coconut fibre slabs, perlite and rockwool slabs to grow cucumber plants, sheep wool had a greater air capacity of 70%, which decreased with use to a comparable 43%, and water capacity that increased from 23% to 44% with use. Using sheep wool resulted in the greatest yield out of the tested substrates, while application

of a biostimulator consisting of humic acid, lactic acid and Bacillus subtilis improved yields in all substrates.

12. **Rock wool:** Rock wool (mineral wool) is the most widely used medium in hydroponics. Rock wool is an inert substrate suitable for both run-to-waste and recirculating systems. Rock wool is made from molten rock, basalt or 'slag' that is spun into bundles of single filament fibres, and bonded into a medium capable of capillary action, and is, in effect, protected from most common microbiological degradation. Rock wool is typically used only for the seedling stage, or with newly cut clones, but can remain with the plant base for its lifetime. Rock wool has many advantages and some disadvantages. The latter being the possible skin irritancy (mechanical) whilst handling (1:1000) Flushing with cold water usually brings relief. Advantages include its proven efficiency and effectiveness as a commercial hydroponic substrate. Most of the rock wool sold to date is a non-hazardous, non-carcinogenic material, falling under Note Q of the European Union Classification Packaging and Labeling Regulation (CLP).

    Mineral wool products can be engineered to hold large quantities of water and air that aid root growth and nutrient uptake in hydroponics; their fibrous nature also provides a good mechanical structure to hold the plant stable. The naturally high pH of mineral wool makes them initially unsuitable to plant growth and requires "conditioning" to produce a wool with an appropriate, stable pH.

13. **Brick shards:** Brick shards have similar properties to gravel. They have the added disadvantages of possibly altering the pH and requiring extra cleaning before reuse

14. **Polystyrene packing peanuts:** Polystyrene packing peanuts are inexpensive, readily available, and have excellent drainage. However, they can be too lightweight for some uses. They are used mainly in closed-tube systems. Note that non-biodegradable polystyrene peanuts must be used; biodegradable packing peanuts will decompose into a sludge. Plants may absorb styrene and pass it to their consumers; this is a possible health risk.

NUTRIENT SOLUTIONS

1. **Inorganic hydroponic solutions:** The formulation of hydroponic solutions is an application of plant nutrition, with nutrient deficiency symptoms mirroring those found in traditional soil based agriculture. However, the underlying chemistry of hydroponic solutions can differ from soil chemistry in many significant ways. Important differences include:

    - Unlike soil, hydroponic nutrient solutions do not have cation-exchange capacity (CEC) from clay particles or organic matter. The absence of CEC means the pH and nutrient concentrations can change much more rapidly in hydroponic setups than is possible in soil.

- Selective absorption of nutrients by plants often imbalances the amount of counter ions in the solution. This imbalance can rapidly affect solution pH and the ability of plants to absorb nutrients of similar ionic charge (see article membrane potential). For instance, nitrate anions are often consumed rapidly by plants to form proteins, leaving an excess of cations in solution. This cation imbalance can lead to deficiency symptoms in other cation based nutrients (e.g. Mg^{2+}) even when an ideal quantity of those nutrients are dissolved in the solution.
- Depending on the pH and/or on the presence of water contaminants, nutrients such as iron can precipitate from the solution and become unavailable to plants. Routine adjustments to pH, buffering the solution, and/or the use of chelating agents is often necessary.

As in conventional agriculture, nutrients should be adjusted to satisfy Liebig's law of the minimum for each specific plant variety. Nevertheless, generally acceptable concentrations for nutrient solutions exist, with minimum and maximum concentration ranges for most plants being somewhat similar. Most nutrient solutions are mixed to have concentrations between 1,000 and 2,500 ppm. Acceptable concentrations for the individual nutrient ions, which comprise that total ppm figure, are summarized in the following table. For essential nutrients, concentrations below these ranges often lead to nutrient deficiencies while exceeding these ranges can lead to nutrient toxicity. Optimum nutrition concentrations for plant varieties are found empirically by experience and/or by plant tissue tests.

2. **Organic hydroponic solutions:** Organic fertilizers can be used to supplement or entirely replace the inorganic compounds used in conventional hydroponic solutions. However, using organic fertilizers introduces a number of challenges that are not easily resolved. Examples include:

- organic fertilizers are highly variable in their nutritional compositions. Even similar materials can differ significantly based on their source (e.g. the quality of manure varies based on an animal's diet and the method of storing).
- organic fertilizers are often sourced from animal byproducts, making disease transmission a serious concern for plants grown for human consumption or animal forage.
- organic fertilizers are often particulate and can clog substrates or other growing equipment. Sieving and/or milling the organic materials to fine dusts is often necessary.
- some organic materials (i.e. particularly manures and offal) can further degrade to emit foul odors.

Nevertheless, if precautions are taken, organic fertilizers can be used successfully in hydroponics.

Organic materials like blood meal, bone ashes, bonemeal, hoof / horn meal, fishmeal, wool waste, wood ashes, cottonseed ashes, cottonseed meal, dried locust or grasshopper, leather waste, kelp meal, liquid seaweed, poultry manure, sheep manure, goat manure, horse manure, cow manure, bat guano, bird guano can be used.

3. **Organically sourced micronutrients:** Micronutrients can be sourced from organic fertilizers as well. For example, composted pine bark is high in manganese and is sometimes used to fulfill that mineral requirement in hydroponic solutions. To satisfy requirements for National Organic Programs, pulverized, unrefined minerals (e.g. gypsum, calcite, and glauconite) can also be added to satisfy a plant's nutritional needs.

ADVANTAGES

1. **No soil needed:** In a sense, you can grow crops in places where land is limited, doesn't exist, or is heavily contaminated. In the 1940s, Hydroponics was successfully used to supply fresh vegetables for troops in Wake Island, a refueling station for Pan American airlines. This is a distant arable area in the Pacific Ocean. Also, Hydroponics has been considered as the farming of the future to grow foods for astronauts in the space (where there is no soil) by NASA.
2. **Make better use of space and location:** Because all that plants need are provided and maintained in a system, you can grow in your small apartment, or a spare bedroom as long as you have some space.

   Plants' roots usually expand and spread out in search for foods, and oxygen in the soil. This is not the case in Hydroponics, where the roots are sunk in a tank full of oxygenated nutrient solution and are directly in contact with vital minerals. This means you can grow your plants much closer, and consequently huge space savings.
3. **Climate control:** Like in greenhouses, hydroponic growers can have total control over the climate - temperature, humidity, light intensification, the composition of the air. In this sense, you can grow foods all year round regardless of the season. Farmers can produce foods at the appropriate time to maximize their business profits.
4. **Hydroponics is water-saving:** Plants grown hydroponically can use only 10% of water compared to field-grown ones. In this method, water is recirculated. Plants will take up the necessary water, while run-off ones will be captured and return to the system. Water loss only occurs in two forms - evaporation and leaks from the system (but an efficient hydroponic setup will minimize or don't have any leaks).

   It is estimated that agriculture uses up to 80% water of the ground and surface water in the US.

   While water will become a critical issue in the future when the food production is predicted to increase by 70% according to the FAQ, Hydroponics is considered a viable solution to large-scale food production.

5. **Effective use of nutrients:** In hydroponics, you have 100% control of the nutrients (foods) that plants need. Before planting, growers can check what plants require and the specific amounts of nutrients needed at particular stages and mix them with water accordingly. Nutrients are conserved in the tank, so there are no loss or change of nutrients like they are in the soil.
6. **pH control of the solution:** All of the minerals are contained in the water. That means you can measure and adjust the pH levels of your water mixture much more easily as compared to the soils. That ensures optimal nutrients uptake for plants.
7. **Better growth rate:** Hydroponically plants grown faster than in soil. You are your own boss that commands the whole environment for your plants' growth - temperature, lights, moisture, and especially nutrients. Plants are placed in ideal conditions, while nutrients are provided at the sufficient amounts, and come into direct contacts with the root systems. Thereby, plants no longer waste valuable energy searching for diluted nutrients in the soil. Instead, they shift all of their focus on growing and producing fruits.
8. **No weeds:** If you have grown in the soil, you will understand how irritating weeds cause to your garden. It's one of the most time-consuming tasks for gardeners - till, plow, hoe, and so on. Weeds are mostly associated with the soil. So eliminate soils, and all bothers of weeds are gone.
9. **Fewer pests & diseases:** And like weeds, getting rid of soils helps make your plants less vulnerable to soil-borne pests like birds, gophers, groundhogs; and diseases like Fusarium, Pythium, and Rhizoctonia species. Also when growing indoors in a closed system, the gardeners can easily take control of most of the surrounding variables.
10. **Less use of insecticide, and herbicides:** Since you are using no soil and while the weeds, pests, and plant diseases are heavily reduced, there are fewer chemicals used. This helps you grow cleaner and healthier foods. The elimination of insecticides and herbicides is a strong point of hydroponics when the criteria for modern life and food safety are more and more placed on top.
11. **Labor and time savers:** Besides spending fewer hours on tilling, watering, cultivating, and fumigating weeds and pests, you enjoy much time saved because plants' growth is proven to be higher in hydroponics. When agriculture is planned to be more technology-based, hydroponics has a room in it.
12. **Hydroponics is a stress-relieving hobby:** This interest will put you back in touch with nature. Tired after a long working day and commute, you return to your small apartment corner, it's time to lay back everything and play with your hydroponic garden. Reasons like lack of spaces is no longer an excuse. You can start growing fresh, tasty vegetables or vital herbs in your small closets, and enjoy the relaxing time with your little green spaces.

DISADVANTAGES

1. **A Hydroponic garden requires a lot of your time and commitment:** Just like any thing worthwhile in life, hard-work and responsible attitude give satisfactory yields. However, in soil-borne counterparts, plants can be left on its own for days and weeks, and they still survive in a short time. Mother nature and soils will help regulate if something is not balancing. That's not the case in Hydroponics. Plants will die out more quickly without proper care and adequate knowledge. Remember that your plants are depending on you for their survival. You must take good care of your plants, and the system upon initial installation. Then you can automate the whole thing later, but you still need to gauge and prevent the unexpected issues of the operations, and do frequent maintenance.
2. **Experiences and technical knowledge:** You are running a system of many types of equipment, which requires necessary specific expertise for the devices used, what plants you can grow and how they can survive and thrive in a soilless environment. Mistakes in setting up the systems and plants' growth ability in this soilless environment and you end up ruining your whole progress.
3. **Organic debates:** There have been some heated arguments about whether hydroponics should be certified as organic or not. People are questioning whether plants grown hydroponically will get microbiomes as they are in the soil. But people around the world have grown hydroponic plants - lettuces, tomatoes, strawberries, etc. for tens of years, especially in Australia, Tokyo, Netherland, and the United States. They have provided food for millions of people. You cannot expect perfection from anything in life. Even for soil growing, there are still more risks of pesticides, pests, etc. compared to Hydroponics. There are some organic growing methods suggested for hydroponic growers. For example, some growers provide microbiomes for plants by using organic growing media such as coco coir and add worm casting into it. Natural-made nutrients are commonly used such as fishes, bones, alfalfas, cottonseeds, neem, etc.

   For this debate for the organic product issue, there will still be researches done currently and in the near future. And we'll know the answer then.
4. **Water and electricity risks:** In a hydroponic system, mostly you use water and electricity. Beware of electricity in a combination of water in close proximity. Always put safety first when working with the water systems and electric equipment, especially in commercial greenhouses.
5. **System failure threats:** You are using electricity to manage the whole system. So suppose you have not taken preliminary measure for a power outage, the system will stop working immediately, and plants may dry out quickly and die within hours. Hence, a backup power source and plan should always be planned, especially for big scale systems.

6. **Initial expenses:** You are sure to spend a lot of money (depending on your garden scale) to purchase equipment for your first installation. Whatever systems you build, you will need containers, lights, a pump, a timer, growing media, nutrients). Once the system is in place, the cost will be reduced to only nutrients and electricity (to keep the water system running, and lightning).
7. **Long return per investment:** If you follow news on agriculture start-up, you may have known that there have been some new indoor hydroponic business started recently. That's a good thing for the agriculture sector and the development of hydroponics as well. However, commercial growers still face some big challenges when starting with hydroponics on a large scale. This is largely because of the high initial expenses and the long, uncertain ROI (return on investment). It's not easy to detail a clear profitable plan to urge for investment while there are also many other attractive high-tech fields out there that seem fairly promising for funding.
8. **Diseases and pests may spread quickly:** You are growing plants in a closed space using water. In the case of plant infection or pests, they can escalate fast to plants on the same nutrient reservoir. In most cases, diseases and pests are not so much of a problem in a small system of home growers. So ignore these issues if you are a beginner.

It's only complicated for big hydroponic greenhouses. So better to have a good disease management plan beforehand. For example, use clean disease-free water sources and growing materials; check the systems periodically, etc.

There are downsides to hydroponics, just like any other things in life. But you can overcome most of them with just some planning and experiences. Meanwhile considering the pros it can offer, hydroponics is worth doing. It is always a part of modern agriculture. Compared with the soil-counterparts, hydroponics is a highly effective growing method.

Below a summary of winning key points of hydroponics over soil:

Hydroponics vs. Soil

- Space saver.
- Effective use of plant nutrients.
- Water-efficient.
- No weeds, fewer pests, and plant diseases.
- Stable and higher yields.
- Control of the whole system.
- Growing media can be chosen and sterilized quickly.
- Fewer fertilizers and insecticides used.
- Easy to transplant.
- Able to grow crops all year round.

FUTURE IN AGRICULTURE

Hydroponics is the fastest growing sector of agriculture, and it could very well dominate food production in the future. As population increases and arable land declines due to poor land management, people will turn to new technologies like hydroponics and vertical farming to create additional channels of crop production. Currently, arable land comprises only around 3 percent of the Earth's surface, and the world population is around 6 billion people, resulting in around 1/5 hectare (2,000 square meters) of arable land per capita. By 2050, scientists estimate that the Earth's population will increase to 9.2 billion, while land available for crop and food production will decline. To feed the increasing population, hydroponics will begin replacing traditional agriculture.

To get a glimpse of the future of hydroponics, we need only to examine some of the early adopters of this science. In Tokyo, Japan, land is extremely valuable due to the surging population. To feed the citizens while preserving valuable land mass, the country has turned to hydroponic rice production. Rice is harvested in underground vaults without the use of soil. Because the environment is perfectly controlled, four cycles of harvest can be performed annually, instead of the traditional single harvest.

Hydroponics also has been used successfully in Israel, which has a dry, arid climate. A company called Organitech has been growing crops in 40-foot (12.19-meter) long shipping containers, using hydroponic systems. They grow large quantities of berries, citrus fruits and bananas, all of which couldn't normally be grown in Israel's climate. The hydroponics techniques produce a yield 1,000 times greater than the same sized area of land could produce annually. Best of all, the process is completely automated, controlled by robots using an assembly line-type system, such as those used in manufacturing plants. The shipping containers are then transported throughout the country.

There has already been a great deal of buzz throughout the scientific community for the potential to use hydroponics in third world countries, where water supplies are limited. While the upfront capital costs of setting up hydroponics systems is currently a barrier, in the long-run, as with all technology, costs will decline, making this option much more feasible. Hydroponics has the ability to feed millions in areas of Africa and Asia, where both water and crops are scarce.

Hydroponics also will be important to the future of the space program. NASA has extensive hydroponics research plans in place, which will benefit current space exploration, as well as future, long-term colonization of Mars or the moon. As we haven't yet found soil that can support life in space, and the logistics of transporting soil via the space shuttles seems impractical, hydroponics could be key to the future of space exploration. The benefits of hydroponics in space are two-fold: It offers the potential for a larger variety of food, and it provides a biological aspect, called a bioregenerative life support system. This simply means that as the plants grow, they will absorb carbon dioxide and stale air and provide renewed oxygen through the plant's natural growing process. This is important for long-range habitation of both the space stations and other planets.

10 ROLE OF CROP RESIDUES IN AGRICULTURE

In India a vast majority of land is used for farming and a wide range of crops are cultivated in its different agro-ecological regions. Rice-wheat (R–W) cropping system is the major production system in the Indo-Gangetic Plains (IGP) covering an area of 10 million ha and has played a significant role in the food security of India. The two North-Western (NW) States i.e. Punjab and Haryana constitute a highly productive rice-wheat cropping system zone in the IGP contributing about 69% of the total food output in the country. High yields of the irrigated rice-wheat system have resulted in the production of huge quantities of crop residues (CRs) (Singh and Sidhu 2014). These crop residues are used for animal feeding, soil mulching, bio- manure making, thatching for rural homes and fuel for domestic and industrial use. Thus crop residues are of tremendous value to the farmers.

The problem of on-farm burning of crop residues is intensifying in recent years due to shortage of human labour, high cost of removing the crop residues by conventional methods and use of combines for harvesting of crops. The residues of rice, wheat, cotton, maize, millet, sugarcane, jute, rapeseed-mustard and groundnut are typically burnt on-farm across different states of the country. The problem is more severe in the irrigated agriculture, particularly in the mechanized rice-wheat system of north west India.

Crop residues, in general, are the parts of the plants left in the field after crops have been harvested and threshed or left after pastures are grazed. These materials have at times been regarded as waste materials that require disposal but it have become increasingly realized that they are important natural resources and are not the wastes. The gravity of the situation demands that an appropriate policy should be evolved to promote multiple uses of crop residues in the context of conservation agriculture and to prevent their on-farm burning.

So, the following steps must be taken towards management of residue:

(i) Quantify the amount of crop residues generated in the country every year and the extent of their on-farm burning

(ii) Assess the environmental impacts of on-farm burning of crop residues

(iii) Identify suitable uses of crop residues and their adoption potential

(iv) Assess the potential of using crop residues for conservation agriculture and constraints involved therein

(v) Develop a model plan for managing crop residues at the local and regional scales

(vi) Identify research and policy issues for safe and sustainable management of crop residues for productive, profitable and sustainable agriculture.

CROP RESIDUE BIOMASS POTENTIAL IN INDIA

Biomass based energy generation is one of the major focus areas of renewable energy programs in India. The strength of India's biomass resources mostly lies in the agricultural sector. A large quantity of crop residue biomass is generated in India. However, crop residue biomass is distributed with variation in spatiotemporal availability and its characteristics. Comparative use of crop residues also vary geographically. Therefore, local biomass databases are important for decentralized bio energy programs. However, in India, State wise crop level biomass database is limited. The present status of crop residue biomass and subsequently bio energy potential in all the 29 states of India is presented in Table 10.1. Overall, India produces 686 MT gross crop residue biomass on annual basis, of which 234 MT is surplus residue (34% of gross) (Table 10.1) for bio energy generation (Hiloidhari *et al* 2014). The highest surplus residue i.e. 55.7 MT is of sugarcane which is followed by cotton (46.9 MT), rice (43.5 MT) and wheat (28.4 MT)

Table 10.1 Gross and surplus crop residue biomass potential in India of different field crops

Crop	Gross potential (MT)	Surplus potential (MT)
Rice	154.0	43.5
Wheat	131.1	28.4
Maize	35.8	9.0
Bajra	24.3	5.1
Barley	1.6	0.2
Small millet	0.6	0.1
Ragi	2.7	0.3
Jowar	17.6	3.5
Mustard and rapeseed	12.7	4.9
Sesame	0.8	0.1
Linseed	0.3	0.0
Niger	0.1	0.0
Safflower	0.6	0.5
Soybean	13.5	4.6
Groundnut	17.0	3.0
Sunflower	3.8	0.6
Tur (arhar)	7.2	1.4
Gaur	2.6	1.8

Gram	6.4	1.6
Lentil	1.7	0.3
Sugarcane	110.6	55.7
Banana	41.9	12.3
Coconut	18.0	9.7
Arecanut	1.5	0.5
Cotton	75.9	46.9
Jute	3.9	0.4
Total (MT)	686.0	234.5

State level potential

Hiloidhari *et al.* 2014 reported that among field crops the highest value of average surplus crop residue (percentage) of different States of India was observed in sugarcane crop (39 %) which was followed by pulses (38%), oilseed (30%) and cereals (29%) on national level (Table 10.2). As far as different States are concerned, the highest percentage of surplus straw was observed in Punjab and Haryana (48%) followed by Uttarakhand and Goa (43%). The lowest value of 21% was observed in Arunachal Pradesh.

Table 10.2 State-wise surplus fraction of residue availability of different crops (% of gross)

State	Cereals	Oilseeds	Sugarcane	Horticulture	Pulses	Others	State avg.
Andhra Pradesh	29	26	40	44	23	38	33
Arunachal Pradesh	27	11	33	25	22	10	21
Assam	25	40	40	45	47	48	41
Bihar	27	32	33	20	23	10	24
Chhattisgarh	29	20	40	25	47	38	33
Goa	26	NA	38	39	NA	70	43
Gujarat	30	25	40	47	53	NA	39
Haryana	34	37	40	NA	40	90	48
H.P.	35	17	38	25	35	48	33
J & K	29	17	40	NA	NA	NA	29
Jharkhand	26	33	40	NA	47	10	31
Karnataka	30	31	38	32	47	33	35
Kerala	30	20	68	38	35	10	32
M. P.	28	23	40	25	43	70	38
Maharashtra	28	33	33	43	47	40	37
Manipur	28	21	40	25	NA	NA	29
Meghalaya	26	15	40	20	35	30	28
Mizoram	29	18	40	32	35	48	34
Nagaland	27	16	40	43	35	38	33
Orissa	29	29	33	52	40	10	32

Punjab	34	30	40	NA	47	90	48
Rajasthan	29	18	40	NA	23	10	24
Sikkim	28	22	NA	NA	NA	NA	25
Tamil Nadu	33	19	40	48	30	29	33
Tripura	34	21	40	45	35	38	37
U. P.	37	23	38	25	25	48	33
Uttarakhand	32	53	38	NA	55	NA	43
West Bengal	24	20	43	50	29	30	29
National avg	29	30	39	42	38	38	34

CROP RESIDUE BURNING

Crop residue burning in Punjab and Haryana are 48% of gross residues and Uttar Pradesh 14%, which causes nutrient loss of organic carbon (3.85 mt), N (59,000 t) P (20,000 t) and K (34,000 t). A large portion of the residues is burnt on-farm primarily to clear the field for sowing of the succeeding crop. The problem of on-farm burning of crop residues is intensifying in recent years due to shortage of human labour, high cost of removing the crop residues by conventional methods and use of combines for harvesting of crops. The residues of rice, wheat, cotton, maize, millet, sugarcane, jute, rapeseed-mustard, etc. are typically burnt on-farm across different States of the country. The problem is more severe in the irrigated agriculture, particularly in the mechanized rice-wheat system of the north-west India.

With IPCC coefficients, the contribution of Uttar Pradesh was maximum, followed by West Bengal, Andhra Pradesh, Punjab, Maharashtra and Haryana. But with our coefficients, the maximum amount of crop residues were burnt in the States of Uttar Pradesh (22.25 Mt) and Punjab (21.32 Mt), followed by Haryana (9.18 Mt) and Maharashtra (6.82 Mt). Highest amount of cereal crop residues were burnt in Punjab followed by Uttar Pradesh and Haryana. Uttar Pradesh contributed maximum to the burning of sugarcane trash followed by Karnataka. Oilseed residues were burnt in Rajasthan and Gujarat while burning of fiber crop residue was dominant in Gujarat (28.6 Mt) followed by West Bengal (24.4 Mt) Maharashtra and Punjab. Among the different crop residues major contribution (93%) was from rice (43%), wheat (21%) and sugarcane (19%) (Jain *et al.* 2014). District wise area under residue burning in Punjab of 2014 and 2015 are given in Table 10.3 (Anonymous, 2015).

Table 10.3 District wise area under residue burning in Punjab 2014 and 2015

District	Burnt area 2014 (000 ha)	Burnt area 2015 (000 ha)
Amritsar	100.95	78.92
Barnala	98.92	89.82
Bathinda	98.31	77.49
Faridkot	87.09	75.35
Fatehgarh Sahib	58.01	57.26

Fazilka	74.21	58.11
Firozpur	165.74	144.99
Gurdaspur	78.59	61.26
Hoshiarpur	46.13	22.23
Jalandhar	82.01	67.16
Kapurthala	81.27	78.69
Ludhiana	180.52	170.18
Mansa	59.79	41.88
Moga	155.83	141.28
Muktsar	105.60	85.18
Pathankot	12.62	9.99
Patiala	165.29	138.55
Rupnagar	14.87	13.59
Sangrur	227.21	202.83
SAS Nagar	23.12	14.45
SBS Nagar	32.08	28.75
Tarn Taran	147.79	134.61
State(Total)	2095.95	26262626

Adverse Consequences of on-farm burning of Crop Residues

- Burning of crop residues leads to release of soot particles and smoke. Many pollutants found in large quantities in biomass smoke are known or suspected carcinogens and could be a major cause of concern leading to various air-borne/lung diseases.
- Adverse effect on soil properties
- Diminish the activity of beneficial microorganism
- It also leads to emission of greenhouse gases. About 70%, 7% and 0.66% of carbon present in rice straw is emitted as carbon dioxide, carbon monoxide and methane, respectively, while 2.09% of nitrogen in straw is emitted as nitrous oxide upon burning.
- It also emits a large amount of particulates that are composed of a wide variety of organic and inorganic species and loss of plant nutrients like N, P, K and S.
- Wastage of valuable resources which could be a source of carbon, bio-active compounds, feed and energy for rural households and small industries.
- Heat generated from the burning of crop residues elevates soil temperature causing death of active beneficial microbial population.
- One tonne of rice straw on burning releases about 3 kg particulate matter, 60 kg CO, 1460 kg CO_2, 199 kg ash and 2 kg SO_2 (Gadi *et al.* 2003).

- Besides other light hydrocarbons, volatile organic compounds and semi-volatile organic compounds including polycyclic aromatic hydrocarbons, polychlorinated biphenyls (PCBs), SO_2 and NO_2 are also emitted. These gases are of major concern for their global impact and may lead to increase in the levels of aerosols, acid deposition, increase in tropospheric ozone and depletion of the stratospheric ozone layer.

Reasons for burning crop residues

- To clear field for sowing of succeeding crop
- Increased crop residue- due to increased mechanization for harvesting
- Unavailability of machinery to handle large crop residue
- Declined number of grazing animals
- Composting requires long time- 4 to 6 month
- For easy field operations- crop residue causes obstruction in field operations
- Large amounts of crop residues result in increased cost of labour requirement for its collection and transportation
- Initial adverse effect of crop residue on soil health and crop yield
- Germination problem
- High cost of cultivation
- Problem of rodents and termite
- Burning kills organisms of some insect pest and diseases
- Price of paddy straw is very low and that is why combines are used for harvesting.

BENEFITS OF CROP RESIDUE

(A) Soil benefits

- Increases soil productivity (yields).
- Sustains soil organic matter.
- Improves soil structure.
- Controls soil erosion by buffering the soil against forces of rain drop impact and wind shear.
- Increases water infiltration rates.
- Conserves soil moistures.
- Recycles plant nutrients.
- Provides habitat and an energy source for soil organisms including earthworms and microorganisms.
- Control weeds
- Reduces pest & disease damage
- Reduction in leaching of fertilizer

(B) Environmental benefits

- Mitigates flooding by holding water on the land rather than allowing it to run off into streams and rivers.
- Reduces surface runoff and decreases sedimentation.
- Improves water quality by denaturing and filtering of pollutants.
- Reduces soil and environmental pollution.

(C) Other benefits: Apart from incorporation, crop residues can be used for the following purposes:

- Mushroom cultivation
- Packaging material
- Biogas production
- Bedding purposes
- As a fodder
- Fuel purposes
- Mulching
- Biomethanation
- Gasification
- Composting- Residue of rice from 1ha give 3.2 tons of manure
- Energy source- 80000 tons of biomass of residue generated 1.5 lakh KW energy/day
- Bio-fuel and Bio-oil production from residue 382 to 471 lit/t of dry matter
- Biochar production-high carbon material produced through slow heating in the absence of oxygen of biomass

Research needs for efficient residue management

- Developing crop varieties to produce more root biomass
- Enhancing decomposition rate of residue for in-situ incorporation
- Designing new generation of long-term experiments
- Improving the implements used for planting, drilling and incorporation
- Assessing life-cycle of residue-based conservation agriculture
- Modifying combine harvester to collect & remove residues from field.
- To study the impact of crop residue as mulching material in different crops.

Laws and legislation

- Monitoring and discouraging burning of crop residues through incentive / punishment
- Supplying suitable machinery for incorporation of crop residue on subsidized rates

- Promoting custom hiring systems for agricultural implements
- Providing soft loans for purchase of implements
- Developing a crop residues management policy for each State

Crop Residues as Source of Plant Nutrients

Crop residues are good sources of plant nutrients and are the primary source of organic matter (as C constitutes about 40% of the total dry biomass) added to the soil, and are important components for the stability of agricultural ecosystems. About 40% of N, 30-35% of P, 80-85% of K, and 40-50% of S absorbed by rice remain in the vegetative parts at maturity (Dobermann and Fairhurst 2002).

On-farm Management of Crop Residues

Recycling of rice residue poses more problems to succeeding wheat than wheat straw to the following rice crop because of the shorter window between rice residue incorporation and wheat sowing, and the slow rate of decomposition of rice straw due to high silica content and low temperature.

(i) **Direct drilling in surface residues:** Direct drilling is a practice that leaves straw residues from a previous crop on the soil surface without any form of incorporation; the residues may or may not be mulched. The following crop is then drilled directly into the soil with residues left on the soil surface. The large volume of residues remaining on the surface often leads to machinery failure, thus affecting the proper sowing of the seeds of the following corps. It is possible with the use of Happy Seeder/Turbo Seeder/Zero Drill etc.

(ii) **Sowing crops:** Residues are spread on the surface after sowing the crop and are maintained on the surface and are not incorporated.

DIMENSIONS OF RICE STRAW AS MULCH

(A) Flourish edaphic characteristics *vis a vis* yield maximization

Ghimire *et al.* (2017) revealed that improved management practices such as reduced- and no-tillage management, N fertilizer and farmyard manure (FYM) application, and crop residues addition can improve soil organic carbon (SOC) accumulation in rice-based cropping systems from South Asia mainly in Bangladesh, Bhutan, India, Nepal, Pakistan and Sri Lanka. Huang *et al.* (2017) reported that both liming and straw retention could significantly enhance grain yield in both early- and late-rice growing seasons. There were interactive effects of liming and straw retention on late-rice yield in both years, whereas no significant interaction was found in the early-rice growing season. Like grain yield, N uptake of both early- and late-rice was significantly increased by both liming and straw retention. Hence, there are synergistic effects of liming and straw retention on late-rice yield in the double rice-cropping system, which may be ascribed to the improvement in rice N uptake. Singh *et al.* (2015) from Ludhiana, Punjab recorded that the optimum N rate for ZT wheat drilled into rice residues after one to three years of residue retention was 120 kg N ha^{-1} in sandy loam soil, the recommended rate for conventional practice (rice residues burning and tillage prior to sowing). The effects of long-term retention of rice residues on fertilizer requirement also need to be determined for the RW cropping system of NW India. Band placement of 20% of the fertilizer N as diammonium phosphate at seeding and top dressing of the remaining 80% as urea in two equal doses before first and second irrigations produced higher grain yield and N use efficiency than other treatments.

A field study was conducted on sandy loam soil at Uttar Pradesh under different rice residue with zero and conventional tillage practices in main plots and four nitrogen (N) levels (0, 50 kg, 100 kg and 150 kg N/ha) in subplots. Residue retained with ZT in wheat recorded significantly higher yield attributes, grain and straw yield, net returns and benefit cost ratio (B: C) over other residue and tillage management practices. The increasing rate of nitrogen at higher dose (150 kg N/ha) produced significantly higher grain weight/spike, grain and straw yield, higher net profit and highest B:C ratio as compared to 50 kg N/ha. Residue retained under ZT with 150 kg N/ha favoured grain weight per spike and was adjudged the best practice in wheat in terms of productivity and profitability (Kumar *et al* 2016). Singh *et al* (2015) studied the field experiment at Ludhiana and showed that mulching significantly increased mean seasonal soil-moisture storage and transpiration under mulching and no mulching treatments. However, mulching significantly reduced mean seasonal soil evaporation compared with no-mulching. Mulching and irrigation significantly increased leaf-area index of spring maize. Mulching significantly increased grain yield, yield attributes and water-use efficiency of spring maize over no-mulching.

The addition of organic materials in sufficient amounts to soil is important for improving soil health as the organic substrates influence the microbial population in the soil, which is responsible for nutrient transformations resulting in the availability of nutrients, particularly N, P, and S. Rice straw can be managed successfully in situ by allowing sufficient time between its incorporation and sowing of the wheat crop in NW India (Singh *et al.* 2004).

(B) Sustainable Weed Management Tool

Crop residue, physically impede seedling growth or inhibit germination and growth by allelopathy.

Brar & Walia (2010) found that under no herbicide application treatments, surface application of rice residues @ 7.5 t/ha significantly reduced the growth and development of *P. minor* as compared to residue removal treatment (Table 10.4). Application of broad spectrum herbicide i.e. sulfosulfuron + metsulfuron under rice straw treatment, provided nearly cent percent control of all weeds and provided highest wheat grain yield (5.48 tonne/ha) than all unsprayed treatments indicating that alone straw mulch cannot control the weeds of wheat crop.

The surface retention of the rice residues 5 and 7 t/ha reduced the weed dry weight in wheat by 23.4 to 30.3 and 35.5 to 44.1 per cent, respectively reported by Chhokar *et al.* (2009) at Karnal (Table 10.5). The less *P. minor* problem under ZT system was due to less soil disturbance as a result *P. minor* seeds present in lower soil layer fail to germinate due to mechanical impedance. Therefore, ZT is a cost effective and sustainable weed management system but continuous use of ZT may shift the weed flora in favour of other weeds such as *Rumex dentatus* and *Malva parviflora* (Chhokar *et al* 2007).

Table 10.4 Effect of residue retention on weeds and yield in wheat under rice-wheat system in India

Treatments	Weed dry matter (g/m^2) at 120 DAS			Yield (t/ha)
	P. minor	*Medicago denticulata*	*Rumex dentatus*	
Residue removal (RR) No herbicide	291.0	215.1	297.3	0.75
2.5 t/ha residue (Surface) No herbicide	304.6	175.4	310.1	1.05
5.0 t/ha residue (Surface) No herbicide	214.8	130.1	214.6	2.58
7.5 t/ha residue (Surface) No herbicide	190.9	123.4	134.7	2.46
Residue Removal (RR)+ herbicide (sulfosulfuron + metoxuron)	2.9	0.0	0.0	5.48
LSD (p=0.05)	83.5	52.5	67.4	0.39

Table 10.5 Influence of ricc residue position and load on population, dry matter and weed control efficiency of *P. minor* in wheat (120 DAS)

Treatments	Population (No./m^2)	Dry Matter (g/m^2)	WCE (%)
ZT+No rice residue (removal)	6.59 (42.2)*	11.23 (125.1)*	–
ZT+Rice residue 5 t/ha (surface)	6.00 (35.0)	10.07 (100.4)	19.74
ZT+Rice residue 6 t/ha (surface)	4.48 (19.1)	8.31 (68.1)	45.56
ZT+Rice residue 7.5 t/ha (surface)	4.27 (17.2)	7.81 (60.0)	52.04
CT+Rice residue 5 t/ha (incorporation)	6.31 (38.8)	10.88 (117.4)	6.16
LSD (p=0.05)	1.81	2.55	–

Figures in parenthesis are original values

(C) Enhancing Crop Water Productivity

Singh *et al.* (2011b) conducted experiment during 2006-07 and 2007-08 in Punjab and found that mulch lowers total soil evaporation over the crop growth season by 35 and 40 mm in relatively high and low rainfall years, respectively. There was a trend for higher biomass production and grain yield (Table 10.6).

Table 10.6 Partitioning of ET (mm) during pre-and post-anthesis periods in wheat

Treatments	Pre-anthesis			Post-anthesis			Total		
	Es	T	ET	Es	T	ET	Es	T	ET
Mulched	67	146	213	33	94	127	100	240	340
Non-mulched	96	135	231	39	75	114	135	210	345
LSD (p=0.05)	14	NS	NS	NS	17	NS	10	26	NS

Es- Soil Evaporation, T- Transpiration, ET- Evapo-transpiration, NS- Non significant

Ram *et al.* (2013) conducted three years experiment at Ludhiana, Punjab and found that straw mulching decreases the soil bulk density and increase organic matter content in 0-15 cm layer and also reported that under limited irrigation water condition rice straw mulching will be beneficial in increasing yield and water use efficiency in wheat (Table 10.7)

Table 10.7 Effect of rice straw mulching on soil volumetric moisture, water use and water use efficiency (WUE) (results of 3rd year)

Treatments	Soil volumetric moisture at harvest (cm)	Water use (cm)	WUE (kg/ha-cm)
No rice straw mulch	0.72	42.2	120.3
Mulch 2 t/ha	0.77	41.5	147.8
Mulch 4 t/ha	0.87	40.4	157.1
Mulch 6 t/ha	0.93	40.1	161.5

(D) Mitigate terminal heat stress in wheat

Singh *et al.* (2011 b) reported that on an average of two years, mulching helps in delay of crop phenological stages starting from germination (2 days), flowering (7.5 days) and physiological maturity (8 days) (Table 10.8), thus helping in improvement in crop yield.

Straw management techniques also influence canopy temperature which was recorded during third week of sowing (Singh *et al* 2011). They reported that wheat sown with Happy Seeder recorded lowest temperature (30.4°C) followed by zero till (31.4°C), conventional tillage (32.0°C) and with rotavator (33.3°C) (Fig. 10.1).

Table 10.8 Wheat crop phenology as affected by rice straw mulch treatments

Growth stage	Mulch treatment	2006-2007	2007-2008
Sowing date		**6 November**	**13 November**
Emergence	Non-mulch	13 November (7)	20 November (7)
	Mulch	15 November (9)	22November (9)
Flowering	Non-mulch	19 February (105)	3 March (113)
	Mulch	25 February (111)	12 March (122)
Physiological maturity	Non-mulch	8 April (153)	10 April (149)
	Mulch	18 April (163)	16 April (155)

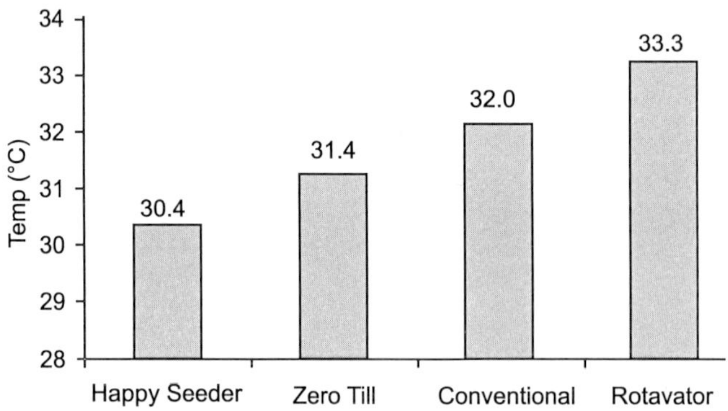

Fig. 10.1. Influence of rice straw mulching on wheat canopy temperature during third week of March 2010 (Source: Singh *et al.* 2011), Farmers' field (n=76)

(E) Environmental prospective

Sidhu *et al.* (2015) concluded that in the extensive rice-wheat system of north-west (NW) India, harvesting is by large combines and the rice residues are normally burnt after harvest, followed by irrigation and intensive tillage prior to sowing wheat. While in field, retention of crop residues can play an important role in replenishing soil quality and reducing environmental pollution from stubble burning, until recently, there has been no suitable technology for seeding wheat in rice residues. To address this need, numerous on-farm trials show that yield of wheat sown into rice residues with the 9-row Turbo Happy Seeder is similar to or higher than yield with straw burning and conventional tillage prior to sowing, while providing many benefits to the farmer. These include greatly reduced fuel consumption and cost of crop establishment, and the ability to sow as soon as desired after harvest, ensuring the possibility of sowing at the optimum time and reducing the need for irrigation. Current options available to farmers for rice residue management before wheat sowing are in situ incorporation in the soil using rotavator and full residue retention using happy seeder machine for sowing of wheat (Jain *et al.* 2014).

CONCLUSION

- Crop residue offer sustainable and ecologically sound alternatives for meeting the nutrient requirements of crops, and improving soil and environmental quality.
- Recent developments in machinery (Happy Seeder) allowing zero-till sowing of wheat with rice residue as surface mulch, while maintaining yield, reduces tillage costs and time savings, avoids the need for burning. Further refinements in the Happy Seeder are needed for fertilizer drilling, reducing its power requirement and improving its ability to work in moist straw.
- The use of crop residues as a mulching material has been found beneficial as it reduces maximum soil temperature and conserves water.
- Nutrient management is more complex with crop residue management techniques because of higher residue levels and reduced options with regard to method and timing of nutrient applications.
- In fact, there is a need to develop complete package of practices (fertilizer, irrigation, weed control, pest management, etc.) for crop residue management systems.
- Genetic variation in Si content in rice straw under different soil and water management situations needs investigation. Long-term studies involving multidisciplinary approaches are needed to study different issues (genotypes with more root biomass, machinery, insects, diseases, weeds, phytotoxicity, soil health, and economics) associated with crop residue management.
- For large-scale spreading of crop residue recycling, burning of crop residues should be discouraged through incentives and punishment.
- Machinery needs to be supplied on subsidized rates and providing soft loans. Subsidies have been an important component in helping make the technology affordable during the awareness raising phase.
- The effects of crop residue management on pests and diseases in rice-wheat cropping system have not received much attention, and systematic work with different management options is needed.
- It is possible that there would be fewer mulch-related disease problems in the rice-wheat rotation than in a rotation of two different upland crops, because the flooded season might inhibit survival of the upland crop pathogens.
- There is a need to support on-farm adaptation of crop residue management technologies in both large and scattered small fields, and developing focused institutional and policy support including appropriate incentives for the widespread dissemination and adoption of crop residue management practices in the Indo Gangetic Plains.
- Modifying combine harvesters to uniformly distribute the residue in the field. Large volume and transport is a major bottleneck in using the residues where those are required. Machinery for volume reduction would facilitate

the process of residue use for conventional tillage. Yield responses to residue retention were affected by both climate and soil conditions.

- Appropriate simulation models should be developed to provide integrated and location-specific practices for crop residue management in order to achieve the multiple goals of increasing rice yield, improving fertilizer use efficiency, and enhancing the profitability of rice production.
- Further research is needed to improve N management in conventional tillage systems. There is a need to develop crop need-based, real-time N management strategies for apply N fertilizer in rice-wheat cropping system amended with crop residues with tools such as SPAD meter, Green Seeker and leaf colour chart.
- Supplying machineries for conventional tillage on subsidized rates and promoting custom hiring systems for agricultural implements will also be needed for increasing the area under direct seeding of wheat into rice residues so as to avoid large scale burning of the residues by the farmers.
- Residue incorporation improves the crop productivity and nutrient uptake.
- Residue retained under zero tillage with 150 kg N ha^{-1} favoured grain weight/ spike and was adjudged the best practice in wheat in terms of productivity and profitability.
- Crop residues should be incorporated into paddy fields along with liming to improve both rice yield and soil fertility under acidic soil condition.
- Band placement of 20% of the fertilizer N as DAP at seeding and top dressing of 80% urea in 2 equal doses in zero till wheat drilled into rice residues produced higher grain yield and N use efficiency.
- Surface application of rice residues @ 6 and 7 t/ha significantly reduced the growth and development of *P. minor* and recorded higher weed control efficiency as compared to corporation and no residue treatment.
- Mulch lower total soil evaporation over the crop growth season by 35 and 40 mm in relatively high and low rainfall years, respectively.
- Straw mulching decreases the soil bulk density and increase organic matter content in 0-15 cm layer and under limited irrigation water condition rice straw mulching will be beneficial in increasing yield and water use efficiency in wheat.

11

SCOPE OF ORGANIC FARMING IN FARMING SYSTEM

What is Organic Farming?

Organic farming system in India is not new and is being followed from ancient times. It is a method of farming system which primarily aims at cultivating the land and raising crops in such a way, as to keep the soil alive and in good health by use of organic wastes (crop, animal and farm wastes, aquatic wastes) and other biological materials along with beneficial microbes (biofertilizers) to release nutrients to crops for increased sustainable production in an eco friendly pollution free environment.

As per the definition of the United States Department of Agriculture (USDA) study team on organic farming, "organic farming is a system which avoids or largely excludes the use of synthetic inputs (such as fertilizers, pesticides, hormones, feed additives, etc.) and to the maximum extent rely upon crop rotation, crop residues, animal manures, off-farm organic waste, mineral grade rock additives and biological system of nutrient mobilization and plant protection".

FAO suggested that "Organic agriculture is a unique production management system which promotes and enhances agro-ecosystem health, including biodiversity, biological cycles and soil biological activity, and this is accomplished by using on-farm agronomic, biological and mechanical methods in exclusion of all synthetic off-farm inputs".

Organic farming is an alternative agricultural system which originated early in the 20th century in reaction to rapidly changing farming practices. Organic farming continues to be developed by various organic agriculture organizations today. It relies on fertilizers of organic origin such as compost manure, green manure, and bone meal and places emphasis on techniques such as crop rotation and companion planting. Biological pest control, mixed cropping and the fostering of insect predators are encouraged. In general, organic standards are designed to allow the use of naturally occurring substances while prohibiting or strictly limiting synthetic substances. For instance, naturally occurring pesticides such as pyrethrin and rotenone are permitted, while synthetic fertilizers and pesticides are generally prohibited. Synthetic substances that are allowed include, for example, copper sulfate, elemental sulfur and Ivermectin. Genetically modified organisms, nanomaterials, human sewage sludge, plant growth regulators, hormones, and antibiotic use in livestock husbandry are prohibited. Reasons for advocation of organic farming include advantages in sustainability, openness, self-sufficiency, autonomy/independence, health, food security, and food safety.

Organic agricultural methods are internationally regulated and legally enforced by many nations, based in large part on the standards set by the International Federation of Organic Agriculture Movements (IFOAM), an international umbrella organization for organic farming organizations established in 1972. Organic agriculture can be defined as an integrated farming system that strives for sustainability, the enhancement of soil fertility and biological diversity whilst, with rare exceptions i.e prohibiting synthetic pesticides, antibiotics, synthetic fertilizers, genetically modified organisms, and growth hormones etc.

Since 1990 the market for organic food and other products has grown rapidly, reaching $63 billion worldwide in 2012. This demand has driven a similar increase in organically managed farmland that grew from 2001 to 2011 at a compounding rate of 8.9% per annum.

As of 2016, approximately 57,800,000 hectares (143,000,000 acres) worldwide were farmed organically, representing approximately 1.2% of total world farmland.

Terminology

Biodynamic agriculturists, who based their work on Steiner's spiritually-oriented anthroposophy, used the term "organic" to indicate that a farm should be viewed as a living organism, in the sense of the following quotation:

"An organic farm, properly speaking, is not one that uses certain methods and substances and avoids others; it is a farm whose structure is formed in imitation of the structure of a natural system that has the integrity, the independence and the benign dependence of an organism".

The use of "organic" popularized by Howard and Rodale, on the other hand, refers more narrowly to the use of organic matter derived from plant compost and animal manures to improve the humus content of soils, grounded in the work of early soil scientists who developed what was then called "humus farming." Since the early 1940s the two camps have tended to merge.

HISTORY

Agriculture was practiced for thousands of years without the use of artificial chemicals. Artificial fertilizers were first created during the mid-19th century. These early fertilizers were cheap, powerful, and easy to transport in bulk. Similar advances occurred in chemical pesticides in the 1940s, leading to the decade being referred to as the 'pesticide era'. These new agricultural techniques, while beneficial in the short term, had serious longer term side effects such as soil compaction, erosion, and declines in overall soil fertility, along with health concerns about toxic chemicals entering the food supply. In the late 1800s and early 1900s, soil biology scientists began to seek ways to remedy these side effects while still maintaining higher production.

Biodynamic agriculture was the first modern system of agriculture to focus exclusively on organic methods. Its development began in 1924 with a series of eight lectures on agriculture given by Rudolf Steiner. These lectures, the first known presentation of what later came to be known as "Organic Agriculture", were held in response to a request by farmers who noticed degraded soil conditions and a deterioration in the health and quality of crops and livestock resulting from the use of chemical fertilizers. The one hundred eleven attendees, less than half of whom were farmers, came from six countries, primarily Germany and Poland. The lectures were published in November 1924; the first English translation appeared in 1928 as *The Agriculture Course*.

In 1921, Albert Howard and his wife Gabrielle Howard, accomplished botanists, founded Institute of Plant Industry to improve traditional farming methods in India. Among other things, they brought improved implements and improved animal husbandry methods from their scientific training; then by incorporating aspects of the local traditional methods, developed protocols for the rotation of crops, erosion prevention techniques, and the systematic use of composts and manures. Stimulated by these experiences of traditional farming, when Albert Howard returned to Britain in the early 1930s he began to promulgate a system of natural agriculture.

In July 1939, Ehrenfried Pfeiffer, the author of the standard work on biodynamic agriculture *Bio-Dynamic Farming and Gardening*, came to the UK at the invitation of Walter James, 4th Baron Northbourne as a presenter at the Betteshanger Summer School and Conference on Biodynamic Farming at Northbourne's farm in Kent. One of the chief purposes of the conference was to bring together the proponents of various approaches to organic agriculture in order that they might cooperate within a larger movement.

In 1940 Howard published his *An Agricultural Testament*. In this book he adopted Northbourne's terminology of "organic farming." Howard's work spread widely, and he became known as the "father of organic farming" for his work in applying scientific knowledge and principles to various traditional and natural methods. In the United States, J.I. Rodale, who was keenly interested both in Howard's ideas and in biodynamics, founded in the 1940s both a working organic farm for trial and experimentation and the Rodale Institute, and the Rodale Press to teach and advocate organic methods to the wider public. These became important influences on the spread of organic agriculture. Further work was done by Lady Eve Balfour (the Haughley Experiment) in the United Kingdom, and many others across the world.

Increasing environmental awareness in the general population in modern times has transformed the originally supply-driven organic movement to a demand-driven one. Premium prices and some government subsidies attracted farmers. In the developing world, many producers farm according to traditional methods that are comparable to organic farming, but not certified, and that may not include the latest scientific advancements in organic agriculture. In other cases, farmers in the developing world have converted to modern organic methods for economic reasons.

METHODS

"Organic agriculture is a production system that sustains the health of soils, ecosystems and people. It relies on ecological processes, biodiversity and cycles adapted to local conditions, rather than the use of inputs with adverse effects. Organic agriculture combines tradition, innovation and science to benefit the shared environment and promote fair relationships and a good quality of life for all involved..."

Organic farming methods combine scientific knowledge of ecology and modern technology with traditional farming practices based on naturally occurring biological processes. (IFOAM) Organic farming methods are studied in the field of agroecology. While conventional agriculture uses synthetic pesticides and water-soluble synthetically purified fertilizers, organic farmers are restricted by regulations to using natural pesticides and fertilizers. An example of a natural pesticide is pyrethrin, which is found naturally in the Chrysanthemum flower. The principal methods of organic farming include crop rotation, green manures and compost, biological pest control, and mechanical cultivation. These measures use the natural environment to enhance agricultural productivity: legumes are planted to fix nitrogen into the soil, natural insect predators are encouraged, crops are rotated to confuse pests and renew the soil, and natural materials such as potassium bicarbonate and mulches are used to control diseases and weeds. Genetically modified seeds and animals are excluded.

While organic farming is fundamentally different from conventional because of the use of carbon based fertilizers compared with highly soluble synthetic based fertilizers and biological pest control instead of synthetic pesticides, organic farming and large-scale conventional farming are not entirely mutually exclusive. Many of the methods developed for organic agriculture have been borrowed by more conventional agriculture. For example, Integrated Pest Management is a multifaceted strategy that uses various organic methods of pest control whenever possible, but in conventional farming could include synthetic pesticides only as a last resort.

Crop Diversity

Organic farming encourages crop diversity. The science of agroecology has revealed the benefits of polyculture (multiple crops in the same space), which is often employed in organic farming. Planting a variety of vegetable crops supports a wider range of beneficial insects, soil microorganisms, and other factors that add up to overall farm health. Crop diversity helps environments thrive and protects species from going extinct.

Soil Management

Organic farming relies heavily on the natural breakdown of organic matter, using techniques like green manure and composting, to replace nutrients taken from the soil by previous crops. This biological process, driven by microorganisms such as mycorrhiza, allows the natural production of nutrients in the soil throughout the growing season, and has been referred to as *feeding the soil to feed the plant.* Organic farming uses a variety of methods to improve soil fertility, including crop rotation, cover cropping, reduced tillage, and application of compost. By reducing tillage, soil is not inverted and exposed to air; less carbon is lost to the atmosphere resulting in more soil organic carbon. This has an added benefit of carbon sequestration, which can reduce green house gases and help reverse climate change.

Plants need nitrogen, phosphorus, and potassium, as well as micronutrients and symbiotic relationships with fungi and other organisms to flourish, but getting enough nitrogen, and particularly synchronization so that plants get enough nitrogen at the right time (when plants need it most), is a challenge for organic farmers. Crop rotation and green manure ("cover crops") help to provide nitrogen through legumes (more precisely, the Fabaceae family), which fix nitrogen from the atmosphere through symbiosis with rhizobial bacteria. Intercropping, which is sometimes used for insect and disease control, can also increase soil nutrients, but the competition between the legume and the crop can be problematic and wider spacing between crop rows is required. Crop residues can be ploughed back into the soil, and different plants leave different amounts of nitrogen, potentially aiding synchronization. Organic farmers also use animal manure, certain processed fertilizers such as seed meal and various mineral powders such as rock phosphate and green sand, a naturally occurring form of potash that provides potassium. Together these methods help to control erosion. In some cases pH may need to be amended. Natural pH amendments include lime and sulfur, but in the U.S. some compounds such as iron sulfate, aluminum sulfate, magnesium sulfate, and soluble boron products are allowed in organic farming.

Mixed farms with both livestock and crops can operate as ley farms, whereby the land gathers fertility through growing nitrogen-fixing forage grasses such as white clover or alfalfa and grows cash crops or cereals when fertility is established. Farms without livestock ("stockless") may find it more difficult to maintain soil fertility, and may rely more on external inputs such as imported manure as well as grain legumes and green manures, although grain legumes may fix limited nitrogen because they are harvested. Horticultural farms that grow fruits and vegetables in protected conditions often rely even more on external inputs.

Biological research into soil and soil organisms has proven beneficial to organic farming. Varieties of bacteria and fungi break down chemicals, plant matter and animal waste into productive soil nutrients. In turn, they produce benefits of healthier yields and more productive soil for future crops. Fields with less or no manure display significantly lower yields, due to decreased soil microbe community. Increased manure improves biological activity, providing a healthier, more arable soil system and higher yields.

Genetic Modification

A key characteristic of organic farming is the rejection of genetically engineered plants and animals. On 19 October 1998, participants at IFOAM's 12th Scientific Conference issued the Mar del Plata Declaration, where more than 600 delegates from over 60 countries voted unanimously to exclude the use of genetically modified organisms in food production and agriculture.

Although opposition to the use of any transgenic technologies in organic farming is strong, agricultural researchers Luis Herrera-Estrella and Ariel Alvarez-Morales continue to advocate integration of transgenic technologies into organic farming as the optimal means to sustainable agriculture, particularly in the developing world, as does author and scientist Pamela Ronald, who views this kind of biotechnology as being consistent with organic principles.

Although GMOs are excluded from organic farming, there is concern that the pollen from genetically modified crops is increasingly penetrating organic and heirloom seed stocks, making it difficult, if not impossible, to keep these genomes from entering the organic food supply. Differing regulations among countries limits the availability of GMOs to certain countries, as described in the article on regulation of the release of genetic modified organisms.

GEOGRAPHIC PRODUCER DISTRIBUTION

The markets for organic products are strongest in North America and Europe, which as of 2001 were estimated to have \$6 and \$8 billion respectively of the \$20 billion global market. As of 2007 Australasia had 39% of the total organic farmland, including Australia's 1,180,000 hectares but 97% of this land is sprawling rangeland (2007:35). US sales are 20x as much. Europe farms 23% of global organic farmland (6,900,000 ha), followed by Latin America with 19% (5.8 million hectares). Asia has 9.5% while North America has 7.2%. Africa has 3 percent.

Besides Australia, the countries with the most organic farmland are Argentina (3.1 million hectares), China (2.3 million hectares), and the United States (1.6 million hectares). Much of Argentina's organic farmland is pasture, like that of Australia (2007:42). Spain, Germany, Brazil (the world's largest agricultural exporter), Uruguay, and England follow the United States in the amount of organic land (2007:26).

In the European Union (EU25) 3.9% of the total utilized agricultural area was used for organic production in 2005. The countries with the highest proportion of organic land were Austria (11%) and Italy (8.4%), followed by the Czech Republic and Greece (both 7.2%). The lowest figures were shown for Malta (0.2%), Poland (0.6%) and Ireland (0.8%). In 2009, the proportion of organic land in the EU grew to 4.7%. The countries with highest share of agricultural land were Liechtenstein (26.9%), Austria (18.5%) and Sweden (12.6%). Sixteen percent of all farmers in Austria produced organically in 2010. By the same year the proportion of organic land increased to 20%.: In 2005 168,000 ha of land in Poland was under organic management. In 2012, 288,261 hectares were under organic production, and there were about 15,500 organic farmers; retail sales of organic products were EUR 80 million in 2011. As of 2012 organic exports were part of the government's economic development strategy.

After the collapse of the Soviet Union in 1991, agricultural inputs that had previously been purchased from Eastern bloc countries were no longer available in Cuba, and many Cuban farms converted to organic methods out of necessity. Consequently, organic agriculture is a mainstream practice in Cuba, while it remains an alternative practice in most other countries. Cuba's organic strategy includes development of genetically modified crops; specifically corn that is resistant to the palomilla moth.

GROWTH OF ORGANIC FARMING IN WORLD AND INDIA

In 2001, the global market value of certified organic products was estimated at USD $20 billion. By 2002, this was USD $23 billion. By 2014, retail sales of organic products reached USD $80 billion worldwide. North America and Europe accounted for more than 90% of all organic product sales.

Organic agricultural land increased almost fourfold in 15 years, from 11 million hectares in 1999 to 43.7 million hectares in 2014.Between 2013 and 2014, organic agricultural land grew by 500,000 hectares worldwide, increasing in every region except Latin America. During this time period, Europe's organic farmland increased 260,000 hectares to 11.6 million total (+2.3%), Asia's increased 159,000 hectares to 3.6 million total (+4.7%), Africa's increased 54,000 hectares to 1.3 million total (+4.5%), and North America's increased 35,000 hectares to 3.1 million total (+1.1%). As of 2014, the country with the most organic land was Australia (17.2 million hectares), followed by Argentina (3.1 million hectares), and the United States (2.2 million hectares).

In 2013, the number of organic producers grew by almost 270,000, or more than 13%. By 2014, there were a reported 2.3 million organic producers in the world. Most of the total global increase took place in the Philippines, Peru, China, and Thailand. Overall, the majority of all organic producers are in India (650,000 in 2013), Uganda (190,552 in 2014), Mexico (169,703 in 2013) and the Philippines (165,974 in 2014)

As per FiBL survey, 2018, there was linear increase in organic area globally from 11.0 mha during 1999 to 57.8 Mha during 2016. (Fig 11.1)

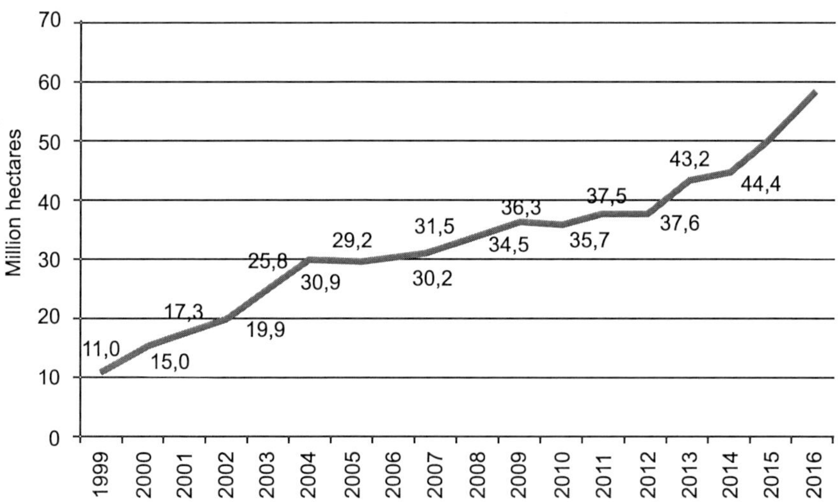

Fig. 11.1 Growth of organic agricultural land 1999-2016.
Source: FiBL-IFOAM-SOEL-Surveys 1999-2018

The largest area of 27.15 mha under organic agriculture during 2016 was in Australia which was followed by Argentina (3.01 mha), China (2.28 mha), USA (2.03 mha) and the minimum area of 1.25 mha was in Germany out of 10 leading countries adopting for organic agriculture. (Fig 11.2)

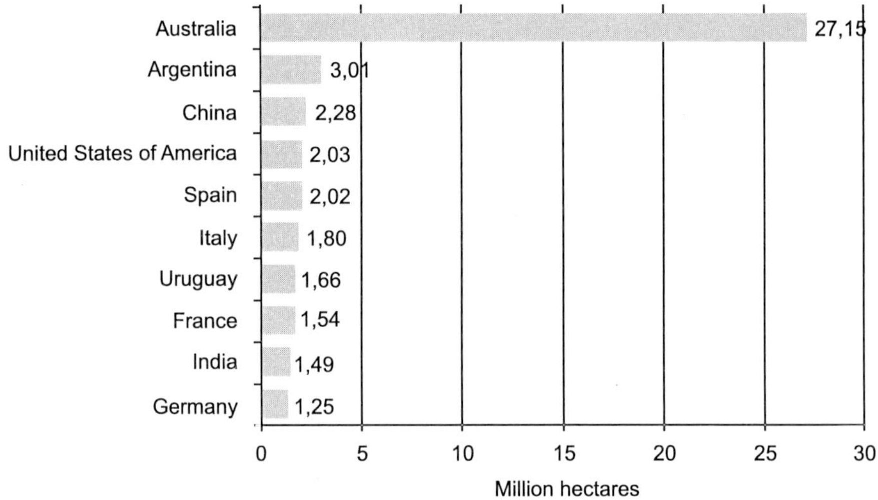

Fig. 11.2 Top 10 countries with the largest areas of organic agriculture land 2016.
Source: FiBL Survey 2018

Among the field crops cereals occupied highest area i.e about 41 lakh hectare during 2016 under organic agriculture land which was followed by green fodder, oilseeds, pulses, textile crops, vegetables etc. (Fig 11.3)

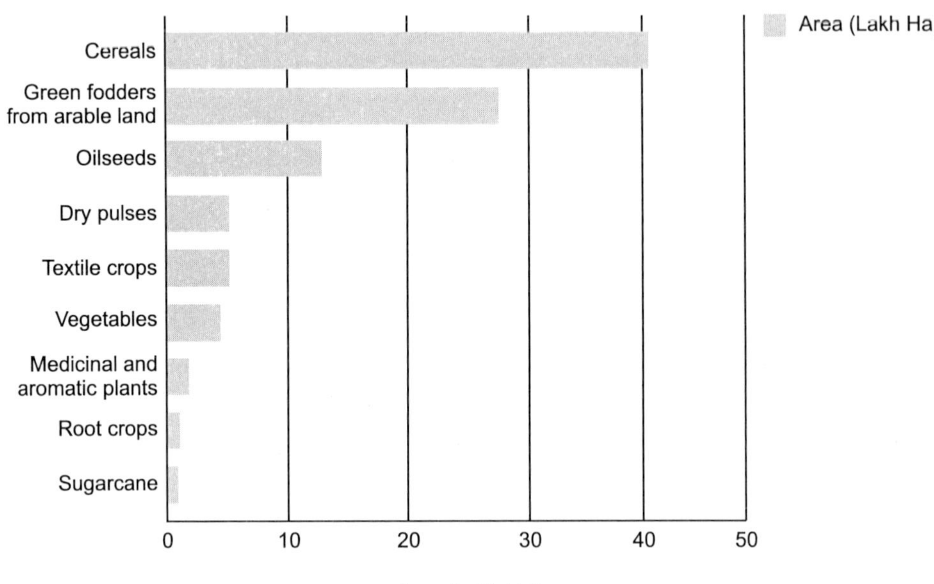

Fig. 11.3 Key crops in organic agriculture during 2016.

Source: FiBL survey 2018

As far as organic agriculture land in different States is concerned, the highest area of 4.62 lakh ha was recorded in Madhya Pradesh, followed by Maharashtra (1.98

lakh ha), Rajasthan (1.55 lakh ha), Telangana (1.04 lakh ha), Orissa (0.96 lakh ha), Karnataka (0.94 lakh ha) etc. (Fig 11.4)

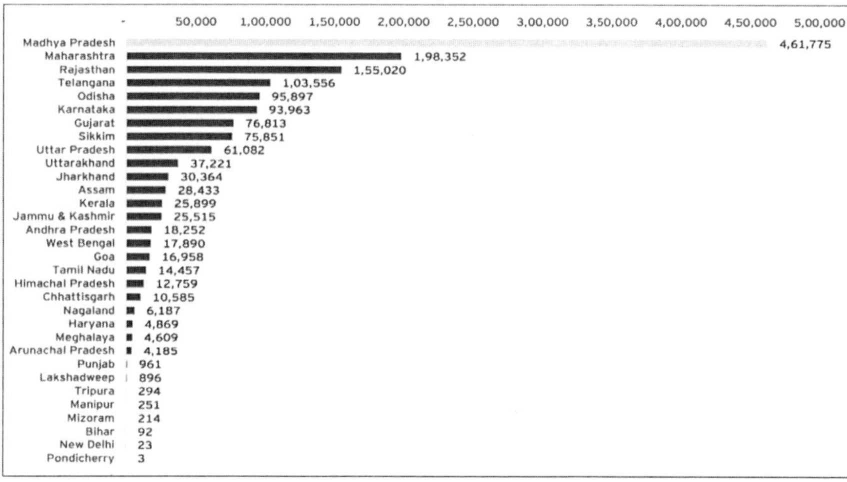

Fig. 11.4 Organic agriculture area by State(india) in ha 2014-2015.

Source: APEDA

The highest production of 338.19 mt was recorded in sugar crops which was followed by oilseeds (228.41 mt), fiber crops (208.93 mt) and cereals (159.50 mt) (Fig 11.5)

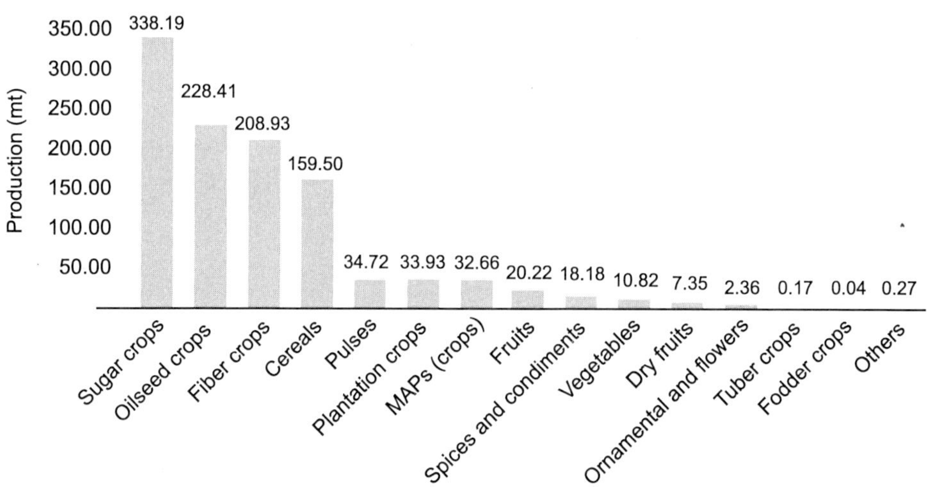

Fig. 11.5 Production of organic crops in india by category (mt) for 2014-15

Source: APEDA

PRODUCTIVITY AND PROFITABILITY

Studies comparing yields have had mixed results. These differences among findings can often be attributed to variations between study designs including differences in the crops studied and the methodology by which results were gathered.

A 2012 meta-analysis found that productivity is typically lower for organic farming than conventional farming, but that the size of the difference depends on context and in some cases may be very small. While organic yields can be lower than conventional yields, another meta-analysis published in Sustainable Agriculture Research in 2015, concluded that certain organic on-farm practices could help narrow this gap. Timely weed management and the application of manure in conjunction with legume forages/ cover crops were shown to have positive results in increasing organic corn and soybean productivity.

Another meta-analysis published in the journal *Agricultural Systems* in 2011 analyzed 362 datasets and found that organic yields were on average 80% of conventional yields. The author's found that there are relative differences in this yield gap based on crop type with crops like soybeans and rice scoring higher than the 80% average and crops like wheat and potato scoring lower. Across global regions, Asia and Central Europe were found to have relatively higher yields and Northern Europe relatively lower than the average.

A 2007 study compiling research from 293 different comparisons into a single study to assess the overall efficiency of the two agricultural systems has concluded that "organic methods could produce enough food on a global per capita basis to sustain the current human population, and potentially an even larger population, without increasing the agricultural land base." The researchers also found that while in developed countries, organic systems on average produce 92% of the yield produced by conventional agriculture, organic systems produce 80% more than conventional farms in developing countries, because the materials needed for organic farming are more accessible than synthetic farming materials to farmers in some poor countries. This study's methodology and results were contested by D.J. Connor of the University of Melbourne, in a short communication published in Field Crops Research. Connor writes that errors in Badgley *et al.* result in "major overestimation of the productivity of OA".

Long Term Studies

A study published in 2005 compared conventional cropping, organic animal-based cropping, and organic legume-based cropping on a test farm at the Rodale Institute over 22 years. The study found that "the crop yields for corn and soybeans were similar in the organic animal, organic legume, and conventional farming systems". It also found that "significantly less fossil energy was expended to produce corn in the Rodale Institute's organic animal and organic legume systems than in the conventional production system. There was little difference in energy input between the different treatments for producing soybeans. In the organic systems, synthetic fertilizers and pesticides were generally not used. As of 2013 the Rodale study was ongoing and a thirty-year anniversary report was published by Rodale in 2012.

A long-term field study comparing organic/conventional agriculture carried out over 21 years in Switzerland concluded that "Crop yields of the organic systems averaged over 21 experimental years at 80% of the conventional ones. The fertilizer input, however, was 34 – 51% lower, indicating an efficient production (Rodale 2012). The organic farming systems used 20 – 56% less energy to produce a crop per unit land area and this difference was 36 – 53%. In spite of the considerably lower pesticide input the quality of organic products was hardly discernible from conventional analytically and even came off better in food preference trials and picture creating methods".

Profitability

In the United States, organic farming has been shown to be 2.9 to 3.8 times more profitable for farmers than conventional farming when prevailing price premiums are taken into account. Globally, organic farming is between 22 and 35% more profitable for farmers than conventional methods, according to a 2015 meta-analysis of studies conducted across five continents.

The profitability of organic agriculture can be attributed to a number of factors. First, organic farmers do not rely on synthetic fertilizer and pesticide inputs, which can be costly. In addition, organic foods currently enjoy a price premium over conventionally produced foods, meaning that organic farmers can often get more for their yield.

The price premium for organic food is an important factor in the economic viability of organic farming. In 2013 there was a 100% price premium on organic vegetables and a 57% price premium for organic fruits. These percentages are based on wholesale fruit and vegetable prices, available through the United States Department of Agriculture's Economic Research Service. Price premiums exist not only for organic versus non organic crops, but may also vary depending on the venue where the product is sold: farmers' markets, grocery stores, or wholesale to restaurants. For many producers, direct sales at farmers' markets are most profitable because the farmer receives the entire markup, however this is also the most time and labor-intensive approach.

There have been signs of organic price premiums narrowing in recent years, which lowers the economic incentive for farmers to convert to or maintain organic production methods. Data from 22 years of experiments at the Rodale Institute found that, based on the current yields and production costs associated with organic farming in the United States, a price premium of only 10% is required to achieve parity with conventional farming. A separate study found that on a global scale, price premiums of only 5-7% percent were needed to break even with conventional methods. Without the price premium, profitability for farmers is mixed.

For markets and supermarkets organic food is profitable as well, and is generally sold at significantly higher prices than non-organic food.

Energy Efficiency

In the most recent assessments of the energy efficiency of organic versus conventional agriculture, results have been mixed regarding which form is more carbon efficient. Organic farm systems have more often than not been found to be more energy efficient, however, this is not always the case. More than anything, results tend to depend upon crop type and farm size.

A comprehensive comparison of energy efficiency in grain production, produce yield, and animal husbandry concluded that organic farming had a higher yield per unit of energy over the vast majority of the crops and livestock systems. For example, two studies - both comparing organically- versus conventionally-farmed apples - declare contradicting results, one saying organic farming is more energy efficient, the other saying conventionally is more efficient.

It has generally been found that the labour input per unit of yield was higher for organic systems compared with conventional production.

Sales and Marketing

Most sales are concentrated in developed nations. In 2008, 69% of Americans claimed to occasionally buy organic products, down from 73% in 2005. One theory for this change was that consumers were substituting "local" produce for "organic" produce.

Distributors

The USDA requires that distributors, manufacturers, and processors of organic products be certified by an accredited State or private agency. In 2007, there were 3,225 certified organic handlers, up from 2,790 in 2004.

Organic handlers are often small firms; 48% reported sales below $1 million annually, and 22% between $1 and $5 million per year. Smaller handlers are more likely to sell to independent natural grocery stores and natural product chains whereas large distributors more often market to natural product chains and conventional supermarkets, with a small group marketing to independent natural product stores. Some handlers work with conventional farmers to convert their land to organic with the knowledge that the farmer will have a secure sales outlet. This lowers the risk for the handler as well as the farmer. In 2004, 31% of handlers provided technical support on organic standards or production to their suppliers and 34% encouraged their suppliers to transition to organic. Smaller farms often join together in cooperatives to market their goods more effectively.

About 93% of organic sales are through conventional and natural food supermarkets and chains, while the remaining 7% of U.S. organic food sales occur through farmers' markets, foodservices, and other marketing channels.

Direct-to-consumer Sales

In the 2012 Census, direct-to-consumer sales equaled $1.3 billion, up from $812 million in 2002, an increase of 60 percent. The number of farms that utilize direct-to-consumer sales was 144,530 in 2012 in comparison to 116,733 in 2002. Direct-to-consumer sales include farmers' markets, community supported agriculture (CSA), on-farm stores, and roadside farm stands. Some organic farms also sell products direct to retailer, direct to restaurant and direct to institution. According to the 2008 Organic Production Survey, approximately 7% of organic farm sales were direct-to-consumers, 10% went direct to retailers, and approximately 83% went into wholesale markets. In comparison, only 0.4% of the value of convention agricultural commodities were direct-to-consumers.

While not all products sold at farmer's markets are certified organic, this direct-to-consumer avenue has become increasingly popular in local food distribution and has grown substantially since 1994. In 2014, there were 8,284 farmer's markets in comparison to 3,706 in 2004 and 1,755 in 1994, most of which are found in populated areas such as the Northeast, Midwest, and West Coast.

Labor and Employment

Organic production is more labor-intensive than conventional production. On the one hand, this increased labor cost is one factor that makes organic food more expensive. On the other hand, the increased need for labor may be seen as an "employment dividend" of organic farming, providing more jobs per unit area than conventional systems. The 2011 UNEP Green Economy Report suggests that "an increase in investment in green agriculture is projected to lead to growth in employment of about 60 per cent compared with current levels" and that "green agriculture investments could create 47 million additional jobs compared with BAU2 over the next 40 years." The United Nations Environment Programme (UNEP) also argues that "by greening agriculture and food distribution, more calories per person per day, more jobs and business opportunities especially in rural areas, and market-access opportunities, especially for developing countries, will be available."

World's Food Security

In 2007 the United Nations Food and Agriculture Organization (FAO) said that organic agriculture often leads to higher prices and hence a better income for farmers, so it should be promoted. However, FAO stressed that by organic farming one could not feed the current mankind, even less the bigger future population. Both data and models showed then that organic farming was far from sufficient. Therefore, chemical fertilizers were needed to avoid hunger. Other analysis by many agribusiness executives, agricultural and ecological scientists, and international agriculture experts revealed the opinion that organic farming would not only increase the world's food supply, but might be the only way to eradicate hunger.

FAO stressed that fertilizers and other chemical inputs can much increase the production, particularly in Africa where fertilizers are currently used 90% less than in Asia. For example, in Malawi the yield has been boosted using seeds and fertilizers. FAO also calls for using biotechnology, as it can help smallholder farmers to improve their income and food security.

Also NEPAD, development organization of African governments, announced that feeding Africans and preventing malnutrition requires fertilizers and enhanced seeds.

According to a 2012 study in *Science Digest*, organic best management practices shows an average yield only 13% less than conventional. In the world's poorer nations where most of the world's hungry live, and where conventional agriculture's expensive inputs are not affordable by the majority of farmers, adopting organic management actually increases yields 93% on average, and could be an important part of increased food security.

Capacity Building in Developing Countries

Organic agriculture can contribute to ecological sustainability, especially in poorer countries. The application of organic principles enables employment of local resources (e.g., local seed varieties, manure, etc.) and therefore cost-effectiveness. Local and international markets for organic products show tremendous growth prospects and offer creative producers and exporters excellent opportunities to improve their income and living conditions.

Organic agriculture is knowledge intensive. Globally, capacity building efforts are underway, including localized training material, to limited effect. As of 2007, the International Federation of Organic Agriculture Movements hosted more than 170 free manuals and 75 training opportunities online.

In 2008 the United Nations Environmental Programme (UNEP) and the United Nations Conference on Trade and Development (UNCTAD) stated that "organic agriculture can be more conducive to food security in Africa than most conventional production systems, and that it is more likely to be sustainable in the long-term" and that "yields had more than doubled where organic, or near-organic practices had been used" and that soil fertility and drought resistance improved.

Millennium Development Goals (MDG)

The value of organic agriculture (OA) in the achievement of the Millennium Development Goals (MDG), particularly in poverty reduction efforts in the face of climate change, is shown by its contribution to both income and non-income aspects of the MDGs. These benefits are expected to continue in the post-MDG era. A series of case studies conducted in selected areas in Asian countries by the Asian Development Bank Institute (ADBI) and published as a book compilation by ADB in Manila document these contributions to both income and non-income aspects of the MDGs. These include poverty alleviation by way of higher incomes, improved farmers' health owing to less chemical exposure, integration of sustainable principles into rural development policies, improvement of access to safe water and sanitation, and expansion of global partnership for development as small farmers are integrated in value chains.

A related ADBI study also sheds on the costs of OA programs and set them in the context of the costs of attaining the MDGs. The results show considerable variation across the case studies, suggesting that there is no clear structure to the costs of adopting OA. Costs depend on the efficiency of the OA adoption programs. The lowest cost programs were more than ten times less expensive than the highest cost ones. However, further analysis of the gains resulting from OA adoption reveals that the costs per person taken out of poverty was much lower than the estimates of the World Bank, based on income growth in general or based on the detailed costs of meeting some of the more quantifiable MDGs (e.g., education, health, and environment).

DISADVANTAGES

Agriculture imposes negative externalities (uncompensated costs) upon society through public land and other public resource use, biodiversity loss, erosion, pesticides, nutrient runoff, subsidized water usage, subsidy payments and assorted other problems. Positive externalities include self-reliance, entrepreneurship, respect for nature, and air quality. Organic methods reduce some of these costs. In 2000 uncompensated costs for 1996 reached 2,343 million British pounds or £208 per ha (£84.20/ac). A study of practices in the US published in 2005 concluded that cropland costs the economy approximately 5 to 16 billion dollars ($30–96/ha – $12–39/ac), while livestock production costs 714 million dollars. Both studies recommended reducing externalities. The 2000 review included reported pesticide poisonings but did not include speculative chronic health effects of pesticides, and the 2004 review relied on a 1992 estimate of the total impact of pesticides.

It has been proposed that organic agriculture can reduce the level of some negative externalities from (conventional) agriculture. Whether the benefits are private or public depends upon the division of property rights.

Several surveys and studies have attempted to examine and compare conventional and organic systems of farming and have found that organic techniques, while not without harm, are less damaging than conventional ones because they reduce levels of biodiversity less than conventional systems do and use less energy and produce less waste when calculated per unit area.

A 2003 to 2005 investigation by the Cranfield University for the Department for Environment, Food and Rural Affairs in the UK found that it is difficult to compare the Global warming potential, acidification and eutrophication emissions but "Organic production often results in increased burdens, from factors such as N leaching and N_2O emissions", even though primary energy use was less for most organic products. N_2O is always the largest global warming potential contributor except in tomatoes. However, "organic tomatoes always incur more burdens (except pesticide use)". Some emissions were lower "per area", but organic farming always required 65 to 200% more field area than non-organic farming. The numbers were highest for bread wheat (200+ % more) and potatoes (160% more).

Environmental Impact and Emissions

Researchers at Oxford University analyzed 71 peer-reviewed studies and observed that organic products are sometimes worse for the environment. Organic milk, cereals, and pork generated higher greenhouse gas emissions per product than conventional ones but organic beef and olives had lower emissions in most studies. Usually organic products required less energy, but more land. Per unit of product, organic produce generates higher nitrogen leaching, nitrous oxide emissions, ammonia emissions, eutrophication, and acidification potential than conventionally grown produce. Other differences were not significant. The researchers concluded that public debate should consider various manners of employing conventional or organic farming, and not merely debate conventional farming as opposed to organic farming. They also sought to find specific solutions to specific circumstances.

Proponents of organic farming have claimed that organic agriculture emphasizes closed nutrient cycles, biodiversity, and effective soil management providing the capacity to mitigate and even reverse the effects of climate change and that organic agriculture can decrease fossil fuel emissions. "The carbon sequestration efficiency of organic systems in temperate climates is almost double (575–700 kg carbon per ha per year – 510–625 lb/ac/an) that of conventional treatment of soils, mainly owing to the use of grass clovers for feed and of cover crops in organic rotations."

Critics of organic farming methods believe that the increased land needed to farm organic food could potentially destroy the rainforests and wipe out many ecosystems.

Nutrient Leaching

According to a 2012 meta-analysis of 71 studies, nitrogen leaching, nitrous oxide emissions, ammonia emissions, eutrophication potential and acidification potential were higher for organic products, although in one study "nitrate leaching was 4.4–5.6 times higher in conventional plots than organic plots".

Excess nutrients in lakes, rivers, and groundwater can cause algal blooms, eutrophication, and subsequent dead zones. In addition, nitrates are harmful to aquatic organisms by themselves.

Land use

The Oxford meta-analysis of 71 studies found that organic farming requires 84% more land for an equivalent amount of harvest, mainly due to lack of nutrients but sometimes due to weeds, diseases or pests, lower yielding animals and land required for fertility building crops. While organic farming does not necessarily save land for wildlife habitats and forestry in all cases, the most modern breakthroughs in organic are addressing these issues with success.

Professor Wolfgang Branscheid says that organic animal production is not good for the environment, because organic chicken requires twice as much land as "conventional" chicken and organic pork a quarter more. According to a calculation by Hudson Institute, organic beef requires three times as much land. On the other hand, certain organic methods of animal husbandry have been shown to restore desertified, marginal, and/or otherwise unavailable land to agricultural productivity and wildlife. Or by getting both forage and cash crop production from the same fields simultaneously, reduce net land use.

In England organic farming yields 55% of normal yields. In other regions of the world, organic methods have started producing record yields.

Pesticides

In organic farming synthetic pesticides are generally prohibited. A chemical is said to be synthetic if it does not already exist in the natural world. But the organic label goes further and usually prohibit compounds that exist in nature if they are produced by chemical synthesis. So the prohibition is also about the method of production and not only the nature of the compound.

A non-exhaustive list of organic approved pesticides with their median lethal doses:

Copper(II) sulfate is used as a fungicide and is also used in conventional agriculture (LD_{50} 300 mg/kg). Conventional agriculture has the option to use the less toxic Mancozeb (LD_{50} 4,500 to 11,200 mg/kg)

Boric acid is used as stomach poison that target insects (LD_{50}: 2660 mg/kg).

Pyrethrin comes from chemicals extracted from flowers of the genus Pyrethrum (LD_{50} of 370 mg/kg). Its potent toxicity is used to control insects.

Lime sulphur (aka calcium polysulfide) and sulfur are considered to be allowed, synthetic materials (LD_{50}: 820 mg/kg)

Rotenone is a powerful insecticide that was used to control insects (LD_{50}: 132 mg/kg). Despite the high toxicity of Rotenone to aquatic life and some links to Parkinson disease the compound is still allowed in organic farming as it is a naturally occurring compound.

Bromomethane is a gas that is still used in the nurseries of strawberry organic farming

Azadirachtin is a wide spectrum very potent insecticide. Almost non toxic to mammals (LD_{50} in rats is $>$ 3,540 mg/kg) but affects beneficial insects.

ADVANTAGES

Food Quality and Safety

While there may be some differences in the amounts of nutrients and anti-nutrients when organically produced food and conventionally produced food are compared, the variable nature of food production and handling makes it difficult to generalize results, and there is insufficient evidence to make claims that organic food is safer or healthier than conventional food.Claims that organic food tastes better are not supported by evidence.

Soil Conservation

Supporters claim that organically managed soil has a higher quality and higher water retention. This may help increase yields for organic farms in drought years. Organic farming can build up soil organic matter better than conventional no-till farming, which suggests long-term yield benefits from organic farming. An 18-year study of organic methods on nutrient-depleted soil concluded that conventional methods were superior for soil fertility and yield for nutrient-depleted soils in cold-temperate climates, arguing that much of the benefit from organic farming derives from imported materials that could not be regarded as self-sustaining.

In *Dirt: The Erosion of Civilizations*, geomorphologist David Montgomery outlines a coming crisis from soil erosion. Agriculture relies on roughly one meter of topsoil, and that is being depleted ten times faster than it is being replaced. No-till farming, which some claim depends upon pesticides, is one way to minimize erosion. However, a 2007 study by the USDA's Agricultural Research Service has found that manure applications in tilled organic farming are better at building up the soil than no-till.

Biodiversity

The conservation of natural resources and biodiversity is a core principle of organic production. Three broad management practices (prohibition/reduced use of chemical pesticides and inorganic fertilizers; sympathetic management of non-cropped habitats; and preservation of mixed farming) that are largely intrinsic (but not exclusive) to organic farming are particularly beneficial for farmland wildlife. Using practices that attract or introduce beneficial insects, provide habitat for birds and mammals, and provide conditions that increase soil biotic diversity serve to supply vital ecological services to organic production systems. Advantages to certified organic operations that implement these types of production practices include: 1) decreased dependence on outside fertility inputs; 2) reduced pest management costs; 3) more reliable sources of clean water; and 4) better pollination.

Nearly all non-crop, naturally occurring species observed in comparative farm land practice studies show a preference for organic farming both by abundance and diversity. An average of 30% more species inhabit organic farms. Birds, butterflies, soil microbes, beetles, earthworms, spiders, vegetation, and mammals are particularly affected. Lack of herbicides and pesticides improve biodiversity fitness and population density. Many weed species attract beneficial insects that improve soil qualities and forage on weed pests. Soil-bound organisms often benefit because of increased bacteria populations due to natural fertilizer such as manure, while experiencing reduced intake of herbicides and pesticides. Increased biodiversity, especially from beneficial soil microbes and mycorrhizae have been proposed as an explanation for the high yields experienced by some organic plots, especially in light of the differences seen in a 21-year comparison of organic and control fields.

Biodiversity from organic farming provides capital to humans. Species found in organic farms enhance sustainability by reducing human input (e.g., fertilizers, pesticides).

The USDA's Agricultural Marketing Service (AMS) published a *Federal Register* notice on 15 January 2016, announcing the National Organic Program (NOP) final guidance on Natural Resources and Biodiversity Conservation for Certified Organic Operations. Given the broad scope of natural resources which includes soil, water, wetland, woodland and wildlife, the guidance provides examples of practices that support the underlying conservation principles and demonstrate compliance with USDA organic regulations. The final guidance also clarifies the role of certified operations (to submit an OSP to a certifier), certifiers (ensure that the OSP describes or lists practices that explain the operator's monitoring plan and practices to support natural resources and biodiversity conservation), and inspectors (onsite inspection) in the implementation and verification of these production practices.

A wide range of organisms benefit from organic farming, but it is unclear whether organic methods confer greater benefits than conventional integrated agri-environmental programs. Organic farming is often presented as a more biodiversity-friendly practice, but the generality of the beneficial effects of organic farming is debated as the effects appear often species- and context-dependent, and current research has highlighted the need to quantify the relative effects of local- and landscape-scale management on farmland biodiversity. There are four key issues when comparing the impacts on biodiversity of organic and conventional farming: (1) It remains unclear whether a holistic whole-farm approach (i.e. organic) provides greater benefits to biodiversity than carefully targeted prescriptions applied to relatively small areas of cropped and/or non-cropped habitats within conventional agriculture (i.e. agri-environment schemes); (2) Many comparative studies encounter methodological problems, limiting their ability to draw quantitative conclusions; (3) Our knowledge of the impacts of organic farming in pastoral and upland agriculture is limited; (4) There remains a pressing need for longitudinal, system-level studies in order to address these issues and to fill in the gaps in our knowledge of the impacts of organic farming, before a full appraisal of its potential role in biodiversity conservation in agroecosystems can be made.[188]

Organic farming is a production system which avoids or largely excludes the use of synthetically compounded fertilizers, pesticides, growth regulators, genetically modified organisms and livestock food additives. To the maximum extent possible organic farming system rely upon crop rotations, use of crop residues, animal manures, legumes, green manures, off farm organic wastes, biofertilizers, mechanical cultivation, mineral bearing rocks and aspects of biological control to maintain soil productivity and tilth to supply plant nutrients and to control insect, weeds and other pests.

Organic methods can increase farm productivity, repair decades of environmental damage and knit small farm families into more sustainable distribution networks leading to improved food security if they organize themselves in production, certification and marketing. During last few years an increasing number of farmers have shown lack of interest in farming and the people who used to cultivate are migrating to other areas. Organic farming is one way to promote either self-sufficiency or food security. Use of massive inputs of chemical fertilizers and toxic pesticides poisons the land and water heavily. The after-effects of this are severe environmental consequences, including loss of topsoil, decrease in soil fertility, surface and groundwater contamination and loss of genetic diversity.

Organic farming which is a holistic production management system that promotes and enhances agro-ecosystem health, including biodiversity, biological cycles, and soil biological activity is hence important. Many studies have shown that organic farming methods can produce even higher yields than conventional methods. Significant difference in soil health indicators such as nitrogen mineralization potential and microbial abundance and diversity, which were higher in the organic farms can also be seen. The increased soil health in organic farms also resulted in considerably lower insect and disease incidence. The emphasis on small-scale integrated farming systems has the potential to revitalize rural areas and their economies.

Advantages of organic farming are summarized as below:

- It helps to maintain environment health by reducing the level of pollution.
- It reduces human and animal health hazards by reducing the level of residues in the product.
- It helps in keeping agricultural production at a sustainable level.
- It reduces the cost of agricultural production and also improves the soil health.
- It ensures optimum utilization of natural resources for short-term benefit and helps in conserving them for future generation.

- It not only saves energy for both animal and machine, but also reduces risk of crop failure.
- It improves the soil physical properties such as granulation, good tilth, good aeration, easy root penetration and improves water-holding capacity and reduces erosion.
- It improves the soil's chemical properties such as supply and retention of soil nutrients, reduces nutrient loss into water bodies and environment and promotes favourable chemical reactions.

NUTRIENT MANAGEMENT

In organic farming, it is important to constantly work to build a healthy soil that is rich in organic matter and has all the nutrients that the plants need. Several methods viz. green manuring, addition of manures and biofertilizers etc. can be used to build up soil fertility. These organic sources not only add different nutrients to the soil but also help to prevent weeds and increase soil organic matter to feed soil microorganisms. Soil with high organic matter resists soil erosion, holds water better and thus requires less irrigation. Some natural minerals that are needed by the plants to grow and to improve the soil's consistency can also be added. Soil amendments like lime are added to adjust the soil's pH balance. However soil amendment and water should contain minimum heavy metals. Most of the organic fertilizers used are recycled by-products from other industries that would otherwise go to waste. Farmers also make compost from animal manures and mushroom compost. Before compost can be applied to the fields, it is heated and aged for at least two months, reaching and maintaining an internal temperature of 130°-140°F to kill unwanted bacteria and weed seeds. A number of organic fertilizers / amendments and bacterial and fungal biofertilizers can be used in organic farming depending upon availability and their suitability to crop. Different available organic inputs are described below:

Organic Manures

Commonly available and applied farm yard manure (FYM) and vermicompost etc. are generally low in nutrient content, so high application rates are needed to meet crop nutrient requirements. However, in many developing countries including India, the availability of organic manures is not sufficient for crop requirements; partly due to its extensive use of cattle dung in energy production. Green manuring with Sesbania, cowpea, green gram etc are quiet effective to improve the organic matter content of soil. However, use of green manuring has declined in last few decades due to intensive cropping and socioeconomic reasons. Considering these constraints International Federation of Organic Agriculture Movement (IFOAM) and Codex Alimentarius have approved the use of some inorganic sources of plant nutrients like rock phosphate, basic slag, rock potash etc. in organic farming systems. These substances can supply essential nutrients and may be from plant, animal, microbial or mineral origin and may undergo physical, enzymatic or microbial processes and their use does not result in unacceptable effects on produce and the environment including soil organisms.

Bacterial and Fungal Biofertilizers

Contribution of biological fixation of nitrogen on surface of earth is the highest (67.3%) among all the sources of N fixation. Following bacterial and fungal biofertilizers can be used as a component of organic farming in different crops.

- ***Rhizobium*** : The effectiveness of symbiotic N_2 fixing bacteria viz. Rhizobia for legume crops eg. *Rhizobium, Bradyrhizobium, Sinorhizobium, Azorhizobium,* and *Mesorhizobium etc.* have been well recognized. These bacteria infecting legumes have a global distribution. These rhizobia have a N_2-fixing capability up to 450 kg N ha^{-1} depending on host- plant species and bacterial strains. Carrier based inoculants can be coated on seeds for the introduction of bacterial strains into soil.
- ***Azotobacter:*** N_2-fixing free-living bacteria can fix atmospheric nitrogen in cereal crops without any symbiosis. Such free living bacterias are: *Azotobacter* sp. for different cereal crops; *Acetobacter diazotrophicus* and *Herbaspirillum* spp. for sugarcane, sorghum and maize crop. Beside fixing nitrogen, they also increase germination and vigour in young plants leading to an improved crop stand. They can fix 15-20 kg/ha nitrogen per year. *Azotobacter* sp. also has ability to produce anti fungal compounds against many plant pathogens. *Azotobacter* can biologically control the nematode diseases of plants also.
- ***Azospirillum:*** The genus *Azospirillum* colonizes in a variety of annual and perennial plants. Studies indicate that *Azospirillum* can increase the growth of crops like sunflower, carrot, oak, sugarbeet, tomato, pepper, cotton, wheat and rice. The crop yield can increase from 5-30%. Inoculum of *Azotobacter* and *Azospirillum* can be produced and applied as in peat formulation through seed coating. The peat formulation can also be directly utilized in field applications.
- ***Plant growth promoting rhizobacteria*** : Various bacteria that promote plant growth are collectively called plant growth promoting rhizobacteria (PGPR). PGPR are thought to improve plant growth by colonizing the root system and pre empting the establishment of suppressing deleterious rhizosphere microorganisms on the roots. Large populations of bacteria established in planting material and roots become a partial sink for nutrients in the rhizosphere thus reducing the amount of C and N available to stimulate spores of fungal pathogens or for subsequent colonization of the root. PGPR belong to several genera viz.*Actinoplanes, Azotobacter, Bacillus, Pseudomonas, Rhizobium, Bradyrhizobium, Streptomyces, Xanthomonas* etc. Bacillus spp. act as biocontrol agent because their endospores are tolerant to heat and desiccation. Seed treatment with *B.subtilis* is reported to increase yield of carrot by 48%, oats by 33% and groundnut upto 37%.
- ***Phosphorus-solubilizing bacteria (PSB):*** Phosphorus is the vital nutrient next to nitrogen for plants and microorganisms. This element is necessary for the nodulation by *Rhizobium* and even to nitrogen fixers, *Azolla* and BGA. The phospho microorganisms mainly bacteria and fungi, make available

insoluble phosphorus to the plants. It can increase crop yield up to 200-500 kg/ha and thus 30 to 50 kg Super Phosphate can be saved. Most predominant phosphorus-solubilizing bacteria (PSB) belong to the genera *Bacillus* and *Pseudomonas*. At present PSB is most widely used biofertilizer in India. PSB can reduce the P requirement of crop up to 25%.

- ***Mycorrhizal fungi:*** Root-colonizing mycorrhizal fungi increase tolerance of heavy metal contamination and drought. Mycorrhizal fungi improve soil quality also by having a direct influence on soil aggregation and therefore aeration and water dynamics. An interesting potential of this fungi is its ability to allow plant access to nutrient sources which are generally unavailable to the host plants and thus plants may be able to use insoluble sources of P when inoculated with mycorrhizal fungi but not in the absence of inoculation.
- ***Blue green algae (BGA):*** BGA are the pioneer colonizers both in hydrosphere and xerosphere. These organisms have been found to synthesize 0.8 x 1011 tonnes of organic matter, constituting about 40 percent of the total organic matter synthesized annually on this planet. BGA constitute the largest, most diverse and widely distributed group of prokaryotic microscopic organisms that perform oxygenic photosynthesis. These are also known as cyanophyceae and cyanobacteria. These are widely distributed in tropics; and are able to withstand extremes of temperature and drought. The significance of the abundance of BGA in Indian rice soils has been well recognized. Multi-location trials conducted under varying agro climatic conditions have indicated that the algal inoculation could save 30 kg N /ha, however, it depends upon the agro ecological conditions. BGA has been reported to reduce the pH of soil and improve upon exchangeable calcium and water holding capacity. The recommended method of application of the algal inoculum is broadcasting on standing water about 3 to 4 days after transplantation. After the application of algal inoculum the field should be kept water logged for about a week's time. Establishment of the algal inoculum can be observed within a week of inoculation in the form of floating algal mats, more prominently seen in the afternoon.
- ***Azolla:*** A floating water fern 'Azolla' hosts nitrogen fixing BGA *Anabaena azollae*. Azolla contains 3.4% nitrogen (on dry wt. basis) and add organic matter in soil. This biofertilizer is used for rice cultivation. There are six species of Azolla viz. *A. caroliniana, A. nilotica, A. mexicana, A.filiculoides, A. microphylla* and *A. pinnata*. Azolla plant has a floating, branched stem, deeply bilobed leaves and true roots which penetrate the body of water .The leaves are arranged alternately on the stem. Each leaf has a dorsal and ventral lobe. The dorsal fleshy lobe is exposed to air and contains chlorophyll. It grows well in ditches and stagnant water. Azolla can be easily grown throughout the year in India if water is not a limiting factor and climatic conditions are favourable for its growth. This fern usually forms a green mat over water. Azolla is readily decomposed to NH_4 which is available to the rice plants.

Field trial have shown that rice yields increased by 0.5-2t/ha due to Azolla application. In India and China, about 20 and 18% increase in rice yield, respectively has been reported due to Azolla application.

PEST MANAGEMENT

Weed Management

Organic weed management promotes weed suppression, rather than weed elimination, by enhancing crop competition and phytotoxic effects on weeds. Organic farmers integrate cultural, biological, mechanical, physical and chemical tactics to manage weeds without synthetic herbicides.

Organic standards require rotation of annual crops, meaning that a single crop cannot be grown in the same location without a different, intervening crop. Organic crop rotations frequently include weed-suppressive cover crops and crops with dissimilar life cycles to discourage weeds associated with a particular crop. Research is ongoing to develop organic methods to promote the growth of natural microorganisms that suppress the growth or germination of common weeds.

Other cultural practices used to enhance crop competitiveness and reduce weed pressure include selection of competitive crop varieties, high-density planting, tight row spacing, and late planting into warm soil to encourage rapid crop germination.

Mechanical and physical weed control practices used on organic farms can be broadly grouped as:

- Tillage - Turning the soil between crops to incorporate crop residues and soil amendments; remove existing weed growth and prepare a seedbed for planting; turning soil after seeding to kill weeds, including cultivation of row crops;
- Mowing and cutting - Removing top growth of weeds;
- Flame weeding and thermal weeding - Using heat to kill weeds; and
- Mulching - Blocking weed emergence with organic materials, plastic films, or landscape fabric. Some critics, citing work published in 1997 by David Pimentel of Cornell University, which described an epidemic of soil erosion worldwide, have raised concerned that tillage contribute to the erosion epidemic. The FAO and other organizations have advocated a 'no-till' approach to both conventional and organic farming, and point out in particular that crop rotation techniques used in organic farming are excellent no-till approaches. A study published in 2005 by Pimentel and colleagues confirmed that 'Crop rotations and cover cropping (green manure) typical of organic agriculture reduce soil erosion, pest problems, and pesticide use.'

In organic farming, chemical herbicides cannot be used. So weeding can be done only manually. Different cultural practices like tillage, flooding, mulching can be used to manage the weeds. Besides, biological (pathogen) method can be used to manage the loss due to weeds. When the ground is fallow, a cover crop can be planted to suppress weeds and build soil quality. Weeds growth can also be limited by using drip irrigation whenever possible, which restricts the distribution of water to the plant line.

Some naturally sourced chemicals are allowed for herbicidal use. These include certain formulations of acetic acid (concentrated vinegar), corn gluten meal, and essential oils. A few selective bioherbicides based on fungal pathogens have also been developed. At this time, however, organic herbicides and bioherbicides play a minor role in the organic weed control toolbox.

Weeds can be controlled by grazing. For example, geese have been used successfully to weed a range of organic crops including cotton, strawberries, tobacco, and corn, reviving the practice of keeping cotton patch geese, common in the southern U.S. before the 1950s. Similarly, some rice farmers introduce ducks and fish to wet paddy fields to eat both weeds and insects.

Insect Pest Management

In organic farming, the presence of pests (where and when) is anticipated in advance and accordingly the planting schedules and locations are adjusted as much as possible to avoid serious pest problems. The main strategy to combat harmful pests is to build up a population of beneficial insects, whose larvae feed off the eggs of pests. The key to building a population of beneficial insects is to establish borders (host crops) around fields planted with blends of flowering plants that the beneficial insects particularly like. Then periodically beneficial insects are released into the fields, where the host crops serve as their home base and attract more beneficial insects over time. When faced with a pest outbreak that cannot be handled by beneficial insects, the used of natural or other organically approved insecticides like neem pesticides is done. The two most important criteria for allowed organic pesticides are low toxicity to people and other animals and low persistence in the environment. These criteria are determined by the National Organic Standards.

Chloroxylon is used for pest management in organic rice cultivation in Chhattisgarh, India

Organisms aside from weeds that cause problems on organic farms include arthropods (e.g., insects, mites), nematodes, fungi and bacteria. Organic practices include, but are not limited to:

- encouraging predatory beneficial insects to control pests by serving them nursery plants and/or an alternative habitat, usually in a form of a shelterbelt, hedgerow, or beetle bank;
- encouraging beneficial microorganisms;
- rotating crops to different locations from year to year to interrupt pest reproduction cycles;
- planting companion crops and pest-repelling plants that discourage or divert pests;
- using row covers to protect crops during pest migration periods;
- using biologic pesticides and herbicides;

- using stale seed beds to germinate and destroy weeds before planting;
- using sanitation to remove pest habitat;
- using insect traps to monitor and control insect populations; and
- using physical barriers, such as row covers.

Examples of predatory beneficial insects include minute pirate bugs, big-eyed bugs, and to a lesser extent ladybugs (which tend to fly away), all of which eat a wide range of pests. Lacewings are also effective, but tend to fly away. Praying mantis tend to move more slowly and eat less heavily. Parasitoid wasps tend to be effective for their selected prey, but like all small insects can be less effective outdoors because the wind controls their movement. Predatory mites are effective for controlling other mites.

Naturally derived insecticides allowed for use on organic farms use include *Bacillus thuringiensis* (a bacterial toxin), pyrethrum (a chrysanthemum extract), spinosad (a bacterial metabolite), neem (a tree extract) and rotenone (a legume root extract). Fewer than 10% of organic farmers use these pesticides regularly; one survey found that only 5.3% of vegetable growers in California use rotenone while 1.7% use pyrethrum. These pesticides are not always more safe or environmentally friendly than synthetic pesticides and can cause harm. The main criterion for organic pesticides is that they are naturally derived, and some naturally derived substances have been controversial. Controversial natural pesticides include rotenone, copper, nicotine sulfate, and pyrethrums Rotenone and pyrethrum are particularly controversial because they work by attacking the nervous system, like most conventional insecticides. Rotenone is extremely toxic to fish and can induce symptoms resembling Parkinson's disease in mammals. Although pyrethrum (natural pyrethrins) is more effective against insects when used with piperonyl butoxide (which retards degradation of the pyrethrins), organic standards generally do not permit use of the latter substance.

Diseases Management in Organic Farming

Plant diseases are major constraints for reductions in crop yield and quality in organic and low input production systems. Proper fertility management to crops through balanced supply of macro and micronutrients and adoption of crop rotation have shown to improve the resistance of crops to certain diseases. Thus one of the biggest rewards of organic farming is healthy soil that is alive with beneficial organisms. These healthy microbes, fungi and bacteria keep the harmful bacteria and fungi that cause disease in check.

Naturally derived fungicides allowed for use on organic farms include the bacteria *Bacillus subtilis* and *Bacillus pumilus*; and the fungus *Trichoderma harzianum*. These are mainly effective for diseases affecting roots. Compost tea contains a mix of beneficial microbes, which may attack or out-compete certain plant pathogens, but variability among formulations and preparation methods may contribute to inconsistent results or even dangerous growth of toxic microbes in compost teas.

Some naturally derived pesticides are not allowed for use on organic farms. These include nicotine sulfate, arsenic, and strychnine.

Synthetic pesticides allowed for use on organic farms include insecticidal soaps and horticultural oils for insect management; and Bordeaux mixture, copper hydroxide and sodium bicarbonate for managing fungi. Copper sulfate and Bordeaux mixture (copper sulfate plus lime), approved for organic use in various jurisdictions, can be more environmentally problematic than some synthetic fungicides dissallowed in organic farming Similar concerns apply to copper hydroxide. Repeated application of copper sulfate or copper hydroxide as a fungicide may eventually result in copper accumulation to toxic levels in soil, and admonitions to avoid excessive accumulations of copper in soil appear in various organic standards and elsewhere. Environmental concerns for several kinds of biota arise at average rates of use of such substances for some crops. In the European Union, where replacement of copper-based fungicides in organic agriculture is a policy priority, research is seeking alternatives for organic production.

Livestock

For livestock, like these healthy cows, vaccines play an important part in animal health since antibiotic therapy is prohibited in organic farming

Raising livestock and poultry, for meat, dairy and eggs, is another traditional farming activity that complements growing. Organic farms attempt to provide animals with natural living conditions and feed. Organic certification verifies that livestock are raised according to the USDA organic regulations throughout their lives. These regulations include the requirement that all animal feed must be certified organic.

Organic livestock may be, and must be, treated with medicine when they are sick, but drugs cannot be used to promote growth, their feed must be organic, and they must be pastured.

Also, horses and cattle were once a basic farm feature that provided labor, for hauling and plowing, fertility, through recycling of manure, and fuel, in the form of food for farmers and other animals. While today, small growing operations often do not include livestock, domesticated animals are a desirable part of the organic farming equation, especially for true sustainability, the ability of a farm to function as a self-renewing unit.

LIMITATIONS AND IMPLICATIONS

There are a few limitations with organic farming such as:

- Organic manure is not abundantly available and on plant nutrient basis it may be more expensive than chemical fertilizers if organic inputs are purchased.
- Production in organic farming declines especially during first few years, so the farmer should be given premium prices for organic produce.
- The guidelines for organic production, processing, transportation and certification etc are beyond the understanding of ordinary Indian farmer.
- Marketing of organic produce is also not properly streamlined. There are a number of farms in India which have either never been chemically managed / cultivated or have converted back to organic farming because of farmers'

beliefs or purely for reason of economics. These thousands of farmers cultivating million acres of land are not classified as organic though they are. Their produce either sells in the open market along with conventionally grown produce at the same price or sells purely on goodwill and trust as organic through select outlets and regular specialized markets. These farmers may never opt for certification because of the costs involved as well as the extensive documentation that is required by certifiers.

In dry lands, covering over 65% cultivated area in India, application of chemical fertilizers and pesticides is always low. So these areas are at least "relatively organic" or "organic by default" and a portion of these lands can be converted easily to an organic one to provide better yields/returns. India can greatly benefit from the export of organic foods, but needs to seriously devote attention to market intelligence regarding which product to grow, where to sell, distribution channels, competition, marketing access etc. Presently, good awareness exists among farmers, researchers and policy makers about organic production but a lot more need to be done to streamline production, certification and marketing of organic produce. Uttaranchal and some other state governments have already declared their states as "Organic" state and created special Export Zones like Basmati Export Zone. A large area of Northeastern states and other states may be developed as commodity based "organic" production areas. With greater political will and investment in research, extension and marketing infrastructure more of this potential could be realized. Therefore to feed the world's hungry and poor and to ensure present and future food security right policies, increased public and private investments and technologies, knowledge and capacity building, grounded in sound ecosystem management and harmony between organic farming and food security goals are required.

Need of Organic Farming

With the increase in population our compulsion would be not only to stabilize agricultural production but to increase it further in sustainable manner. The scientists have realized that the 'Green Revolution' with high input use has reached a plateau and is now sustained with diminishing return of falling dividends. Thus, a natural balance needs to be maintained at all cost for existence of life and property. The obvious choice for that would be more relevant in the present era, when these agrochemicals which are produced from fossil fuel and are not renewable and are diminishing in availability. It may also cost heavily on our foreign exchange in future.

The key characteristics of organic farming include

- Protecting the long-term fertility of soils by maintaining organic matter levels, encouraging soil biological activity, and careful mechanical intervention
- Providing crop nutrients indirectly using relatively insoluble nutrient sources which are made available to the plant by the action of soil micro-organisms
- Nitrogen self-sufficiency through the use of legumes and biological nitrogen fixation, as well as effective recycling of organic materials including crop residues and livestock manures

- Weed, disease and pest control relying primarily on crop rotations, natural predators, diversity, organic manuring, resistant varieties and limited (preferably minimal) thermal, biological and chemical intervention
- The extensive management of livestock, paying full regard to their evolutionary adaptations, behavioural needs and animal welfare issues with respect to nutrition, housing, health, breeding and rearing
- Careful attention to the impact of the farming system on the wider environment and the conservation of wildlife and natural habitats

Standards

Standards regulate production methods and in some cases final output for organic agriculture. Standards may be voluntary or legislated. As early as the 1970s private associations certified organic producers. In the 1980s, governments began to produce organic production guidelines. In the 1990s, a trend toward legislated standards began, most notably with the 1991 EU-Eco-regulation developed for European Union, which set standards for 12 countries, and a 1993 UK program. The EU's program was followed by a Japanese program in 2001, and in 2002 the U.S. created the National Organic Program (NOP). As of 2007 over 60 countries regulate organic farming (IFOAM 2007:11). In 2005 IFOAM created the Principles of Organic Agriculture, an international guideline for certification criteria. Typically the agencies accredit certification groups rather than individual farms.

Economics

The economics of organic farming, a subfield of agricultural economics, encompasses the entire process and effects of organic farming in terms of human society, including social costs, opportunity costs, unintended consequences, information asymmetries, and economies of scale. Although the scope of economics is broad, agricultural economics tends to focus on maximizing yields and efficiency at the farm level. Economics takes an anthropocentric approach to the value of the natural world: biodiversity, for example, is considered beneficial only to the extent that it is valued by people and increases profits. Some entities such as the European Union subsidize organic farming, in large part because these countries want to account for the externalities of reduced water use, reduced water contamination, reduced soil erosion, reduced carbon emissions, increased biodiversity, and assorted other benefits that result from organic farming.

Traditional organic farming is labor and knowledge-intensive whereas conventional farming is capital-intensive, requiring more energy and manufactured inputs.

Organic farmers in California have cited marketing as their greatest obstacle.

12

FARMING SYSTEM OF DRYLAND AGRICULTURE

Dryland agriculture refers to cultivation of crops entirely under natural rainfall without irrigation. It is a form of subsistence farming in the regions where deficit of the soil moisture retards the growth of water consuming crops like rice *(Oryza sativa)*, sugarcane etc. Dryland areas are characterized by low and erratic rainfall and no assured irrigation facilities. Dryland agriculture is important for the economy as most of the coarse grain crops, pulses, oilseeds, and raw cotton are grown on these lands. Dryland areas receive annual rainfall between 500 and 1200 mm.

Types of dryland agriculture

Depending on the amount of rainfall received, dryland agriculture has been grouped into three categories:

- ***Dry farming:*** It is production of crops without irrigation in areas where annual rainfall is less than 750 mm. Crop failures are more frequent under dry farming condition owing to prolonged dry spells during crop period. The growing season is less than 200 days. It is generally practiced in arid regions of the country
- ***Dryland farming:*** Cultivation of crops in areas receiving rainfall above 750 mm is known as dryland farming. Dry spell during crop duration occurs, but crop failures are less frequent. Semi-arid regions are included under this category.
- ***Rainfed farming:*** It is practice of crop cultivation without irrigation in areas receiving 1150 mm rainfall, mostly in sub-humid and humid areas. Here chances of crop failure and water stress are very less.

AREA UNDER DRYLANDS

Globally the area under drylands is about 6150 m.ha. In India out of the total cultivated area of 143 m.ha, the area under drylands is about 85 m.ha, which comes to 60% . It is estimated that even after creating entire irrigation potential for irrigation about 55% of total cultivated area remains as rainfed. Except in the States of Punjab, Haryana and Pondichery the percentage of area under drylands is high in all other

Dryland area in different regions of India

Region	States	Per cent of rainfed area
Cold and northern region	Jammu and Kashmir, Uttaranchal and Himachal Pradesh	60 to 81
Arid western region	Rajasthan and Gujarat	66 to 88
Semi arid to arid central and southern region	Madhya Pradesh, Maharastra, Andhra Pradesh, Andhra Pradesh, Karnataka, Tamilnadue	76 to 82
Sub humid to humid eastern region	Eastern Uttar Pradesh, Bihar, Jharkhand, Orissa, West Bengal	33 to 73
Humid to per humid north eastern region	Assam and north eastern hill States	Up to 90

Distribution of drylands

Our country has fertile cultivable land and receives the highest rainfall on per unit area basis anywhere in the world due to short duration of rainfall in a year. One hundred and twenty eight districts in India have been recognized as dryland farming areas. Of these, 91 districts are spread in the states of Madhya Pradesh, Chhattisgarh, Uttar Pradesh and Tamil Nadu, representing typical dry farming tracts. Rest of the districts belongs to Central Rajasthan, Saurashtra region of Gujarat and rain shadow region of the Western Ghats.

India has about 108 million hectares of rainfed area which constitutes nearly 75% of the total 143 million hectares of arable land. In such areas crop production becomes relatively difficult as it mainly depends upon intensity and frequency of rainfall. The crop production, therefore, in such areas is called rainfed farming as there is no facility to give any irrigation, and even protective or life saving irrigation is not possible.

Major dry farming crops are millets such as *jwar, bajra, ragi*, oilseeds like mustard, rapeseed, and pulse crops like pigeon pea , gram and lentil. Almost 80% of maize and *Jwar*, 90 per cent of *Bajra*and approximately 95% of pulses and 75% of oilseeds are obtained from dryland agriculture. In addition to these, 70% of cotton is produced through dryland agriculture. Dryland areas also contribute significantly to wheat and rice production. Thirty three per cent of wheat and 66% of rice are still rainfed.

Prospects of dryland areas

More than 75% of the peasants involved in dry farming are small and marginal. Therefore, improvement in dry farming would raise the economic status of farmers thus helping in poverty elimination. Dryland farming holds immense significance especially in the context of fluctuating food grain production and expanding population in our country. The biggest employer in our country, the cotton mills are fed by raw cotton grown mostly in dryland areas. Increasing production of cotton subsequently leads to increase in exports of cotton good. The expanding import of oilseeds is a cause of concern to Indian nation. The improvement of production of oilseeds in these regions will save valuable foreign exchange reserves. By enhancing the productivity of crops like *jowar, bajra* and *ragi* which are mainly grown in dryland farming would increase the nutrient consumption levels of our nation.

Marginal lands in the semi-arid regions offer potential for fodder production to feed the cattle population which is an integral component of farming practice of this region. Providing importance to these areas can solve the problems of pulses, oilseeds and cotton. The dryland areas have also tremendous potentiality of increased food grain production. Thus enhanced agricultural production in these areas would boost the agriculture dependent economy of India. Moreover it would also be helpful in eliminating the problem of hunger and malnutrition prevailed in below poverty line society of the country.

Constraints of drylands

Drylands are characterized by low and uncertain rainfall therefore, crop failure is common feature. The various constraints of drylands includes:

1. In dryland areas in general, the rainfall is low and highly variable which results in uncertain crop yields. The distribution of rainfall during the crop period is uneven, receiving high amount of rain when it is not required and lack of it when crop need it.
2. Generally in dryland areas when the monsoon sets in late, the sowing of crops are delayed resulting in poor yields. At times, the rains may cease very early in season exposing the crop to drought and during flowering and maturity stages which reduces the crop yields considerably.
3. Soils of the drylands are not only dry but also deficient in macronutrients like nitrogen and phosphorous. Thus in other words drylands are not only thirsty but they are hungry too.
4. The temperature in dryland varies greatly. During the period of moisture stress and drought, the temperatures accelerate the crop development resulting into forced maturity. Chilling or frost injury at flowering results in poor grain setting and deteriorates the grain quality.
5. Dryland areas suffer from various process of soil degradation especially soil erosion.
6. Small size of land holdings (less than 2 hectares) usually fragmented and scattered, lack of market facilities, frequent crop failure, poor economic condition and other socio-economic problem related to drylands.
7. Extremely poor condition of farmers, lack of infrastructure to boost production.

Major areas of concern

Major areas of concern in dryland agriculture are:

- Proper marketing and price policy to cover crops and animal products.
- Conservation of soil and water resources.
- Need to evolve high yielding and drought resistant crop varieties.
- Low cost and locally suited agricultural implements.
- Judicious and balance use of costly chemicals.
- Proper financial availability to purchase inputs; and
- Extension education.

DRYLAND TECHNOLOGIES

Reducing evaporation losses

Soil moisture is the most limiting factor in dryland agriculture. It is lost as evaporation from the soil surface and as transpiration from the plant surfaces. Evaporation has to be arrested as it is not directly related to productivity whereas transpiration can be reduced to some extent without affecting productivity of plants. The evaporation losses can be reduced by:

1. Mulches
2. Antitranspirants
3. Wind breaks
4. Weed control

Crop Planning

1. Mulches

About 60 to 75% of the rainfall is lost through evaporation. These evaporation losses can be reduced by applying mulches. Mulch is any material applied on the soil surface to check evaporation and improve soil water. Application of mulches results in additional benefits like soil conservation, moderation of temperature, reduction in soil salinity, weed control and improvement of soil structure. Following types of mulches can be used.

(i) **Soil mulch or dust mulch:** If the surface of the soil is loosened, it acts as a mulch for reducing evaporation. This loose surface soil is called soil mulch or dust mulch. Intercultivation creates soil mulch in a growing crop.

(ii) **Stubble mulch:** Crop residues like wheat straw or cotton stalks etc., are left on the soil surface as a stubble mulch. The advantages of stubble mulch farming are protection of soil from erosion and reduction of evaporation losses.

(iii) **Straw mulch:** If straw is used as mulch, it is called as straw mulch. Mostly, by-products / waste materials of crops are used as straw mulch.

(iv) **Plastic mulch:** Plastic materials like polyethylene, polyvinyl chloride are also used as mulching materials.

(v) **Vertical mulching:** To improve infiltration and storage of rainwater in these soils, vertical mulches are formed. It consists of digging narrow trenches across the slope at intervals and placing the straw or crop residues in these trenches. The pruned plant material is placed in contour trenches formed between rows or in trenches around the plants in concentric circles each year in one circle.

2. Antitranspirants

About 99 per cent of the water absorbed by the plants is lost in transpiration. If transpiration is controlled, it may help in maintenance of favourable water balance.

Antitranspirant is any material applied to transpiring plant surfaces for reducing water loss from the plant. These are of four types:

- Stomatal closing
- Film forming
- Reflective
- Growth retardant

(i) **Stomatal closing type:** Most of the transpiration occurs through the stomata on the leaf surface.

Fungicides like phenyl mercuric acetate (PMA) and herbicides like atrazine in low concentrations serve as antitranspirants by inducing stomatal closing.

These might reduce the photosynthesis also simultaneously. PMA was found to decrease transpiration to a greater degree than photosynthesis in a number of plants.

(ii) **Film forming type:** Plastic and waxy materials which form a thin film on the leaf surface retard the escape of water due to formation of physical barrier.

Mobileaf, hexadeconol, silicone are some of the film forming type of antitranspirants. The success of these chemicals is limited since they also reduce photosynthesis.

The desirable characteristics of film forming type of antitranspirants are:

- They should form a thin layer,
- They should be more resistant to the passage of water vapour than carbon dioxide and the film should maintain continuity and should not break

(iii) **Reflectant Type:** These are white materials which form a coating on the leaves and increase the leaf reflectance (albedo). By reflecting the radiation, they reduce leaf temperatures and vapour pressure gradient from leaf to atmosphere and thus reduce transpiration.

- Application of 5% kaolin spray reduces transpiration losses.
- A diatomaceous earth product (celite) also increases reflection of solar radiation from crop canopy

(iv) **Growth Retardants:** These chemicals reduce shoot growth and increase root growth and thus enable the plants to resist drought. They may also, induce stomatal closure.

- Cycocel is one such chemical useful for improving water status of the plant
- Antitranspirants generally reduce photosynthesis. Therefore, their use is limited to save the crop from death under severe moisture stress. If the crop survives, it can utilize the rainfall that is received subsequently. Antitranspirants are also useful for reducing the transplantation shock of nursery plants. They have some practical use in nurseries and horticultural crops.

3. Wind breaks and shelterbelts

(i) **Wind breaks** are any structures that obstruct wind flow and reduce wind speed while shelterbelts are rows of trees planted for protection of crops against wind. The direction from which wind is blowing is called windward side and direction to which wind is blowing is called leeward side.

(ii) **Shelterbelts** are planted across the direction of wind. They do not obstruct the wind flow completely. Depending upon their porosity, certain amount of wind passes through the shelterbelts while the rest deflects and crosses over the shelterbelts. It thus reduces wind speed without causing turbulence. The protection offered by the shelterbelts is dependent on the height of central tree row in the shelterbelts. Generally, shelterbelts give protection from desiccating winds to the extent of 5 to 10 times their height on windward side and up to 30 times on leeward side. Due to reduction in wind speed, evaporation losses are reduced and more water is available for plants. The beneficial effect of shelterbelts is seen more clearly in drought years. In addition, shelterbelts reduce wind erosion.

4. Weed control

- Prompt weed control eliminates the competition of weeds with crops for limited soil moisture.
- Transpiration rate from weeds is more compared to crops.
- Effective weed control in dryland agriculture leads to increasing availability of soil moisture to crops.

  This is the most useful measure to reduce transpiration losses.

5. Crop Planning

Crop varieties for dryland areas should be of short duration through resistant tolerant and high yielding which can be harvested within rainfall periods and have sufficient residual moisture in soil profile for post-monsoon cropping.

(i) Planning for weather: Variation in yields and output of the dryland agriculture is due to the observation in weather conditions especially rainfall. An aberrant weather can be categorised in three types viz.

- Delayed onset of monsoon.
- Long gaps or breaks in rainfall and
- Early stoppage of rains towards the end of monsoon season.
- Farmers should make some changes in normal cropping schedule for getting some production in place of total crop failure.

(ii) Crop substitution: Traditional crops/varieties which are inefficient utilizer of soil moisture, less responsive to production input and potentially low producers should be substituted by more efficient ones.

(iii) Cropping systems: Increasing the cropping intensities by using the practice of intercropping and multiple cropping is the way of more efficient utilization of resources. The cropping intensity would depend on the length of growing season, which in turn depends on rainfall pattern and the soil moisture storage capacity of the soil.

(iv) Fertilizer use: The availability of nutrients is limited in drylands due to the limiting soil moisture. Therefore, application of the fertilizers should be done in furrows below the seed. The use of fertilizers is not only helpful in providing nutrients to crop but also helpful in efficient use of soil moisture. A proper mixture of organic and inorganic fertilisers improves moisture holding capacity of soil and increase during tolerance.

Nutrient solution spray is recommended in the event of revival of rain and release of moisture stress. Urea or DAP spray (2% solution) is useful for quicker regeneration of crops like legumes and castor after rain.

(v) Rain water management: Efficient rainwater management can increase agricultural production from dryland areas. Application of compost and farm yard manure and raising legumes add the organic matter to the soil and increase the waterholding capacity. The water, which is not retained by the soil, flows out as surface runoff. This excess runoff water can be harvested in storing dugout ponds and recycled to donar areas in the server stress during rainy season or for raising crops during winter.

(vi) Water shed management: Water shed management I s a approach to optimize the use of land, water and vegetation in a area and thus, to provide solution drought, moderate floods, prevent soil erosion, improve water availability and increase fuel, fodder and agricultural production on a sustained basis.

(vii) Alternate land use: All drylands are not suitable for crop production. Same land may be suitable for range/ pasture management and for tree farming and let farming, dryland horticulture, agro-forestry systems including alley cropping. All these systems which are alternative to crop production are called as alternate land use systems. This system helps to generate off-season employment in mono-cropped dryland and also, minimizes risk, utilizes off-season rains, prevents degradation of soils and restores balance in the ecosystem. The different alternate land use systems are alley cropping, agri-horticultural systems and silvi-pastoral systems, which utilizes the resources in a better way for increased and stabilized production from drylands.

To summarize following farming technologies are needed to enhance agricultural production in dryland areas.

- Timely preparatory and seeding operations including conservation of stored soil moistures.
- The use of improved crop varieties should be done which can withstand stress. For moisture conservation in the soil, deep tillage, surface cultivation and stubble mulching need to be practiced. Deep tillage is required to break plough soles and layers because repeated ploughing over centuries has resulted in the growth of hard compacted layers which restrict infiltration and movement of water and penetration of water.
- Conjunctive use of rainfall, surface and groundwater.
- Harvesting of water for use in dry periods. Watershed a natural hydrological unit is a good device for water harvesting. Proper watershed management can stop not only further degradation of ecosystem, but degraded lands can also be restored.
- Soil conservation by contour bunding, terracing, land sloping and land levelling and also by practicing conservational tillage (zero tillage and minimum tillage).
- Practice of drip irrigation to save water.
- Lining of canals to minimize water loss.
- Agronomic practices like mixed cropping and crop rotation which increase the yield of crops need to be practiced.
- Integrated nutrient management need to be practiced with special emphasis on use of bio-fertilizers to maintain the soil fertility.
- Integrated weed management and integrated pest management need to be adopted to control weeds and pests, respectively.
- For the non-farm operation dryland areas have to be supplemented with non-form occupation like animal husbandry, fisheries, poultry, social forestry and cottage for the development of these areas.
- Alley cropping, pasture management, tree farming, silvi-pastoral management systems and agro-horticultural system which are more relevant to dryland situations have to be adopted for successful dryland farming system.

Dryland areas constituting more than two-third of total arable lands in India are the chief contributor of pulses, oilseeds, coarse grain crops and cotton. Drylands also contribute significantly to wheat and rice production. Therefore, it is the need of the hour to adopt and practice the available dryland technology to maximum extent for the enhancement of agricultural production in these areas which would not only boost the food grain production of the country but would also improve the economic status of farmers in these areas.

PROBLEMS OF CROP PRODUCTION IN DRYLAND

Dry farming crops are characterized by very low and highly variable and uncertain yields. Crop failures are quite common. These are mainly due to the following causes.

- **Inadequate and uneven distribution of rainfall:** In general, the rainfall is low and highly variable which results in uncertain crop yields. Besides its uncertainty, the distribution of rainfall during the crop period is uneven, receiving high amount of rain, when it is not needed and lack of it when crop needs it.
- **Late onset and early cessation of rains:** Due to late onset of monsoon, the sowing of crop are delayed resulting in poor yields. Sometimes the rain may cease very early in the season exposing the crop to drought during flowering and maturity stages which reduces the crop yields considerably
- **Prolonged dry spells during the crop period:** Long breaks in the rainy season is an important feature of Indian monsoon. These intervening dry spells when prolonged during crop period reduces crop growth and yield and when unduly prolonged crops fail.
- **Low moisture retention capacity:** The crops raised on red soils, and coarse textured soil suffer due to lack of moisture whenever prolonged dry spells occur due to their low moisture holding capacity. Loss of rain occurs as runoff due to undulating and sloppy soils.
- **Low fertility of soils:** Drylands are not only thirsty, but also hungry too. Soil fertility has to be increased, but there is limited scope for extensive use of chemical fertilizers due to lack of adequate soil moisture.

Importance of dryland farming in India

In India, the importance of dryland farming is increasing year by year. With the continuous growth in the size of population in India the gap between the requirements and the supply of agricultural output is increasing gradually leading to a crisis in respect of both food and non-food crops.

The World Bank in a study had observed that at the turn of the century, the total requirements of food would increase to about 235 million tonnes which would again require an additional production of food grains to the force of 65 million tonnes over the present level of production.

Prospects of dryland farming

With the present initiatives, the production of food grains would increase at best by another 35 million tonnes, leaving a gap of 30 million tonnes. Under this situation, the only way open to the country is to introduce and promote better dry-land farming which has a huge potential throughout the country.

Until recently, the development of technology for dry-land farming remained totally neglected. Considering the potential and also its importance, the Indian Council of Agricultural Research (ICAR) very recently evolved dry farming technology for its application in those areas which are experiencing inadequate rainfall even in the rainy season.

In India, a huge area of land remained fallow for want of assured water supply required for agriculture. Recently, the Planning Commission has made an estimate of such arid lands where dryland farming technology can be adopted.

As per this estimate, a total of about 37 million hectares of such arid lands, spread over 92 districts in different regions of the country can be brought under dry-land farming technology which could serve the twin purpose of controlling soil erosion, as well as increasing agriculture production and productivity.

In the mean time, the Eighth Plan has given much thrust on the improvement of dry-land farming more particularly for the production of pulses and oilseeds in which the country is deficient. Accordingly about 5,000 micro watersheds have been identified by the Government for adopting dry-farming technology.

In order to improve dry-land farming in the country, the Government proposed to follow an integrated and multi- disciplinary approach for the adoption and development of dry-land farming in the country. NABARD and cooperative banks have already expressed their willingness to extend financial help for running those viable dry-land farming projects in the country.

Concept

Indian agriculture is predominantly a rainfed agriculture under which both dry farming and dryland agriculture is included. Dry faring was the earlier concept for which amount of rainfall (less then 500 mm annually) remained the deciding factor for more then 50 years. In modern concept, dryland areas are those where the balance of moisture is always on the deficit side. In other words, annual evapotranspiration exceeds precipitation. In dryland agriculture, there is no consideration of amount of rainfall. It may appear quiet strange to a layman that even those areas which receive 1100 mm or more rainfall annually fall in the category of dryland agriculture under this concept. To be more specific, the average annual rainfall of Varanasi is around 1100 mm and the annual potential evapotranspiration is 1500 mm. thus the average moisture deficit so created comes to 400 mm. This deficit in moisture is bound to affect the crop production under dryland situation ultimately resulting into total or partial failure of the crops. Accordingly the production is either low or extremely uncertain and unstable which are the real problems of dryland in India.

The success of crop production in these areas depends on the amount and distribution of rainfall, as these influences the stored soil moisture and moisture used by crops. The amount of water used by the crop and stored in the soil is governed by the water balance equation: ET = P-(R+S). When the balance of the equation shifts towards right, precipitation (P) is higher then ET, so that there may be water logging or it may even lead to run off (R) and flooding. On the other hand, if the balance shifts to the left, ET becomes higher then the precipitation, resulting in drought in the various severity. Taking the country as a whole, as per meteorological report, severe drought as large area is experienced once in 50 years and partial drought in five years while folds are expected every year in one part of the country or the other, especially during rainy season. In fact the balance of the equation is controlled by the weather, season, crops and cropping pattern.

STATUS OF DRYLAND FARMING

Out of 14.3 million ha of net sown area in the country, rainfed agriculture is practiced in 85 million ha (67%). Nearly 60 m ha of rainfed area falls in the mean annual precipitation range of 500-1500 mm.

The average annual rainfall of the country is 1200 mm amounting to 400 million ha meter of rain water over the country's geographical area (329 m ha). However, the distribution across the country varies from less than 100 mm in extreme arid areas of western Rajasthan to greater than 3600 mm in NE states and 1100 mm from east coast 2500-3000 mm in the west coast. The broad area of the summer monsoon activity extends between 30^0 N to 30^0 S and from 30^0 W to 16.5^0 E. the detail information on rainfall and monsoonal pattern in India has been summarized in the following table:

Rainfall pattern in fall

Season/Period	m ha m	Percent
Winter (Jan-Feb)	12	3
Pre-monsoon (Mar-May)	52	13
South-west monsoon (Jun-Sept)	296	74
North-east monsoon (Oct-Dec)	40	10
Total for the year	400	100

Rainfed farming comprises about 91% area of coarse cereals (sorghum, pearl millet, maize and finger millet), 91% pulses (chickpea and pigeon pea), 80% of oilseeds (groundnut, rape seed, mustard and soybeen), and 65% of cotton. Also, about 50% area under rice and 19% area under wheat is rainfed.

During the last 25 years there had occurred significant changes in the area and yield of imported crops of rainfed farming areas. The area under coarse cereals decreased by about 10.7 million ha and most of this was under sorghum. The area under oilseeds increased by 9.2 million ha and most of this increase was due to irrigated rapeseed and mustard and soybeen. The total area under pulses and cotton remained constant but more of cotton became irrigated and shifts in the area under occurred from one agro-ecological region to others. Area under chickpea in the northern belt decreased but increase in the central belt. This change occurred due to increase in area under rice-wheat cropping system which displaced chickpea and also pearl millet to a great extent and maize to a small extent.

According to the present concept, there are 128 districts in the country which face the problems of dryland, of these 25 districts covering 18 m. ha of net area sown with 10 % irrigation receive 375-750 mm rainfall annually spread over Central Rajasthan, Saurashtra region of Gujarat and rain shadow region of Western Ghats in Maharashtra and Karnataka. Twelve districts have irrigation covering 30-50% of the cropped area and do not pose serious problems. The remaining 91 districts covering mainly Madhya Pradesh, Gujarat, Maharashtra, Andhra Pradesh, Karnataka, Uttar Pradesh, parts of Haryana, Tamil Nadu etc., represent typical dryland area. The total net sown area in these districts is estimated to be 42 million hectares of which 5 m ha are irrigated. Rainfall in these districts varies from 375 to 1125 mm. therefore, more and more efforts are to be made for enhanced and stable production in these areas so that the recurring droughts do not stand in the way of meeting the growing food demands.

It is not that no attention has been paid in the country towards the development of dryland farming. Efforts were made right from 1923 to improve crop yields with the establishment of a research projects at Manjari in Maharashtra and later at Solapur, Bijapur, Raichur and Hagari in Deccan and Rohtak (Haryana) in the north. An All India Coordinated Research Project for Dryland Agriculture was launched by ICAR in 1970 in collaboration with Government of Canada and later Central Research Institute for Dryland Agriculture (CRIDA) was established in 1985 at Hyderabad. These projects generated technology, which, if followed, can bring marked improvement in cropping intensity, productivity and stability in production.

Limitations of dryland farming

In dryland agriculture, scarcity of water is the main problem. Apart from the low and erratic behavior of rainfall, high evaporative demand and limited water holding capacity of the soil constitute the principle constraint in the crop production in dryland area. Yield fluctuations are high mainly due to vagaries of weather, often much behind the risk bearing capacity of the farmers. It is surprising to a layman that even humid areas with 2000 mm of annual rainfall not only suffer from moisture stress, but also face drinking water scarcity. Monsoon starts in the month of June and ends in last week of September or sometimes in the first week of October. Most of the rainfall is received during this period. With undulating topography and low moisture retention capacity of the soil, major portion of the rain water is lost through runoff, causing erosion and adding to the water logging of low lying areas. After the rain stops, very little moisture is left in the profile to support plant growth and grain production.

In dryland area deficiency and uncertainty in rainfall of high intensity causes excessive loss of soil through erosion which leaves the soil infertile. Owing to erratic behavior and improper distribution of rainfall, agriculture is risky, farmers lack resources, tools become inefficient and ultimately productivity is low.

Vertisoles have high clay content and high moisture retention capacity. Owing to its swelling and shrinking characteristics, permeability is low and hence the rate of infilteration of water is minimum. This causes more surface and high soil loss from the top layer owing to surface erosion. It is estimated that 68.5 tones/ha per year soil is lost from *vertisoles*. Due to high clay content it develops cracks during Rabi season at flowering stage of crops.

Alfisols are, by and large, light textured soils which have low moisture holding capacity but high water intake. The rain water falling in such areas gets soaked up and saturates the profile. The soil water percolation is more and therefore, is lost for crop use. Owing to faster intake of water in the profile the surface runoff is limited and soil loss from erosion is low (3.05 t/ha/year). Soil crusting is a common problem in low rainfall areas.

Entisols are generally loamy sand or sandy loam. Depth in these soils is not a constraint. These soils have very low clay content and hold water up to 200 mm per meter of soil profile. Its nutrient holding capacity is poor. In low rainfall areas monsoon cropping is practiced and in high rainfall areas double cropping is possible.

Submontane soils are medium in texture and depth is medium to deep as well as moderate in clay content. Moisture retention capacity is high (300 mm/m. profile). These soils are poor in nitrogen but in other nutrients. Phosphorous may be limiting in high production system. Due to high rainfall double cropping is possible in these soils.

Sierozems are extremely light soils, effectively depth being influenced by the $CaCo_3$ concentration in soil profile. Its moisture holding capacity is low (150 mm water.m). Sierozemic soils are low in nitrogen and sometimes inadequate in phosphorous. Subsoil salinity is common. These soils are mostly monsoon cropped, except in deep sandy loams where post-monsoon cropping is also possible. Crusting is very frequent.

IMPROVED AGRONOMIC PRACTICES FOR DRYLAND

The improved techniques and practices, which have so far been generated and recommended for achieving the objective of increased and stable crop production in dryland areas, have been summarized as follows.

Crop Planning

The farmers of the dryland areas, prior to the development of dryland techniques, were growing a crop either on rainwater in kharif or on conserved soil moisture during the winter. The crop varieties grown when moisture is sufficiently available. Such varieties have low genetic potential for yield. Selecting suitable crops and varieties capable of maturing with in actual rainfall periods will not only help in enhancing production of a single crop but in intensifying the cropping intensity. Many criteria have been laid out for selecting a crop variety for drylands. The capacity to produce a fairly good yield under limited soil moisture conditions is the most desirable criteria. The duration of kharif crops/varieties should not normally exceed the number of rainy days. In other words, crop varieties for dryland areas should be of short duration, through resistant tolerant and high yielding which can be harvested with in rainfall periods and have sufficient residual moisture in soil profile for post-monsoon cropping.

Under dryland agriculture determination of length of growing period (LPG) i.e., moisture availability of a given soil type, provides better index than total rainfall based crop planning. LPG is defined as the period when the moisture and temperature regious are suitable for crop growth and the period is determined by the FAO method (1983). The LPG is computed as the sum of the period when P is more than 0.5 PET plus time taken to utilize stored soil moisture (assured 100 mm) after P falls short of PET. For example 'Nagpur' and Ratnagiri in Maharashtra receive mean annual rainfall of 1120 mm and 2500 mm, respectively but LPG determination indicates that both the places have LPG of 210 days in deep black soils. Therefore, both the places are suitable for single long duration on a short duration crop with a relay **rabi** crop.

Planning for aberrant weather

Dryland agriculture is subject to high variability in areas sown, yields and output. These variations are the results of aberrations in weather conditions, especially rainfalls. Delays in normal monsoonal pattern cause problems of timing and the organization of preparatory tillage and other initial activities for commencing cultivation processes for the season. Such monsoonal delays have repercussions on the programme of activities for the entire agricultural year. Even after the onset of monsoon and the commencement of planting, there may be monsoonal withdrawal causing moisture stress on plants and creating difficulties in the adoption and timing of approval cultural practices ultimately causing reactions in yields and outputs. Some crops are highly susceptible to such mid-season variations in moisture availability such as at the flowering stage in rice. Major crops like rice and maize get seriously affected if monsoonal rains cease early.

The need for modifying and introducing new technology for increasing and sustaining yield in dryland areas can hardly be overemphasized. Equally urgent is the need to decelerate and ultimately eliminate the process of damage to agricultural assets which are proceeding unbated in dryland areas. Erratic rainfall results in fluctuating production. This in turn leads to frequent scarcities, like the ones experienced in Indonesia and Vietnam in 1977 which created severe food shortages. Droughts in China in 1972, 1974 and 1985 brought depression of food grain production by up to about 25 million tones. Frequent droughts in India during 1966, 1968, 1972, 1974, 1979, 1982 and 1987 seriously affected the food and fodder production in the country. Hence, it is necessary to understand the distribution of South-West monsoon within the season to determine the extent to which the crop productions are likely to be affected by the vagaries of monsoon.

Several attempts have been made to understand the behavior of South-west monsoon rainfall in different agro-climatic regions on the basis of historical rainfall records. These studies (Singh, 1987, Ramanna Rao, 1988) have brought out that (i) there is large variation in dates of commencement of South-west monsoon from year to year in different parts of the country, (ii) the monsoon rainfall is of sequential nature with long dry spells or breaks extends sometimes to the period of even one month or more, (iii) there is large year to year variation in dates of withdrawal of South-west monsoon, (iv) there is variation in quantum of rainfall received from year to year and (v) high intensity rainfall occurs in association with movement of cyclones or depression resulting in sizeable loss of rainwater through run-off and deep drainage.

Thus, crop production in drylands fluctuates widely from year to year due to vagaries of weather. An aberrant weather can be categorized under three heads i.e. (i) delayed onset of monsoon, (ii) long gaps or breaks in rainfall, and (iii) early stoppage of rains towards the end of monsoon season. Therefore, to mitigate such weather situations, farmers should make some changes in normal cropping schedule for getting some production in place of total crop failure.

Crop Substitution

Alternate crop strategies have been worked out for important regions of the country for *vertisols, alfisols, entisols, submontane* and *sierozemic* soils. Strategy has also been evolved for normal onset of rains, breaks in rains, early withdrawal, its uneven distribution; through selection of uneven crops/varieties which are inefficient utilize of the soil moisture, less responsive to production input and potentially low producers should be substituted by more efficient ones. Appropriate crops, suiting varying rainfall situations, have been identified for most of the dryland regions of India which are presented below.

Traditional and alternate efficient crops in different dryland regions of India

S. no	Region	Traditional crop	Alternate efficient crop
1.	Deccan *Rabi* season	Cotton, wheat	safflower
2.	Malwa Plateau	wheat	Safflower, Chick pea
3.	Uplands of Bihar Plateau and Orissa	Rice	Ragi, Black gram, Groundnut
4.	South-east Rajasthan	Maize	Sorghum
5.	North Madhya Pradesh	Maize	Soybean
6.	Eastern UP	Kalitur	Chick pea
7.	Sierozems of North-west India	Wheat	Mustard, Taramira (Eruca sativa)

Crops which do not produce yield under normal rainfall years may not do so under abnormal years. Studies conducted in agro climatic conditions of Varanasi (eastern U.P.) revealed that under normal monsoon crops like short duration upland rice, maize, pearl millet, blackgram, greengram, sesame, pigeon pea etc. should be taken up on the basis of needs. These crops should be followed by chickpea, lentil, barley, mustard, safflower, linseed etc. on residual moisture during winter season.

If monsoon sets in as late as second week of July, short duration upland rice (variety - NDR-97 and NDR-118) may be included in place of Akashi and Cauvery. If the rains are delayed beyond the period but start somewhere in last week of July or first week of August and growing season is reduced to 60-70 days, then cultivation of hybrid pearl millet (NHB 3-4, B.J. 104), blackgram (Type 9), greengram (var-Jagriti and Jyoti) may be included in place of T-44 and k-851 etc. should be grown. Yet another alternative could be to harvest a fodder of either pearl millet, maize, sorghum or a mixture of cowpea, blackgram and one of the above fodder crops.

In case monsoon rains stop early towards the end of season, normal sowing of short duration upland rice, blackgram and sesame may be taken up. If the rain stops very early, i.e. by the end of August or first week of September, only fodder crops or grain legumes could be harvested. Depending upon the soil moisture condition, relay sowing of crops like chickpea, lentil, mustard, linseed and barley could be done in rabi season.

During the drought, it was found that farmers in Karnataka, Andhra Pradesh and Maharashtra who went in for safflower cultivation were in gainers. Safflower succeeded where other crops failed. In other dryland regions, alternative efficient crops can profitably substitute the traditional ones (Table 12.1).

Table 12.1. Relative yield of traditional and efficient crops in dryland areas

Region	Traditional	Yield (q/ha)	Efficient crops	Yield (q/ha)
Bellary	Cotton	2.0	Sorghum	26.7
Varanasi	Wheat	8.6	Chickpea	28.5
Ranchi	Upland Rice	28.8	Maize	33.6
Indore	Green gram, Wheat	11.8 11.0	Soybean Safflower	33.3 24.2
Agra	Wheat	10.3	Mustard	20.4
Hisar	Wheat	3.0	Taramira	16.0
Udaipur	Maize	18.0	Hybrid sorghum	29.0

Dryland research has remained confined to important traditional crops such as sorghum, millet, pulse and oilseeds and has not explored the possibility of growing non-traditional crops such as dye-providing crops {e.g. Henna (*Lawsonia inermis*: mehadi) and jaffra(*Bixa ovellana*) species (e.g. cumin), and medicinal value crops (e.g. citronella, lemongrass, senna and isabgol)}. These crops need to find an important place in research agenda of dryland farming.

Time has come for the relevant researchers to plan a joint integrated research programme for maximizing the profitability, productivity and sustainability of learning systems of rainfed areas. Sericulture offers great promise in rainfed farming strategy, particularly of the watershed approach in peninsular India.

Efficient cropping system

Besides putting various measures to increase the productivity levels of dryland crops, efforts would also be needed to increase the cropping intensity in dryland areas which was generally 100%, implying that a single crop was taken during the year.

Cropping intensities of these areas could be increased by practice of inter cropping and multi cropping (sequential) by way of more efficient utilization of resources. The cropping intensity would depend on the length of growing season which in turn depends on rainfall pattern and the soil moisture storage capacity of the soil. For example in Indore region, receiving 1000 mm annual rainfall, only single crop can be taken on shallow soils, inter cropping in medium depth soils and double cropping on deep soils. Similar crop combinations have been identified for different regions of the country. In dryland of Varanasi region upland rice-chickpea/lentil sequence can be practices with advantage.

Inter cropping of vegetables with grain crops was pursued vigorously in some centers such as Varanasi and Phulbani. At both the places long duration pigeon pea was inter cropped with vegetables such as okra, radish and chilli. Such inter cropping systems would be very useful to get maximum returns from rainfall agriculture. Even at solapur, leafy vegetables and some short duration beans were grown as intercrops during the rainy season.

Fertilizer use

Soils of drylands in the country are not only thirsty but hungry also because these soils are severely eroded horizontally as well as vertically. Whenever efforts are made towards bunding and levelling of the fields in dryland areas, it is the surface soil which is removed. The resultant effect is that the fields are rendered shallow in depth and completely deprived of plant nutrients, particularly nitrogen, phosphorus and potassium. It is, therefore, necessary to apply all the three major nutrients in adequate amounts. Since soil moisture is limiting in drylands, the availability of nutrients becomes limited, attempt should always be made to apply fertilizers in furrows below the seed. If seed-cum-fertilizer drills drawn by bullocks or tractors are available, this objective can be fulfilled. There has been a belief among the farmers of dryland areas that use of fertilizer increases the chances of crop failure but recent findings have shown that the use of fertilizer is not only helpful in providing nutrients to crop but also helpful in efficient use of profile soil moisture (Table 12.2). If dryland farmers are shown such results, they will be convinced to use more and more fertilizers.

Studies on the management of legumes in crop sequences for their residual effect indicated that in alluvial soils an advantage of 25-30 kg N/ha could be obtained in barley or mustard grown after black gram or green gram. Another possibility for nitrogen management in cropping system is to use legumes as green manures either at flowering stage or after one picking. Studies conducted at Varanasi centers clearly showed that general yield levels of barley and mustard were greater when legumes raised in the previous season was incorporated in soil after first picking as compared to that harvested at normal maturity (Table 12.3).

Table 12.2. Effect of N-levels on Yield and Moisture Use Efficiency (MUE) of Barley and Wheat (Varanasi Centres)

Nitrogen levels (kg/ha)	Grain yield (q/ha)	Total moisture use (mm)	MUE (kg/mm)
Barley			
0	14.05	133.7	10.5
30	20.45	136.3	15.0
60	30.00	142.3	21.0
90	37.20	141.6	26.3
Wheat			
0	9.55	145.5	6.6
30	13.55	144.4	9.3
60	18.35	153.6	119
90	24.15	155.1	13.6

In dryland areas, a proper mixing of organic and inorganic would be desirable. Organics have low nutrient content, but help to improve the moisture holding capacity of soils. In addition to yield advantage, nutrients like potassium help to increase drought tolerance by affecting plant-soil relationship. Transpiration losses are reduced and productivity per unit water increases.

Table 12.3. Nitrogen economy to legume-cereal system (4 years average)

Nitrogen level (kg/ha)	Crop yield (q/ha) Incorporated	Crop yield (q/ha) Unicorporated
Green Gram	1.89	2.23
Barley		
0	16.98	13.65
30	21.30	18.64
60	24.43	21.84
90	27.27	25.20

Rainwater management

Efficient management of rain water can boost agricultural production from drylands. The broad bed and furrow system of the **International Crop Research Institute for the Semi Arid Tropics (ICRISAT)** for managing rainwater in vertisols made it possible to increase crop yields four to five times as compared to normal practice. However, this method could not be adopted widely by the farmers in India because it is costly and labour intensive. The vertical mulching developed at Bellary centres increases the infiltration of water in soil profiles and improves in situ moisture conservation. The scope for managing profile moisture is limited in alfisols but the surface run off in such soils can be reduced by ridge-and-furrow technique. Alternatively, application of compost and farm yard manure as well as raising legumes will add the organic matter to the soil and increase the water holding capacity.

The water which is not retained by the soil flows out as surface runoff. The run-off-recycling holds immense prospects in deep black soils where the seepage losses are much less. This runoff water, if not permitted to drain out safely, causes erosion. Therefore, safe disposal of excess water from the field drains to the disposal system should be planned properly. This excess runoff water can also be harvested in storing in dug out ponds and recycled and donated as in the event of severe moisture stress during rainy season or for raising crops during the winter.

Watershed approach for resource improvement and utilization

Watershed management is a holistic approach arrived at optimizing the use of land, water and vegetations in an area and thus, providing solution to alleviate drought, moderate folds, prevents soil erosion, improves water availability and increases fuel, fodder and agricultural production on a sustained basis. On the basis of the experiences of ICAR Operational Research Projects, which attracted the attention of our farmers, State departments, administrators and scientists, 47 model watersheds were established during the year 1983 for development, jointly by the Ministry of Agriculture, ICAR and various State Government Departments and Agricultural Universities, in 16 states and then the Department of Agriculture and Co-operation launched the National Watershed Development Project for Rainfall Areas (NWDPRA) covering almost the same states. Out of these 47 model watersheds, the Central research Institute for dryland Agriculture (CRIDA). Hyderabad has been entrusted with 30 watersheds. These activities were in micro and mini-watersheds covering 500-2000 ha. Major components in these model watersheds are: (i) Improvement of water resources, (ii) In situ soil and water conservation : rain water harvesting for safe disposal of surface runoff, (iii) increase in cropping intensity and (iv) alternate land use system for efficient use of lands as per land capability to provide stability in productivity.

The model watersheds in operation have provided a fruitful experience of how development can lead to all round improvement in food and fodder production, economic condition of the farmers. Sakho-majori model, where creation of eater source worked as a catalyst and triggered the development process can be repeated under similar situations. Similar experiences have been gained a Tejpura (Jhansi), Ariel (Bareilly District) and Tejpura watersheds which have been awarded the First and Second Prizes respectively by the President of India on 14-11-1988 based on the recommendation of National Productivity Council.

CROPPING SYSTEMS OF DRYLAND

All drylands are not suitable for crop production. Some lands may be suitable for range/ pasture management, while others for tree farming, ley farming, dryland horticulture, agroforestry systems including alley cropping. All these systems which are alternatives to crop production are called as alternate land use systems. This system not only helps in generating much needed off-season employment in mono crop dryland but also minimizes risk, utilizes off season rains which may otherwise go waste as runoff, prevents degradation of soils and restores balance in the ecosystem.

Crop production may be disastrous in the years of drought, whereas drought resistant grasses and trees could be remunerative. Scientists of dryland have developed many alternate land use systems which may suit different agro ecological situations. These are alley cropping, agri-horticultural system and silvi-pastoral system which utilize the resources in better way for increased and stabilized production from drylands.

Alley Cropping: For imparting stability and providing sustainability to the farming system, a tree-cum-crop system will be one most appropriate for such situations. One such system called 'alley cropping' - a version of agro-forestry system, could meet the multiple requirements of food, fodder, fuel, fertilizer etc. Alley cropping is a system in which food crops are grown in alleys formed by hedge rows of trees or shrubs. The essential feature of the system is that hedgerows are cut back at planting and kept pruned during cropping to prevent shading and to reduce competition with food crops.

For example, fast growing leguminous trees such as *Leucaena leucocephala* or *liliricidia spp.* are planted in rows. During the cropping season, trees are lopped at about 0.5 metre height. These loppings are used as much to reduce moisture loss and improve the nutrient status of soil. Arable crops like maize, rice, pearl millet, legumes, oilseeds etc. are planted in the alleys formed by the two rows of threes. This is also known as agrisilviculture. Alley cropping is also a form of conservation farming which enhances soil fertility and prevents erosion.

One very strong argument in favour of alley cropping is its ability to produce usable material even in years of severe drought. At Rajkot in 1985, rainfall received during the season was only 30% of the normal. There was total failure of grain production of the three legume crops tried in the system. In sole crop plots production was limited to 5.0 q/ha to 17.0 q/ha of green fodder. However, in alley cropped plots, *Leucaena* hedge-rows produced over 50.0 q/ha of green fodder.

Agri-horticultural system: Agri-horticultural system palys an important role in dryland areas, especially in semi-arid regions where production of annual crops is not only low but also highly unstable. Fruit trees if suitably integrated in dryland farming system could add significantly to overall agricultural production including food, fuel and fodder, conservation of soil and water and stability in production and income. Dryland fruit trees being deep rooted and hardy, can better tolerated monsoonal aberrations than short duration seasonal crops. Hence, in drought year when annual crops usually fail or their production is highly depressed, fruit trees species yield considerable food, fodder and fuel.

A suitable example of agri-horti-system is growing of cow pea/green gram/horse gram in inter space of *ber (Zizyphus mauritiaria)* at 6 x 6 m spacing at Hyderabad. Phalsa (*Grewia asiatica*) may be planted in between two ber plants in a row with a view to intensify the system. A well managed dryland orchard of *ber* should give 50 kg fruits per tree/year. There should be 250 plants/ha for optimized production. The grow income would touch around Rs. 50,000/ha (250 x 50 x 4), assuming that one kg *ber* fetches Rs. 4. One could get an additional income of Rs. 800-Rs. 1000 from green gram/cow pea (2.5-3.0 q/ha).

Silvi-Pastoral System: This system is suitable for marginal drylands and is most preferable where fodder shortages are experienced frequently. The system essentially consists of a top feed tree species carrying grasses on legumes (preferable perennial) as understorey crops. Dryland farmers having larger holdings and keeping a land follow for a longer period for one reason or the other, should go in for this system which could provide both fodder and fuel. In a survey carried out in Andhra Pradesh, Karnataka and Maharashtra by CRIDA scientists, it was revealed that after food it is fodder which is of paramount importance for sustaining animal wealth in rural areas. In years to come, fuel will assume greater importance.

In August 1981 *Leucaena leucocephalla* was planted in contour trenches 7.5 m apart, the plant to plant spacing being maintained at 2.0 m at CRIDA. Four strips at upper reaches of plot (2% slope) were put under *Cenchrus ciliaris,* while lower four strips were seeded with *Stylosanthes hamata.* The system has come up very well.

Efficient Implements

In order to take full advantage of annual precipitation in dryland agriculture, higher doses of energy input is essential. Farmers in drylands have been using traditional and outdated farm equipments which not only perform poorly but also demand a lot of energy and time and post-harvest operations. Farm implements can help to conserve as much rain water in situ as possible and to harvest rain water. Shallow off season tillage with pre-monsoon showers ensures better moisture conservation and lesser weed intensity. It has resulted in 20% yield increase in sorghum in Andhra Pradesh. Deep tillage helps in increasing water in soils having textural profiles and hard pan. This has resulted in 10% yield increase in sorghum and 9% yield increase in case of caster. For in-situ moisture conservation, land has to be opened so that it can cause hurdle to flow of rainwater. Tillage machines of appropriate size and type matching the power sources need to be used. Location specific seeders have been developed for dryland areas and these have shown good prospects and promise. A feature of these machines is that the seeds and fertilizers are placed in the moist zone of the soil resulting in a high percentage of seed germination and good crop vigour. In deciding farm mechanization in dryland areas, where farmers are generally poor, and their socio-economic condition should always be kept in mind.

The foregoing discussions show that technology of crop production in dryland areas have been generated to a great extent. What is important now is to view it in socio-economic context of the farmers. Once the technology is adopted by the farmers, the contribution of dryland areas to the total production can be sizably improved and the living standards of the farmers of these areas can be improved. This has been clearly shown in selected watershed areas and what is needed is to have more watersheds identified, proper technology to be developed and implemented.

Choice of crops and varieties in drylands

Traditional cropping pattern in the dry farming areas is dominated by food grains viz., millets and pulses. In a predominantly subsistence type of farming system, such dominance of food crops is natural. The choice of crops for drylands is affected by

- Rainfall quantity and distribution
- Time of onset of rainy season
- Duration of rainy reason
- Soil characters including amount of rain water stored in the soil
- Farmer's requirements

The major focus of research under AICRPDA has been on the identification of most efficient crops for each dry farming region. The criteria for choice of crops for dry farming regions comprise the following

- Tolerance to drought
- Fast growth during initial period to withstand harsh environment
- Genetic potential for high yield
- Short or medium duration to escape terminal drought
- Adaptability to wide climatic variations
- Responsive to fertilizers.

For many dry farming regions of India, more suitable crops than existing ones have been identified. However, the acceptance and adoption of the practice of crop substitution by dryland farmers is poor since in most instances the new crops replace food crops (Table 12.4).

Table 12.4. Suggested crops in place of traditional crops

Region	Traditional crop	Yield (q/ha)	More suitable crop	Yield (q/ha)
Agra	Wheat	10.3	Mustard	20.4
Bellary	Cotton	2.0	Sorghum	26.7
Bijapur	Wheat	9.4	Safflower	18.8
Varanasi	Upland rice	28.0	Maize	33.8

In most crops of dry farming regions, traditional local varieties still dominate. The preference for these local varieties is based on their pronounced drought tolerance. But they are susceptible to moisture stress at maturity. They have low yield potential even under favourable rainfall. They do not respond significantly to improved management practices such as nutrient supply. The criteria now adopted for selection of crop varieties for drylands include drought tolerance, short or medium duration, high yield potential, response to nutrient supply, high water use efficiency, moderate resistance to pest and diseases. Suitable varieties for all dryland crops have been developed in all the dry farming regions and have proved their high yield potential.

Choice of Cropping System

Choice of suitable cropping system must aim at maximum and sustainable use of resources especially water and soil. Cropping systems depend on rainfall quantity, length of rainy reason and soil storage capacity. The broad guidelines in choosing a cropping system for drylands are given below (Table 12.5).

Table 12.5 Potential cropping systems based on rainfall and soil characters

Rainfall (mm)	Soil type	Growing season (weeks)	Profile storage capacity (mm)	Suggested cropping system
350-600	Alfisols, shallow vertisols	20	100	Single rainy season cropping sorghum/maise/ soybean
350-600	Deep aridisols, Entisols (alluvium)	20	100	Single cropping sorghum/maise/ soybean in *kharif/rabi*
350-600	Deep vertisols	20	100	Single post rainy season cropping sorghum
600-750	Alfisols, vertisols, entisols	20-30	150	Intercropping Sorghum + Pigeon pea Cotton + Black gram
750-900	Entisols, deep vertisols, deep alfisols, inceptisols	30	200	Double cropping with monitoring Maize-safflower Soybean-chick pea Groundnut-horsegram
> 900	As above	> 30	> 200	Assured double cropping Maize-chick pea Soybean-safflower

Monocropping

In general a single crop is being taken in dry farming areas where the annual rainfall is below 500 mm. Rainy season crop is taken in light soils like alfisols, inceptisols and oxisols. Crops are being grown on residual moisture in vertisols in the post rainy season in winter season. The most common monoc ropping systems are groundnut in alfisols and associated soils in Rayalaseema region of Andhra Pradesh, sorghum or chickpea or wheat in black soils of Madhya Pradesh, Andhra Pradesh, and Karnataka

Intercropping

Intercropping refers to growing two or more crops in the same field with specific row arrangement during the same season. In areas where the annual rainfall is in the range of 600-850 mm, intercropping is being recommended and practiced. In such areas, at least one of the component crops succeeds in drought years and there is every chance to get two crops in good rainfall years. The different intercropping systems are followed in different regions of the country based on soil, Category Effective cropping season(weeks) in different areas Potential system < 20 weeks Bellary (8), Hissar (17), Jodhpur (11), Rajkot (17), Anantapur (13), Bijapur (17), Sole cropping 20-30 weeks Jhansi (21), Udaipur (22), Anand (25), Kovilpatti (21), Sholapur (23), Akola (27), Hyderabad (22), Agra (24), Intercropping > 30 weeks Bhubaneswar (32), Bangalore (36), Rewa (36), Samba (44), Varanasi (32), Hoshiarpur (35), Ranchi (45), Indore (36), Dehradun (44), Sequence cropping rainfall, market price and marketing facilities. Intercropping systems are subject to change due to change in market price of component crops. Intercropping is widely practiced in dry farming since it offers many advantages as outlined below.

- Intercropping is a risk minimization strategy and provides an insurance against complete crop failure due to rainfall abnormalities.
- It provides more yield and income per unit area per unit time than sole cropping
- Stability in production is achieved
- Multiple products for home consumption as well as for marketing are made available
- When legumes are included in intercropping, soil fertility is enriched
- Intercrop canopy suppresses weed growth
- Resource use efficiency is increased viz., light, water and nutrients efficiently used are
- One crop provide physical support to other crop.

Disadvantages are

- Yield may be decreased due to competition effect
- Allelopathy effect
- Build up of certain diseases and pests
- May not be suitable for large farms with adequate resources

The different intercropping systems suitable for drylands are furnished in Table 12.6 and 12.7.

Table 12.6. Inter cropping systems suitable for dryland

Crops	Geometry	Base crop duration (days)	Intercrop duration (days)
Sorghum + Lablab	6-8:2	100–120	150–180
Sorghum + Redgram	6-8:1	100–120	180
Sorghum + Cowpea	2:1	100–120	80
Cotton + Black gram	2:1	150–185	65–75
Groundnut + Redgram	6-8:1	105	180
Groundnut + Castor	6-8:1	105	150–180
Bengalgram + Coriander	4:1	100	80
Maize + Cowpea	2:1	100–110	75–80
Ragi + Cowpea + Redgram	6:1:	100	75 + 180

Table 12.7. Dominant intercropping systems in different regions of India

Location	Soil	Rainfall (mm)	Intercropping	Row ratio
Anantapur	Alfisols	550	Groundnut + pigeon pea	11 : 1
Akola	Vertisols	825	Sorghum + pigeon pea	2 : 1
Rajkot	Vertisols	590	Groundnut + caster	4 : 1
Sholapur	Vertisols	560	Chickpea + safflower	3 : 1
Bangalore	Alfisols	890	Finger millet + pigeon pea	8 : 1

Double cropping in Drylands

Double cropping either by sequential cropping or relay cropping is possible in places with high rainfall (> 900 mm), extended rainy season and high soil moisture storage capacity. Double cropping is also possible with rain water harvested in farm ponds which is used for establishing winter crop. Some of the important double cropping systems of different regions are presented in Tables 12.8 and 12.9.

Table 12.8. Important double cropping systems of different location

Location	Rain fall (mm)	Soils	Double cropping
Ranchi	1,370	Oxisols	Rice-linseed or chickpea
Indore	960	Vertisols	Soybean-Wheat
Pune	1,050	Vertisols	Rice-Chickpea
Bangalore	890	Alfisols	Cowpea-Finger millet

Table 12.9. Efficient double cropping systems for drylands of India

Soil type	Region	Water availability 9days)	Double cropping system
Vertisols	MP	210–230	Maize–chickpea Soybean–wheat
	Maharastra	190–210	Sorghum–safflower
	Karnataka	130–150	Cowpea–sorghum Greengram–saf-flower
Oxisols	Bihar	160–180	Maize–chickpea Groundnut–barley
Alfisols	Karnataka	190–220	Cowpea–ragi Soybean–ragi
Alfisols and aridisols		<120	No double cropping

13

ROLE OF AGROFORESTRY IN FARMING SYSTEM

Agroforestry is the management and integration of trees, crops and/or livestock on the same plot of land and can be an integral component of productive agriculture. It may include existing native forests and forests established by landholders. It is a flexible concept, involving both small and large-sized land holdings.

Scientifically speaking, agroforestry is derived from ecology and is one of the three principal land-use sciences, the other two being agriculture and forestry. Agroforestry differs from the latter two principals by placing emphasis on integration and interactions among a combination of elements rather than just focussing on each element individually.

Agroforestry has a lot in common with intercropping (the practice of planting two or more crops on the same plot) with both practices placing an emphasis on interaction between different plant species. Generally speaking, both agroforestry and intercropping can result in higher overall yields and reduced operational costs.

History of Agroforestry

Agroforestry was formally outlined in the early 20th century by American economic geographer J. Russell Smith in his book *Tree Crops: A Permanent Agriculture* (1929). Smith viewed tree-based "permanent agriculture" as a solution to the destructive erosion that often followed the cultivation of sloping lands. However, his contributions were largely overlooked during the Green Revolution of the 1960s and the subsequent and more-inclusive farming systems research/extension (FSR/E) development approach of the early 1970s that sought sustainable agricultural alternatives. Those endeavours failed to address many basic needs of small-scale resource-limited farmers in the tropics, such as the essential timber and non-timber products derived from trees, including food, fuel, fodder, building materials, medicine, and income, as well as the ecosystem sustainability that trees maintain.

BENEFITS OF AGROFORESTRY

Over the past two decades, a number of studies have been carried out analysing the viability of agroforestry. The combined research has highlighted that agroforestry can reap substantial benefits both economically and environmentally, producing more output and proving to be more sustainable than forestry or agricultural monocultures.

Agroforestry systems have already been adopted in many parts of the world. According to the Agroforestry Research Trust, agroforestry systems can include the following benefits:

1. They can control runoff and soil erosion, thereby reducing losses of water, soil material, organic matter and nutrients.
2. They can maintain soil organic matter and biological activity at levels satisfactory for soil fertility. This depends on an adequate proportion of trees in the system- normally at least 20% crown cover of trees to maintain organic matter over systems as a whole.
3. They can maintain more favourable soil physical properties than agriculture, through organic matter maintenance and the effects of tree roots.
4. They can lead to more closed nutrient cycling than agriculture and hence to more efficient use of nutrients. This is true to an impressive degree for forest garden/farming systems.
5. They can check the development of soil toxicities, or reduce exiting toxicities. Both soil acidification and salinization can be checked and trees can be employed in the reclamation of polluted soils.
6. They utilize solar energy more efficiently than monocultural systems. Different height plants, leaf shapes and alignments all contribute.
7. They can lead to reduced insect pests and associated diseases.
8. They can be employed to reclaim eroded and degraded land.
9. Agroforestry can augment soil water availability to land use systems. In dry regions, though, competition between trees and crops is a major problem.
10. Nitrogen-fixing trees and shrubs can substantially increase nitrogen inputs to agroforestry systems.
11. Trees can probably increase nutrient inputs to agroforestry systems by retrieval from lower soil horizons and weathering rock.
12. The decomposition of tree and pruning can substantially contribute to maintenance of soil fertility. The addition of high-quality tree prunings leads to large increase in crop yields.
13. The release of nutrients from the decomposition of tree residues can be synchronized with the requirements for nutrient uptake of associated crops. While different trees and crops will all have different requirement, and there will always be some imbalance. The addition of high quality prunings to the soil at the time of crop planting usually leads to a high degree of synchrony between nutrient release and demand.
14. In the maintenance of soil fertility under agroforestry, the role of roots is as important as that of above-ground biomass.
15. Agroforestry can provide a more diverse farm economy and stimulate the whole rural economy, leading to more stable farms and communities. Economic risks are reduced when systems produce multiple products.

As well as building on practices used in forestry and agriculture, agroforestry also works towards land protection and conservation through more effective protection of stock, control of soil erosion, salinity and water tables and a higher quality control of timber.

A denser, more dependable tree covering can provide shelter to livestock during the warmer months allowing the animals can conserve energy. That same tree covering helps block out wind, helping to boost water retention levels that can help produce a more robust crop yield.

According to the Central Queensland Forest Association (based in rainforest-rich Northern Australia), agroforestry can improve land protection in the following areas:

- **Salinity and water table control:** Salinity is mainly caused by rising water tables. Trees help to lower water tables, acting as pumps to take up water from the soil and then evaporating it into the atmosphere.
- **Soil erosion control:** Soil erosion or loss results from the action of wind and water on unprotected soils. The forest canopy, roots and leaf litter all have a role in controlling soil erosion.
- **Water logging:** Through water removal, established trees can substantially reduce water logging in their immediate area, which may result in improved land uses, e.g. pasture or crop. Agroforestry can have immense benefits for the environment and the farmer (a detailed breakdown of agroforestry's main benefits can be found on AgriInfo's site). For farmers, the ability to maintain some sort of control over land and production in the face of climate change means agrofrestry could hold huge promise for the agricultural sector. On an environmental level, agroforestry's ability to help prevent soil erosion while simultaneously aiding water retention and promoting soil fertility could help provide a solution for areas where rainfal is irregular or might become irregular due to climate change while dense plantations of trees would also help absorb CO_2 and regulate local temperature

AGROFORESTRY PRACTICES

Agroforestry systems are intensively managed to maintain their productive and protective functions through cultivation, fertilization, irrigation, pruning, and thinning. Ideally, components are structurally and functionally combined and actively managed to optimize the positive biophysical interactions between them. In some systems, for example, the trees are regularly coppiced (severely cut back), and the cuttings are applied as mulch to the soil. Such management not only encourages new tree growth but also augments the light levels reaching shaded crops, reduces weeds, and helps to maintain soil moisture.

Although American and Canadian temperate-zone agroforestry nomenclature differs from that used in the tropics and Europe, five temperate-zone agroforestry practices are generally recognized worldwide.

- **Riparian and upland buffers:** Strips of permanent vegetation, consisting of trees, shrubs, grasses, and forbs planted and managed together
- **Windbreaks:** Trees or shrubs planted and managed as barriers to reduce wind speed as part of a crop or livestock operation
- **Alley cropping** (known as silvoarable agroforestry in Europe): Trees planted in multiple rows combined with crops cultivated in the alleyways between the tree rows
- **Silvopasture** (also known as agrosilvopastoral agroforestry): Trees combined with forage (pasture) and livestock production
- **Forest farming:** Cultivation of high-value specialty crops under the protection of a forest overstory that provides an appropriate microclimate

In addition to many of the temperate-zone practices, tropical agroforestry systems often include a variety of other agroforestry methods. Home gardens and *taungya*, in which food crops are grown between tree seedlings as they mature (often used in teak or mahogany production), are useful to supplement the nutritional needs of landowners growing tree crops. Terrace cultivation, living fences, multipurpose trees e.g., *Albida acacia / Faidherbia albida*, fodder trees, and multi-strata systems (featuring trees of different heights, as in shade-grown coffee) are also used in tropical systems to combine trees with other crops and livestock.

Key traits of agroforestry practices

Agroforestry practices are intentional combinations of trees with crops and/or livestock which involve intensive management of the interactions between the components as an integrated agroecosystem. These four key characteristics - intentional, intensive, interactive and integrated - are the essence of agroforestry and are what distinguish it from other farming or forestry practices. To be called agroforestry, a land use practice must satisfy all of the following four criteria:

- **Intentional:** Combinations of trees, crops and/or animals are intentionally designed and managed as a whole unit, rather than as individual elements which may occur in close proximity but are controlled separately.
- **Intensive:** Agroforestry practices are intensively managed to maintain their productive and protective functions, and often involve annual operations such as cultivation, fertilization and irrigation.
- **Interactive:** Agroforestry management seeks to actively manipulate the biological and physical interactions between the tree, crop and animal components. The goal is to enhance the production of more than one harvestable component at a time, while also providing conservation benefits such as non-point source water pollution control or wildlife habitat.
- **Integrated:** The tree, crop and/or animal components are structurally and functionally combined into a single, integrated management unit. Integration may be horizontal or vertical, and above- or below-ground. Such integration utilizes more of the productive capacity of the land and helps to balance economic production with resource conservation.

Research and Development

To advance agroforestry in the North America, research is needed both on basic, process-level questions and on applied management techniques that are appropriate for commercial farm or forest operations. While basic research may, for example, investigate the long-term biological interactions between the components of an agroforestry practice, applied research should seek to maximize the tangible short and intermediate term benefits. Agroforestry practices should be tailored to readily integrate into existing farming or forestry enterprises, minimize the displacement of existing crops, use equipment and technical skills that are readily available, and allow some harvesting of products within conservation agroforestry practices (e.g., hardwood timber from riparian buffer strips). There is the potential to expand the participation of State, community and junior colleges, through their agriculture and forestry programs, in agroforestry research.

The greatest research need is to develop farm-level analyses of the potential economic costs, benefits, and risks associated with agroforestry practices. This information is a vital prerequisite to the objective comparison of both production and conservation-driven agroforestry practices with alternative land use options. Furthermore, attention should be given to evaluations of future price trends in regional, national and international markets for commodities that can be produced using agroforestry (e.g., hardwood lumber or high-value, wind-sensitive crops). Research on tree-crop-animal-environment interactions should be pursued to provide a scientific basis for optimizing agroforestry designs.

Information and technology transfer

Technical information must be developed locally or regionally for application within that region. Information which is too general or which is based on studies conducted in dissimilar regions or climate zones is not likely to convince landowners to adopt agroforestry practices, or provide relevant skills and knowledge to ensure their success. on-farm demonstrations and field days are key to the understanding and appreciation of agroforestry practices by landowners. Education and training in agroforestry are needed both for natural resource professionals and college students.

In addition to the traditional model for the transfer of technology from researcher to extension agent to practitioner, landowners should have greater involvement in all phases of this process. With the assistance of research and extension personnel, local groups of landowners may analyze their own needs for agroforestry development, conduct on-farm experiments under real-life conditions, and then choose the practices most appropriate for their individual properties. Rather than accusing landowners of causing environmental degradation, they should be approached from a "win-win" perspective. Emphasis should be placed on participatory decision-making including landowner advisory groups. Research and information development should focus on agroforestry practices that afford economic opportunities, increase production efficiency, and provide cost-effective and pro-active solutions to conservation problems.

Landowner adoption

Farmers, ranchers, and foresters may not recognize particular practices as "agroforestry" even though they make use of them (e.g., field windbreaks to protect crops). Agroforestry is a set of practices integrated into larger land use systems. It is not a product or commodity. As a part of integrated land use management systems, agroforestry is relevant to the sustainable production of a wide variety of agricultural commodities, as well as the production of high-value specialty products.

Economic gain is the primary motivating factor in the adoption of agroforestry in the North America. The decision whether or not to adopt an agroforestry practice depends on the decision maker's perception of how that practice compares with alternative land use options. To be acceptable, agroforestry practices must offer:

- At least as much income potential, without significantly greater risk, compared to other market-driven land uses
- Better prospects for solving a particular conservation problem compared to other practices that do not involve tree planting.

While economics are often paramount in the decision to adopt one land use practice over another, social and aesthetic considerations may also be important to the landowner. The relative weighting of economic, social and other factors will vary among landowners depending on the size of the farming or forestry enterprise, the level of production intensity, proximity to markets, and whether it is a full or part-time activity.

TYPES OF AGROFORESTRY SYSTEMS

1. Structural basis:

A. Nature of components

(i) **Agrisilvicultural systems:** In this system, agricultural crops are intercropped with tree crops in the interspace between the trees. Under this system agricultural crops can be grown upto two years under protective irrigated condition and under rainfed farming upto four years. The crops can be grown profitably upto the above said period beyond which it is uneconomical to grow grain crops. However fodder crops, shade loving crops and shallow rooted crops can be grown economically. Wider spacing is adopted without sacrificing tree population for easy cultural operation and to get more sunlight to the intercrop. Performance of the tree crops is better in this system when compared to monoculture.

(ii) **Silvopastora systems:** The production of woody plants combined with pasture is referred to as Silvipasture system. The trees and shrubs may be used primarily to produce fodder for livestock or they may be grown for timber, fuelwood, fruit or to improve the soil.

This system is classified into three categories

(a) Protein bank

(b) Live fence of fodder trees and hedges

(c) Trees and shrubs on pasture

(a) Protein bank: In this Silvipastoral system, various multipurpose trees (protein rich trees) are planted in or around farmlands and range lands for cut and carry fodder production to meet the feed requirement of livestock during the fodder deficit period in winter.

Example: *Acacia nilotica, Albizia lebbeck, Azadirachta indica, Leucaena leucocephala, Gliricidia sepium, Sesbania grandiflora*

(b) Live fence of fodder trees and hedges: In this system, various fodder trees and hedges are planted as live fence to protect the property from stray animals or other biotic influences.

Example: *Gliricidia sepium, Sesbania grandiflora, Erythrina sp, Acacia sp.*

(c) Trees and shrubs on pasture: In this system, various tree and shrub species are scattered irregularly or arranged according to some systemic pattern to supplement forage production.

Example: *Acacia nilotica, Acacia leucophloea ,Tamarindus indica, Azadirachta indica.*

(iii) Agrosilvopastoral Systems: The production of woody perennials combined with annuals and pastures is referred to as Agrisilvopastural system.

This system is grouped into two categories.

(a) Home gardens

(b) Woody hedgerows for browse, mulch, green manure and soil conservation

(a) Home gardens: This system is found extensively in high rainfall areas in tropical South and South east Asia. This practice finds expression in the states of Kerala and Tamil Nadu with humid tropical climates where coconut is the main crop. Many species of trees, bushes , vegetables and other herbaceous plants are grown in dense and in random or spatial and temporal arrangements. Most home gardens also support a variety of animals. Fodder grass and legumes are also grown to meet the fodder requirement of cattle. In India, every homestead has around 0.20 to 0.50 ha land for personal production.

Home gardens represent land use systems involving deliberate management of multipurpose trees and shrubs in intimate association with annual and perennial agricultural crops and livestock within the compounds of individual houses. The whole tree- crop- animal units are being intensively managed by family labour. Home gardens can also be called as multitier system or multitier cropping.

Home gardens are highly productive, sustainable and very practicable. Food production is the primary function of most home gardens.

Home gardens are characterized by high species diversity and usually 3-4 vertical canopy strata. The layered configuration and compatible species admixture are the most conspicuous characteristics of all home gardens. Generally all home gardens consist of an herbaceous layer near the ground, a tree layer at the upper levels and an intermediate layer. The lower layer can be partitioned into two, the lowermost being at less than 1.0 m in height, dominated by different vegetables and the second layer of 1.0-3.0 m height comprising food crops such as banana, papaya and so on. The upper tree layer can also be divided into two, consisting of emergent , full grown timber and fruit trees occupying the upper most layer of 25 m height and medium size trees of 10-20 m occupying the next lower layer. The intermediate layer of 5-10 m height is dominated by various fruit trees.

(b) Woody hedgerows for browse, mulch, green manure and soil conservation

- **Woody species:** *Anacardium occidentale,Artocarpus heterophyllus, Citrus spp, Psiduim guajava, Mangifera indica, Azadirachta indica, Cocus nucifera,*
- **Herbaceous species:** Bhendi, Onion, Cabbage, Pumpkin, Sweet potato, Banana, Beans, etc.
- **Woody Hedgerows:** In this system various woody hedges, especially fast growing and coppicing fodder shrubs and trees are planted for the purpose of browse, mulch, green manure, soil conservation etc. The following species viz., *Erythrina sp, Leucaena luecocephala, Sesbania grandiflora* are generally used.

(iv) Other systems

(a) **Apiculture with trees:** In this system various honey (nector) producing trees frequently visited by honeybees are planted on the boundary of the agricultural fields

(b) **Aquaforestry:** In this system various trees and shrubs preferred by fish are planted on the boundary and around fish ponds. Tree leaves are used as feed for fish. The main role of this system is fish production and bund stabilization around fish ponds

(c) **Mixed wood lots:** In this system, special location-specific Multi Purpose Trees (MPTs) are grown mixed or separately planted for various purposes such as wood, fodder, soil conservation , soil reclamation etc.

B. Arrangement of components

(a) **Spatial arrangement:** Spatial arrangement of plants in an agroforestry mixture may result in dense mixed stands (as in home gardens) or in sparse mixed stands (as in most systems of trees in pastures).

(b) **Temporal arrangement:** Temporal arrangements of plants in Agroforestry may also take various forms. An extreme example is the conventional shifting cultivation cycles involving 2-4 years of cropping and more than 15 years of fallow cycle, when a selected woody species or mixtures of species may be planted. Similarly, some silvipastoral systems may involve grass leys in rotation with some species of grass remaining on the land for several years. These temporal arrangement of components in agroforestry are termed coincident, concomitant, overlapping, separate and interpolated.

2. Functional Basis

All agroforestry systems have two functions:

(a) Productive functions

(b) Protective functions

(a) Productive functions

The productive functions are:

- Food
- Fodder
- Fuel wood
- Cloths
- Shelter
- NTFPs

(b) Protective functions

The Protective functions are:

- Wind breaks
- Shelterbelts
- Soil conservation
- Soil improvement

3. Socio-economic Classification

Based on socioeconomic criteria as scale of production and level of technology input and management, agroforestry systems have been grouped into three categories:

(a) Commercial Agroforestry systems

(b) Intermediate Agroforestry systems

(c) Subsistence Agroforestry systems

(a) **Commercial Agroforestry Systems:** The term commercial is used whenever the scale of the production of the output is the major aim of the system.

Examples:

(i) Commercial production of plantation crops such as rubber, oilpalm, and coconut with permanent underplanting of food crops, pastures

(ii) Commercial production shade tolerating plantation crops such as coffee, tea and cocoa under overstorey of shade trees

(b) Intermediate Agroforestry systems: Intermediate systems are those between commercial and subsistence scale of production and management.

Examples: Production of perennial cash crops and subsistence food crops undertaken on farms wherein the cash crops fulfill the cash needs and the food crops meet the family's food needs.

(c) Subsistence Agroforestry systems: Subsistence Agroforestry systems are those wherein the use of land is directed towards satisfying basic needs and is managed mostly by the owner and his family.

4. Ecological Classification

(a) Humid / sub humid

(b) Semiarid / arid

(c) Highlands

(a) Agroforestry systems in humid / Subhumid lowlands

Examples: Homegardens, Trees on rangelands and pastures, improved fallow in shifting cultivation and Multipurpose woodlots.

(b) Agroforestry systems in Semiarid and arid lands

Examples: Various forms of silvopastoral systems, windbreaks and shelterbelts.

(c) Agroforestry systems in Tropical highlands

Examples: Production systems involving plantation crops such as coffee, tea, use of woody perennials in soil conservation and improved fallow.

5. Miscellaneous types

Common types of agroforestry systems applied by landowners and land managers across the United States are silvopasture, alley cropping, forest farming, riparian forest buffers, and windbreaks. There are also several special applications worth consideration. Most agroforestry practices can be applied almost anywhere in the United States, climate, soil type, and rainfall remain limiting factors. A professional can help you decide which practices are most appropriate in your particular situation.

(a) Silvopasture: Silvopasture is a form of agroforestry that combines trees with forage and livestock production. The trees in a silvopasture system are typically managed for high-value sawlogs and, at the same time, provide shade and shelter for livestock and forage. The partial shade throughout a silvopasture can reduce stress on the animal, and in some cases, it can increase forage production and quality. In plantations of conifers or hardwoods for timber or Christmas trees, managed grazing provides additional annual income from hay or livestock production. Silvopasture is a particularly popular agroforestry system in the Southeast, but it is becoming more popular in other areas across the country where coniferous trees exist. Some nut (e.g. walnut and pecan) and fruit orchards may also be managed as a silvopasture. Silvopasture is successful when the tree, forage, and livestock components are all compatible.

(b) **Alley cropping:** Alley cropping is a type of agroforestry that involves growing an agricultural crop simultaneously with a long-term tree crop. It is broadly defined as widely-spaced rows of trees and/or shrubs (single or multiple), that create alleyways within which agricultural crops or horticultural crops are produced. Alley cropping is usually done with the specific purpose of providing annual income while the tree crop matures. Fine hardwoods, like walnut, oak, and pecan, are favored species in alley cropping systems and can potentially provide high-value lumber or veneer logs. Nut crops can be an intermediate product. In addition to cash crops such as corn, squash, and melon, fruit bearing shrubs such as blueberry and ornamentals can be grown in the alleyways.

(c) **Forest farming:** Forest farming is a specific form of agroforestry that involves the cultivation of high-value non-timber crops under the protection of a forest canopy that has been modified to provide the shade level appropriate for a specific crop. Ginseng, shiitake mushrooms, and decorative ferns are all crops that are typically cultivated under forest cover and are sold for medicinal, culinary, and ornamental uses. Forest farming provides annual income while high-quality trees are being grown on a longer rotation for wood products. Additionally, forest farming can promote biodiversity by reestablishing previously exploited, naturally occurring plants. The diversity created with forest farming can also attract a variety of wildlife species.

(d) **Riparian forest buffers:** Riparian forest buffers are a form of agroforestry that involves the natural or re-established streamside forests made up of trees, shrubs, and grasses. They intercept and reduce the impact of non-point source pollution associated with agricultural operations on land adjacent to waterways. Riparian forest buffers also reduce bank erosion, protect aquatic environments from excess nutrients and sedimentation, enhance wildlife, and increase biodiversity.

(e) **Windbreaks/Shelterbelts:** Windbreaks are linear plantings of trees and shrubs designed to enhance crop production and protect people, livestock, soil, and water. There are several types of windbreaks. Field windbreaks protect a variety of wind-sensitive crops, control wind erosion, and increase bee pollination and pesticide effectiveness. They can also spread snow evenly across a field, increasing spring soil moisture. Livestock windbreaks help reduce animal stress and mortality, reduce feed consumption, and help reduce visual impacts and odors. Living snowfences keep roads clear of drifting snow and increase driving safety. All properly designed windbreaks provide protection for wildlife from harsh winds. Over 50 bird species are known to use windbreaks during the breeding season.

Special Applications in Agroforestry

1. **Carbon removal:** Growing trees remove carbon from the earth's atmosphere. Several agroforestry applications can lead to substantial removal of carbon

dioxide from the atmosphere and carbon storage. These include windbreaks, riparian forest buffers, silvopasture, and short rotation woody crops (SRWC). Properly planned field windbreaks can provide adequate crop and soil protection when only 2-3% of the farmed acres are planted to windbreaks. To illustrate this point, there are 185 million acres of tilled land in the Great/ Central Plains. Planting only 2.5 percent of this field area to 30-foot-wide tree windbreaks would span 1.3 million miles and cover 4.6 million acres. As trees continue to grow, more carbon is stored. Carbon dioxide removal, based on 20-year-old plantings, would exceed: 80 million metric tons. In another example, USDA has committed to planting 2,000,000 miles of conservation buffers. If one-fourth of these buffers were 100 foot wide forested riparian buffers, carbon dioxide removal would exceed: 110 million metric tons.

2. **Wildlife management:** Having trees and shrubs on farms in corridors makes for good wildlife habitat and encourages wildlife movement. Every landscape has, at some point, had a dominant land cover such as woodland or grassland. However, human influences like agriculture, fence rows, roads, and urbanization have dissected this dominant cover making the landscape more suitable for some wildlife species and less suitable for others. Planting trees and shrubs on farms in corridors helps offset fragmentation by providing food, water quality, cover, and territory. All agroforestry applications, from forest farming to special applications, can benefit a wide variety wildlife species.
3. **Waste management:** Excess nutrients and other chemicals from agricultural, municipal, and industrial operations impact surface and ground water quality. Trees can be a natural solution to compromised water quality. Fast growing tree crops, in particular hybrid poplar, hybrid willow, sweetgum, sycamore, yellow poplar, and loblolly pine, can use the water and excess nutrients found in agricultural, municipal, and industrial waste. At the same time, they can break down and filter harmful chemicals found in the waste. By applying tailwater, animal waste, and municipal and industrial waste, to trees, you are turning waste into a useful product.
4. **Biomass/Energy:** As Americans search for cleaner, more sustainable energy options, there will be an opportunity in integrating food or energy crops with fast growing trees such as hybrid poplar, hybrid willow, sweetgum, sycamore, and locust. On tree rotation lengths as short 6 and 9 years, farmers and landowners could apply an alley cropping system, with the goals of producing an annual energy crop (e.g., switchgrass) and a longer rotation energy crop (e.g., willow) or producing an annual cash crop (e.g., corn) and a longer term energy crop (e.g., poplar).

SOME IMPORTANT CROPS IN RELATION TO AGROFORESTRY

Maize (*Zea mays*)

- **Origin:** Maize originated in Mexico and was introduced into East Africa as early as 1643. The early varieties came from the Caribbean and were only suited to the coastal strip. The spread of maize into the highlands and medium-altitude areas, with a consequent decline in most places of the indigenous cereals, was largely due to the introduction by European settlers of varieties from South Africa.

- **Ecology:** Maize has a higher yielding potential than the indigenous cereals where water and drainage are favourable. The young maize plant is moderately drought resistant, but from five weeks onwards it is less so. Maize requires well-drained soil and a good supply of nutrients.
- **Features in agroforestry:** Maize is light demanding, so shade may reduce crop yields. It is sensitive to competition for moisture, so intercropped trees must be of a non-competitive type in drier areas. For example, Leucaena in hedgerows is not recommended in areas with less than 800 mm rain. Wind damage can be a problem, so trees are useful as windbreaks.

Beans (*Phaseolus vulgaris*)

- **Origin:** South America.
- **Ecology:** Beans are annual legumes and have the capacity to fix atmospheric nitrogen. The degree of nodulation is very variable, however: in some areas large nodules are formed while in others none are produced. Beans are not drought resistant and ideally they need moist soil throughout the growing period. High temperatures cause a poor fruit set. Beans demand free-draining soils with a reasonably high nutrient content.
- **Features in agroforestry:** Beans are shade tolerant, and since they prefer lower temperatures they normally perform well under maize, bananas or other trees.

Cassava (*Manihot esculenta*)

- **Origin:** Indigenous to Africa.
- **Ecology:** Cassava is widespread in areas below 1,500 m, occasionally higher. It is very drought-resistant and is able to give good yields on poor soils. Cassava is suitable for areas with erratic rainfall where other crops may fail in bad years. Its main requirement is well-drained soil. As a result, Cassava is one of the few crops that fits into a definite rotational pattern on small holdings and owing to its undemanding nature is often planted as the last crop in the arable period before the land reverts to fallow. Weeding is only required during the first year; later weeds are tolerated. Cassava has been demonstrated to be a soil improver.
- **Features in agroforestry:** Cassava appears to be sensitive to shade but not to other forms of competition from trees.

Sorghum (*Sorghum vulgare*)

- **Origin:** Indigenous to Africa. It has been cultivated for more than 4,000 years in Egypt.

- **Ecology:** Sorghum is drought-resistant and often out-yields maize under dry conditions. It is reasonably tolerant of waterlogging and yields comparatively well on infertile soils. Sorghum has twice as many roots as maize in a given volume of soil. The roots do not die back during drought and the leaves roll up to minimize transpiration. In this way the plant can remain dormant when other crops would be killed, and when the rains start again it recovers rapidly.
- **Features in agroforestry:** Sorghum is sensitive to shade, as are all cereals. Birds, particularly Quelea, are one of the main causes of crop loss in sorghum: bird-susceptible varieties sometimes give no yield at all. Weavers, starlings, bishop birds and many other kinds of bird also attack sorghum but are less devastating than quelea. The exact role trees may play in attracting such birds has not been determined, and neither do we know if increasing perching sites for birds of prey, by pollarding trees for example, can reduce the damage by grain-eating birds.

Finger millet (*Eleusine coracana*)

- **Origin:** Some millets are believed to be indigenous to eastern Africa, but they are now widely grown in both Africa and Asia.
- **Ecology:** Finger millet tolerates dry spells in the early stages of growth, but after the first month it requires a steady supply of moisture if good yields are to be obtained. It is usually grown between 900 and 2,400 m in areas with at least 900 mm rainfall. It only yields well on fertile soils with good drainage.
- **Features in agroforestry:** Similar to sorghum with regard to shade and bird problems.

Rice (*Oryza sativa*)

- **Origin:** Southeast Asia is regarded as the centre of origin for rice from where it has spread to many parts of the world.
- **Ecology:** Rice is an unusual crop in that it can grow not only under waterlogged conditions but also in standing water. It prefers high temperatures and grows best below 1,00 m. Sandy soils can support rice, but only if there is a permanent high watertable. Otherwise heavy soils are needed to retain irrigation water.
- **Features in agroforestry:** There are practically no trees on the irrigation schemes in Kenya where rice is grown. Paddy fields are special environments and few trees can thrive there. Rice is also a shade-sensitive crop and wood supplies for irrigation schemes are best met from woodlots planted nearby.

Tea (*Camellia sinensis*)

- **Origin:** From Burma tea has been taken to many other parts of the world. The first tea was planted in Kenya in 1903 near Limuru, but most tea plantations were established after 1945.
- **Ecology:** Tea tolerates short dry spells but only produces well with adequate rainfall. The lowest altitude at which it will grow well, 1,200 m, is set by

scarcity of rainfall, whereas the upper limit, 2,200 m, is set by the incidence of frost. For good growth, soils should be deep, well drained and slightly acidic.

- **Features in agroforestry:** Tea growing areas frequently experience strong winds and windbreaks are recommended. A tree species particularly well suited to use in tree growing areas is Hakea saligna. Armillaria root rot, Armillaria mellea, is a serious disease of tea in east Africa. Armillaria is promoted by the presence of old stumps and roots of trees which have been infected, so such stumps should always be carefully removed before new tea is established.

In other parts of the world shade trees are interplanted with tea, but this is not normally done in Kenya. Shade trees in tea plantations are only beneficial in hot and humid areas, e.g. Assam, but due to the higher altitudes of the tea plantations in Kenya the soil temperatures are sub-optimal for the growth of tea. Shade trees serve to moderate the soil-temperature differences between day and night, but there is no need for this in the cool tea-growing areas in Kenya. The overall impact of shade trees on the micro-environment has been found to result in lower tea yields.

Drying of tea leaves requires a lot of energy. Currently the energy source for this purpose in Kenya is largely imported oil, and it has been argued that this could be substituted with fuelwood produced near the tea estates. Increased wood production for this purpose would best be achieved through establishment of more woodlots.

Arabica coffee (*Coffea arabica*)

- **Origin:** Ethiopia. Coffee was introduced into Kenya in 1897 by missionaries. Coffee is one of the most valuable commodities in international world trade and is very important for Kenya's foreign exchange earnings.
- **Ecology:** The best coffee areas in Kenya are between 1,400 and 1,900 m altitude, with rainfall of 1,500-2,250 mm. The best soils are well-drained volcanic soils, deep, fertile and slightly acidic.
- **Features in agroforestry:** Coffee is shade tolerant, which enables farmers to grow trees in many spatial arrangements with the crop. Grevillea was originally introduced as a shade tree for coffee and is still the most popular tree in coffee-growing areas. Other common trees with coffee are Cordia abyssinica and Albizia spp. Armillaria root rot attacks coffee, so all old roots of trees that may harbour Armillaria should be removed when new coffee is established. In large plantations where there are no shade trees windbreaks should be considered.

Bananas (*Musa spp.*)

- **Origin:** Bananas originally came from Southeast Asia, but were introduced to Africa many hundreds of years ago.
- **Ecology:** For good yields bananas require a constant supply of moisture. They are thus most important in areas with well-distributed rainfall, e.g. Kisii and Kaka-mega. Bananas also require a warm climate, so the crop performs best

at altitudes below 1,800 m. Bananas also require rich, deep and well-drained soils.

- **Features in agroforestry:** Bananas are often used as a middle- or upper-storey crop with other crops being grown underneath. Sometimes they are mixed with taller trees. Occasionally bananas are also grown in monoculture. Bananas are shallow rooted, most roots being found in the top 15 cm of the soil. Therefore they are sensitive to wind damage and windbreaks or support stakes to prevent the banana plants from falling over are essential. Bananas are sensitive to high nematode populations in the soil. If the nematode population is too high banana cultivation must be halted temporarily and other crops grown until the nematode population has decreased.

Wheat (*Triticum spp.*)

- **Origin:** South-western Asia. Wheat was introduced into Kenya early this century by the white settlers.
- **Ecology:** The best wheat-growing areas are above 1,800 m with at least 750 mm rainfall. The soils should be well drained and have a high nutrient content.
- **Features in agroforestry:** In some wheat-growing areas Acacia trees have been left in the fields, apparently with little or no reduction in yields. Strong winds may cause lodging, hence trees are useful for windbreaks. Some bird damage may occur, but it is not as serious as for sorghum and millet. Wheat is a light-demanding crop, so trees in fields must either have a natural light shade or be managed to reduce the intensity of the shade.

Sweet potato (*Ipomoea batatas*)

- **Origin:** South America, but widespread in the tropics.
- **Ecology:** Sweet potato is a crop which grows under a very wide range of ecological conditions from sea level up to 2,400 m, in a wide variety of soils, and in areas receiving more than 750 mm rainfall. Little weeding is required since the vines cover the soil very effectively and suppress most weeds.
- **Features in agroforestry:** Sweet potatoes are not normally grown with an upper storey of trees, and in general trees do not have a big role to play in association with sweet potatoes.

Irish potato (*Solarium tuberosum*)

- **Origin:** South America, but now widespread in temperate regions and in the tropical and subtropical highlands.
- **Ecology:** The Irish potato grows best in cool climates with a rainfall of 25 mm per week during the growing season and at altitudes ranging from 1,500 m to 2,900 m. The soils should be well drained and rich in nutrients. Sandy soils are better than heavy soils.
- **Features in agroforestry:** Irish potatoes are relatively shade tolerant and can sometimes be seen growing under bananas with good results. Irish potatoes

are badly damaged by high nematode populations, and if nematode infestation is severe potato cultivation may have to be discontinued.

Yams (*Dioscorea spp.*)

- **Origin:** Different species have different origins. Now widespread all over the tropics.
- **Ecology:** Tropical areas which have a long rainy season are best for this crop. Most types of yams require at least six months of rain, and the average annual rainfall should be at least 1,500 mm. Temperatures of 25-30°C are best for yam cultivation, and the soils should be fertile, loose and well drained. Intercropping and crop rotation are recommended since yams are sensitive to nematode build-up. In Kenya, yams are common in Central Province and in Eastern Province around Mount Kenya.
- **Features in agroforestry:** In Kenya, yams are often grown in association with Commiphora eminii, subsp. zimmermanni. A cutting of Commiphora is planted as a stake for the yam, and since Commiphora cuttings take root easily eventually a Commiphora tree provides support to the yams. Both grow well together without any signs of competition and there are several other benefits from the Commiphora beside being a support to the yams, e.g. emergency fodder, timber, making of mole traps and firewood.

Tobacco (*Nicotiana tabacum*)

- **Origin:** South America from where it was brought to Europe and later by the Portuguese to Africa. Tobacco growing was initiated by the British American Tobacco Company (BAT) in the late 1930s, but it was in the 1970s that the cultivation of tobacco expanded considerably.
- **Ecology:** Tobacco requires about 750 mm of rain and the altitude range is from sea level up to 1,260 m. A high light intensity is required for good production and the relatively low light intensities in Kenya result in low yields compared to those in some other countries. Crop rotation is required to avoid build-up of nematode populations.
- **Features in agroforestry:** Curing tobacco requires fuelwood and BAT has long promoted tree growing among small-scale tobacco producers in Kenya. Recommended tree species are Grevillea robusta, Cassia siamea, Markhamia lutea, Eucalyptus spp., and to a lesser extent Terminalia brownii, Gmelina arborea, Acacia albida, Cordia abyssinica, Sesbania sesban and Azadirachta indica. Eucalyptus spp. cannot be used for fire-curing tobacco because of the odour the wood imparts to the smoke. Woodlots with a spacing of 2 x 2 m should be established, preferably on marginal land where little or no other production takes place. Planting of trees is also recommended on farm lands, e.g. on boundaries and terrace risers. In dry areas, water-harvesting methods are adopted. The light-demanding nature of tobacco limits its scope for intercropping with trees.

Sugarcane (*Saccharum officinarum*)

- **Origin:** New Guinea. Sugar-cane growing expanded in Kenya during the 1920s with establishment of the first sugar factories.
- **Ecology:** Sugarcane is a perennial grass which grows well at altitudes from sea level up to 1,600 m in areas with not less than 1,500 mm rainfall. It prefers heavy fertile soils.
- **Features in agroforestry:** Sugarcane is strongly light demanding, so intercropping trees and Sugarcane cannot be recommended when Sugarcane is commercially grown. Sugarcane growing areas are often those where severe shortages of wood are experienced, so establishment of woodlots and intensified tree growing around homesteads are recommended to meet the needs for wood. Under small-scale irrigation in the Kerio Valley Sugarcane is sometimes grown with Ficus sycomorus and Acacia tortilis.

MEDICINAL AND AROMATIC PLANTS IN AGROFORESTRY

Medicinal and aromatic plants (MAPs) play an important role in the healthcare of people around the world, especially in developing countries. Until the advent of modern medicine, man depended on plants for treating human and livestock diseases. Human societies throughout the world have accumulated a vast body of indigenous knowledge over centuries on medicinal uses of plants, and for related uses including as poison for fish and hunting, purifying water, and for controlling pests and diseases of crops and livestock. About 80% of the population of most developing countries still use traditional medicines derived from plants for treating human diseases (De Silva 1997). China, Cuba, India, Sri Lanka, Thailand, and a few other countries have endorsed the official use of traditional systems of medicine in their healthcare programs. For example, the Indian systems of medicine "Ayurveda," "Sidha" and "Unani" entirely, and homeopathy to some extent, depend on plant materials or their derivatives for treating human ailments (Prajapati et al. 2003). People in villages and remote areas primarily depend on traditional medicines as the modern system is out of reach and expensive. Many among the educated in Asian and African countries use traditional medicines for reasons of firm belief that they are more effective than modern medicine for certain chronic diseases, they do not have side effects of some of the modern medicines, and/or for economic reasons. Thus, in many societies, traditional and modern systems of medicines are used independently.

About 12.5% of the 422,000 plant species documented worldwide are reported to have medicinal value; the proportion of medicinal plants to the total documented species in different countries varies from 4.4% to 20% (Schippmann et al. 2002). About 25% of drugs in modern pharmacopoeia are derived from plants and many others are synthetic analogues built on prototype compounds isolated from plants. Up to 60% of the drugs prescribed in Eastern Europe consist of unmodified or slightly altered higher plant products. These drugs carry important therapeutic properties including contraceptives, steroids and muscle relaxants for anesthesia and abdominal surgery (all made from the wild yam, *Dioscorea villosa*); quinine and artemisinin against malaria; digitalis derivatives for heart failure; and the anti-cancer drugs vinblastin, etoposide and taxol. These compounds cannot be synthesized cost-effectively, which means that their production requires reliable supplies of plant material.

The global importance of MAP materials is evident from a huge volume of trade at national and international levels. During the 1990s, the reported annual international importation of MAPs for pharmaceutical use amounted on average to 350 000 Mg valued at over USD 1 billion. A few countries dominate the international trade with over 80% of the global import and export allotted to 12 countries each. Whereas Japan and Korea are the main consumers of medicinal plants, China and India are the world's leading producing nations. Hong Kong, United States and Germany stand out as important trade centers. It is estimated that the total number of MAPs in international trade is around 2500 species worldwide (Schippmann et al. 2002).

Medicinal trees in traditional agroforestry systems

Many plants in traditional agricultural systems in the tropics have medicinal value. These can be found (either planted or carefully tended natural regenerations) in homegardens, as scattered trees in croplands and grazing lands, and on field bunds. Many Acacia species found in Africa such as *Acacia nilotica, A. seyal, A. senegal* and *A. polyacantha*, as well as several species in *African croplands* (e.g. *Faidherbia albida, Vitellaria paradoxa, Adansonia digitata, Markhamia lutea* and *Melia volkensii*) have medicinal value (ICRAF 1992). Similarly, "arjun" (*Terminalia arjuna*) in India, chinaberry (*Melia azederach*) in Asia and *Erythrina species* in Latin and Central America combine many uses including medicinal. Holy Basil or "tulsi" (*Ocimum sanctum*), drumstick (*Moringa oleifera*) and curry leaf (*Murraya koenigii*) are backyard plants in many Indian households and they are routinely used for common ailments or in food preparations. A number of plants used as live fences around home compounds such as henna (*Lawsonia inermis*) in India (Singh *et al.* 1996), *Ipomoea carnea* ssp. fistulosa in Bolivia and Asian countries (Frey 1995) and *Euphorbia tirucalli* around crop fields in Africa (ICRAF 1992) have medicinal values. Although the medicinal value of these plants is "exploited" locally, they are seldom used for commercial purposes (except in the case of commercially exploited neem). In fact, many of these species are valued for poles, fuel wood, fodder, fruit, shade, and/or boundary demarcation and their medicinal value is secondary.

Forests and forest plantations

MAPs growing in forests require (or tolerate) partial shade, moist soils high in organic matter, high relative humidity, and mild temperatures. Cultivation of such MAPs can be taken up in thinned forests and cleared forest patches, and as intercrops in new forest plantations . In China, cultivation of medicinal plants has been an age-old practice under the name of "silvo-medicinal" systems. In northeast China, ginseng (Panax ginseng) and other medicinal plants are grown in pine (*Pinus spp.*) and spruce (*Picea spp.*) forests; in central China, many medicinal plants are planted with Paulownia tomentosa and in southern China medicinal herbs are often planted in bamboo (*Bambusa spp.*) and Chinese fir (*Cunninghamia lanceolata*) forests (Zou and

Sanford 1990). In Yunnan province, China, traditional "Dai and Jinuo" agroforestry systems involve the medicinal crop *Amomum villosum* in the forest areas cleared of undergrowth. The forest is thinned to give 30% to 40% shade and seedlings or cuttings are planted, which produce an average dried fruit yield of 375 kg ha^{-1} per annum (Zhou 1993). Gupta (1986) listed a number of indigenous understory herbs and shrubs that can be produced as part of forest farming or in new forest plantations to improve economic returns from forests in India. Indigenous people living in the Himalayan forest margins in Uttaranchal, India, are known to have conserved and cultivated several medicinal species for centuries (Kumar et al. 2002).

A farmers' cooperative in the northern lowlands of Costa Rica has successfully demonstrated cultivation of the medicinal herb "raicilla" (Cephaelis ipecacuanha) in natural forests for export to the Netherlands and Germany (Hager and Otterstedt 1996). American ginseng (Panax quinquefolium), a medicinal herb exported to China from the United States and Canada is grown as an understorey in red maple (Acer rubrum) forests (Nadeau et al. 1999) or deciduous hardwoods such as black walnut (Juglans nigra) and sugar maple (Acer saccharum), instead of growing under artificial shade with considerable expense (Hill and Buck 2000). Indeed, cultivation of ginseng and several other medicinal plants in the forests is a common and growing form of forest-farming practice of agroforestry in North America. Light demanding understory species (e.g. Echinacea sp.) may be intercropped initially to provide early returns from plantations and after canopy closure, shade-tolerant species such as ginseng and goldenseal can be intercropped (Teel and Buck 2002). Studies in New Zealand have indicated that the American ginseng can be successfully grown under *Pinus radiata* with best growth under a tree stand of 130 stems ha^{-1} (Follett 1997). In addition to providing shade, the trees may also benefit the understory component from hydraulically lifted water. Fungal diseases are a major concern in forest farming but the application of fungicide can be detrimental to the forests' health, therefore proper spacing and mixed cropping is recommended. Mechanical cultivation may not be feasible under forested conditions so labor availability needs to be considered as a constraining factor (Hill and Buck 2000).

As in the taungya system, newly established forest plantations can be intercropped with MAPs similar to food crops until the trees cover the ground. The participation of the local people with right to share benefits of the plantations, especially ownership to crops, has helped governments to establish and protect large-scale tree plantations without conflict with the local people in many Asian countries (Nair 1993). The same approach can be employed for the cultivation of MAPs in the new plantations. In the rehabilitation of degraded forestlands, participatory planning and implementation with local communities and economic benefits from an early stage onwards will ensure commitment of the people (Rao et al. 1999).

Homegardens

Homegardens are complex agroforestry systems involving many plant species characterized by different morphology, stature, biological function and utility, practiced mostly in the humid and subhumid tropics (Kumar and Nair 2004). Food, fruit and timber species may dominate the homegardens and occupy the middle and upper strata, but medicinal plants, spices and vegetables occupy the lower stratum. Three categories of medicinal plants could be noted in homegardens: species used exclusively for medicine, horticultural or timber species with complementary medicinal value, and "weedy" medicinal species. While the first two categories are deliberately planted the latter group is part of spontaneous growth. The species composition, plant density and level of management vary considerably depending on the soil, climate and market opportunity and cultural background of the people. Homegardens of individual holdings generally cover small parcels of land and are established around homesteads. Although these systems in the past were mostly seen to meet the home needs of small-scale farmers in the forest margins, increased urbanization, transport and market opportunities in recent times are helping to produce cash value crops. Multistrata systems involve fewer species (three to 10 species) than in homegardens in definite planting arrangement and can be designed for home consumption as well as commercial production.

Medicinal plants are an invariable component of homegardens, whether they are in the Peruvian Amazon (Lamont *et al.* 1999), on the slopes of the Mt. Kilimanjaro in Tanzania (O'Kting'ati *et al.* 1984), or in the humid and semiarid Cuba (Wezel and Bender 2003). The species composition differs depending on cultural background, distance from markets and influence of tourism. Medicinal plants accounted for about 27% of total plant species in the homegardens in Amazon (Padoc and de Jong 1991), 56% in northern Catalonia (Iberian Peninsula) (Agelet *et al.* 2000) and 45% in the floodplains of the river Jamuna in Bangladesh (Yoshino and Ando 1999). In the Soqotra Island, Yemen, endemic medicinal plants such as *Aloe perryi, Jatropha unicostata* and *Commiphora ornifolia* are cultivated in homegardens (Ceccolini 2002). On St. Croix and St. Thomas, U.S. Virgin Islands, the medicinal trees neem, moringa, and noni (*Morinda citrifolia*) have become popular in homegardens (Palada and Williams 2000).

Riparian buffer zones

An agroforestry system that has received considerable attention in North America is riparian buffers zones (Schultz et al. 2004). Riparian buffer zones can improve water quality and protect streams and rivers from degradation by nutrient loading and chemical pollutants from agriculture and urban areas, from erosion by attenuating peak flows and provide habitat for wildlife. NTFP production can help defray the cost of buffer zone installation and maintenance. Slippery elm (Ulmus rubra), harvested for its aromatic and medicinal inner bark, is commonly found in riparian areas in North America (Teel and Buck 2002). Riparian buffer zones are an ideal location for the production of this species, which suffers from commercial over-exploitation and the Dutch elm disease.

Intercropping of Medicinal, aromatic and spice crops (MAP's)

Two types of intercropping systems can be distinguished involving MAPs:

(1) Medicinal plants as upperstory trees and (2) MAPs as intercrops in other tree crops.

1. **Medicinal plants as overstorey trees:** Coffee (*Coffea arabica*), cacao (*Theobroma cacao*) and tea (*Camellia sinensis*) are traditionally grown under shade offered by multipurpose trees that produce timber, fruit, flowers, nuts, palms etc. Medicinal tree species that grow tall and develop open crown at the top can also be used for this purpose, for example yongchak (*Parkia roxburghii*) in India, the protein rich seeds of which are used to treat stomach disorders (Balasubramanian 1986) and *Ginkgo biloba* in China, the nuts of which are used in Chinese medicine and fetch high value (Shen 1998). In Ivory Coast, 19 of the 41 tree species planted as shade trees in coffee and cacao provide pharmaceutical products for traditional medicine (Herzog 1994). New plantations of coffee, tea, and cacao offer scope for cultivation of forest medicinal trees that are under demand. However, research needs to identify the medicinal trees that can be grown in association with these plantation crops and develop management practices for them. Tall and perennial medicinal trees that need to be planted at wider spacing such as *Prunus africana, Eucalyptus globulus* (for oil), sandalwood (*Santalum album*), ashok (*Saraca indica*), bael (*Aegle marmelos*), custard apple (*Anona squamosa*), amla (*Emblica officinalis*), drumstick or moringa (*Moringa oleifera*) and soapnut tree (*Sapindus mukorossi*) can be intercropped with annual crops in the early years until the tree canopy covers the ground. Some of the medicinal trees may allow intercropping for many years or on a permanent basis depending on the spacing and nature of the trees. The intercrops give some income to farmers during the period when the main trees have not started production.
2. **Medicinal plants as intercrops:** Many tropical MAPs are well adapted to partial shading, moist soil, high relative humidity and mild temperatures (Vyas and Nein 1999), allowing them to be intercropped with timber and fuel wood plantations, fruit trees and plantation crops. Some well known medicinal plants that have been successfully intercropped with fuel wood trees (e.g. *Acacia auriculiformis, Albizia lebbeck, Eucalyptus tereticornis, Gmelina arborea,* and *Leucaeana leucocephala*) in India, include safed musli (*Chlorophytum borivilianum*), rauvolfia (*Rauvolfia serpentina*), turmeric (*Curcuma longa*), wild turmeric (*C. aromatica*), *Curculigo orchioides*, and ginger (*Zingiber officinale*) (Prajapati et al. 2003). Only 10 out of 64 herbaceous medicinal plants tried in intercropping with two-year old poplar (*Populus deltoides*) spaced 5 m apart gave poor performance (Kumar and Gupta 1991), indicating that many medicinal plants can be grown in agroforestry systems. The trees may benefit from the inputs and management given to the intercrops. Short stature and short cycle MAPs and culinary herbs are particularly suited for short-term intercropping during the juvenile phase of trees. Wherever markets are established, MAPs are remunerative alternative intercrops to the traditionally grown annual crops (Maheswari et al. 1985; Zou and Sanford 1990). The number of years MAPs can be intercropped with a given tree species depends on the size and intensity of its canopy shade, tree spacing and management, especially pruning of branches and nature of the MAPs.

Shade-tolerant and rhizomatic MAPs can be grown on a longer-term basis in widely spaced plantations.

Intercropping of medicinal plants in coconut (*Cocos nucifera*) and arecanut (*Areca catechu*) stands is an age-old practice in India and other parts of south- and southeast Asia. These palms allow 30% to 50% of incident light to the underneath, which is ideal for some MAPs, including cardamom (*Elettaria cardamomum*). Kacholam or galang (*Kaempferia galanga*) - a medicinal herb - is traditionally intercropped in mature coconut gardens in Kerala, India. Kacholam intercropped in a 30 year-old coconut plantation produced 6.1 Mg ha^{-1} of rhizomes compared with 4.8 Mg ha^{-1} as a sole crop (Maheswarappa *et al.* 1998). Twelve-year old coconut trees did not adversely affect the growth and yields of a number of medicinal species grown as intercrops compared to the yields in the open (Nair *et al.* 1989). In Karnataka and Kerala states, India, arecanut palm is commonly intercropped with ginger, turmeric, black pepper (*Piper nigrum*) and cardamom (Korikanthimath and Hegde 1994). Some of these intercrops may cause small reduction in arecanut yields but the combined returns from both the components are greater than from arecanut alone. Another plantation crop intercropped with MAPs is rubber (*Hevea brasiliensis*), for example with *Dioscorea floribunda* in the state of Assam in India (Singh et al. 1996) and with *Amomum villosum* in Yunnan province of China (Zhou 1993). In Sikkim, India, large cardamom (*Amomum subulatum*) is grown under 30 different shade tree species (Patiram *et al.* 1996). In Fujian Province, China, *Cunninghamia lanceolata* - an important timber tree - is intercropped with a variety of cereals, cash and medicinal and oil-producing crops (Chandler 1994). Many of the medicinal herbs commonly grown in thinned forests can also be grown intercropped with trees (Zhou 1993).

In the Caribbean islands, there has been increased interest on alternative crops that have better economic potential than traditional crops. For example, in the U.S. Virgin Islands, a number of farmers are now opting for specialty crops such as the West Indian hot peppers (*Capsicum chinense*), thyme (*Thymus vulgaris*) and chives (*Allium schoenosprasum*) instead of vegetables (Crossman *et al.* 1999). The prospects of growing indigenous MAPs such as "japana" (*Eupatorium triplinerve*), worrywine (*Stachytarpheta jamaicensis*), inflammation bush (*Verbersina alata*) and lemongrass (*Cymbopogon citratus*) in association with the medicinal trees noni (*Morinda citrifolia*) and moringa have been explored at the University of the Virgin Island, St. Croix, (Palada and Williams 2000). These local herbs are commonly used as bush teas and very popular in the Caribbean. Medicinal plants and herbs in intercropping produced similar yields to those in sole cropping at the first harvest, but they tended to be lower than in sole cropping at subsequent harvests (Palada and Williams 2000).

3. **Intercropping ginger and turmeric:** Jaswal *et al.* (1993) reported that performance of two rhizomatous crops i.e. ginger (*Zingiber officinale* Rosc.) and turmeric (*Curcuma Ionga* L.) was investigated under rainfed conditions in

pure stands and as intercrops with 5-year-old poplars planted at three spacings, viz., 5 x 5 m, 5 x 4 m and 5 x 3 m. The average illumination below the canopies was 53, 46 and 38% of incident radiation, respectively. Both crops performed better as intercrops than as pure stands. Survival was inversely correlated to light intensity. Plant height, tillers per plant and leaves per plant in ginger and leaf length and leaf breadth besides plant height in turmeric were significantly enhanced when intercropped. The rhizome length, rhizome breadth, yield per plant and yield per ha in ginger exceeded under poplars but showed a drastic reduction under the closest poplar spacing. In turmeric, the trend for the first two characters was the same, whereas yield per plant as well as yield per ha were slightly greater in the open than under 5 x 3 m spacing. Dry matter content varied significantly with spacing. For quality parameters, only oil content in ginger and oleoresin in turmeric showed significant differences. The cultivation of turmeric proved more remunerative than ginger. Among the poplar spacings, 5 × 4 m for ginger and 5 × 5 m for turmeric were delineated as the best spacings.

Pandey *et al.* (2016) conducted field experiments under support irrigated conditions at NAU, Navsari (Gujarat) during 2013 and 2014. The nine year old plantation of Sapota (*Manilkara acharas* (Mill.) Fosberg) with seven year old plantation of Jatropha (Jatropha curcas L.) was used for the present intercropping study. Two crops with two varieties viz. ginger (*Zingiber officinale* L var. Navsari local and *Zingiber officinale* L. var. Udaipur local) and turmeric (*Curcuma longa* L. var. Sugandham and *Curcuma longa* L. var. Guj. Navsari Turmeric- 1) was selected for the study. Significantly higher B: C ratio was noted in intercropping of ginger with Sapota + Jatropha or Jatropha based agro-forestry systems as compared to sole cropping or intercropping under Sapota, whereas in intercropping of turmeric with Sapota or Sapota + Jatropha agro-forestry systems as compared to Intercropping under Jatropha or sole cropping. Overall economic analysis states that the total cost of production found higher in case of ginger but net income as well as BCR observed higher in case of turmeric

Conclusion

Traditional systems of medicine in most developing countries depend primarily on the use of plant products either directly or indirectly. Besides serving the healthcare needs of a large number of people, medicinal plants are the exclusive source of some drugs even for modern medical treatment. The use of plant products as nutrition supplements and in the cosmetic and perfume industry has increased the value of medicinal and aromatic plants in recent years. The over dependence on forests, natural woodlands and long-term fallows for extraction of MAPs is threatening the survival of many valuable plant species. It is imperative therefore that such endangered species are cultivated outside their natural habitats to ensure their regular supply for human needs as well as to preserve the genetic diversity. Cultivation is an important strategy for conservation and sustainable maintenance of natural stocks, but few MAPs are actually cultivated. Lack of basic knowledge of biology, ecology, propagation methods and cultural practices for the concerned species is an important constraint.

AGRI HORTICULTURE

Cropping systems in Horticultural crops

1. **Cover crops:** Cover crops are raised mainly in the beginning or advanced rainy season so that these may provide necessary protection to the soil when damage by soil erosion is greatest. Retention of cover crops during summer is undesirable as it will compete with the main crop of fruit trees for soil moisture. During monsoon cover crops help in drawing off excess moisture. Experiment on cover cropping in citrus orchard reported by Singh and Nauriyal showed that when berseem and methi were grown as cover crops there was a significant increase in nitrogen content of the soil.

   The cultural operations aim at:

   (i) Reducing the prevalence of weeds and

   (ii) Conservation of moisture.

   To meet these objects, artificially mulches of saw dust, alkathene black film, dried leaves etc. are becoming popular. The experiments conducted at Pune on effect of different mulching materials on yield and grows of grapes have shown that the yield of grapes was significantly higher under bluck alkathene film and sawdust, mulches.

   Similarly giving of mulch of dried leaves and grass has been found to be beneficial for Cashewnut under Kokum conditions.

2. **Inter crop:** The crop which are growing between the rows of fruit trees in pre-bearing stage with a view to get some income are known as intercrops. The commonly grown inter crops are vegetable gram, groundnut, mung, black gram, etc.

   The following factors should be kept in mind while selecting the inter crop to be grown in an orchard.

   (i) The inter crops should get occupy the space for more the 5 to 9 months.

   (ii) The water requirement of the inter crop should be minimum.

   (iii) The inter crops should adjust with main crop.

   (iv) Vegetables like ground vines should not be taken as intercrops as the creepers are likely to dimb on the trees.

   The following set of inter crops may be good for orange orchards. In the first and second year papaya may be taken as intercrop.

   (v) After papaya until the plants becomes 3 to 4 years old vegetables could be profitably grown in between the rows.

   (vi) When the orange trees are 4 to 5 years old instead of growing vegetables, an inter crop of groundnut can be taken, when there is no trouble of pigs. In areas where there is such trouble an inter crop of peas may be taken.

3. **Mulching:** Mulching is a system of soil management in which hay, straw, or cut grass is placed on the soil between the tree rows to conserve the soil moisture by checking surface evaporation,. It is considered that cultivation and the maintenance of a soil mulch are important as means of preventing the evaporation of moisture. If seeds are removed without disturbing the surface of the soil, very little after is lost by evaporation. Mulching checks germination of weeds.

Merits:

- Mulching helps in conserving soil moisture.
- Mulching helps in moderising of soil temperature.
- Adds organic matter in the field

.Demerits:

- Mulching is liable to fire out break.
- Mulching also facilitates the hibernation of rodents and pests.

FRUIT-BASED AGROFORESTRY SYSTEMS FOR DRYLANDS

Crop production on drylands in general and marginal rainfed lands in particular results in low and unstable, and often times uneconomical yields because of aberrant monsoon behavior. Poor management of marginal lands results in land degradation. It is estimated that nearly 65 m.ha out of 98 m.ha under rainfed cultivation are facing some kind of land degradation or the other. These marginal lands are not able to sustain arable crops particularly during the drought years. Therefore there has been thinking to develop some alternate land use systems for these lands. Tree component in dryland agriculture increases production and income, besides imparting stability to the farming system. Fruit trees, apart from the above advantages also yield valuable byproducts like fodder, and fuel wood, through their annual prunings and fruits, which are supposed to improve and maintain good health of human beings.

Among the alternate land use system developed, dryland horticulture and agri-horticulture system (fruit based agroforestry systems) are readily picked up by the farmers due to cash benefits derived from these system.

Agri-horticulture system or horti-pastoral system is one form of agroforestry where the tree component is a fruit tree. Agroforestry is a collective term for systems of land use in which woody plants (trees and shrubs) are deliberately combined on the same land management unit with herbaceous crops in some form of spatial arrangement.

Advantages of Fruit-based agroforestry systems

This system mainly focuses on higher income per unit area. Well-maintained and established orchards bring better returns than field crops from the same piece of land. The farmer can practice intercropping during the early stages of the fruit trees. The system is helpful in generating more employment especially during the off-season.

Fruit growing also provides scope for ancillary industries like fruit processing, canning, preservation, dehydration, essential oils, package transport and refrigeration. Fruits are protective foods, necessary for the maintenance of human health. Many fruits are a rich source for energygiving carbohydrates. Besides, fruits are a valuable source of minerals, vitamins and enzymes, which are required for better functioning of human body and for fighting diseases. Many fruits are also known to possess specific medicinal values. Pectin and cellulose found in several fruits stimulates intestinal activity in human body.

Fruit trees are deep rooted and hardy. These provide minimum guaranteed returns even in drought years (atleast fodder and wood) when arable crops fail. Trees utilize off-season rainfall in addition to the rainfall during the cropping season.

Some of the key elements for successful raising of dryland fruit trees are as follows.

Selection of fruit species

Success of fruit trees on rainfed land depends on selection of fruit species for a given soil and climate. Fruit tree species like

1. **Custard apple (*Annona squamosa*) (Variety, Balanagar):** It prefers dry climate. It can withstand mild frost. The root system is confined to relatively shallow layers and therefore, these do not require deep soils. However, it needs well-drained soil. The trees withstand high amounts of lime found in calcareous soils. It is one of the most drought tolerant fruit trees in India. It is grown in rocky soils. Fruits contains up to 16.5% sugar. Custard apple is free from pests and diseases except mealy bugs.
2. **Amla (*Emblica officinalis*):** Amla is becoming a commercially important fruit crop. It is also known as amla, amlaki and usiri. It is rich in vitamin C. The fruits are made into murrabba, sauce, candy, dried chips, jellies, pickles and powder. Used extensively in Ayurvedic and Unani medicines. Important varieties are NA 10 (Balwant); NA 7 (Neelam), NA 6 (Amrit), NA 4 (Kanchan), Chakaiya, & Banarasi.
3. **Ber (*Zizyplus mauritiana*):** It is a drought hardy tree and can be grown purely as rainfed crop. It can even tolerate alkalinity and slight water logging. It has very good nutritious value, comparable to that of apple. It yields sufficient fodder and fuel every year. Processed products viz. candy, juice, squash, jam & jellies. Varieties: Gola, Seb, Umran. Umran has better keeping quality over others.
4. **Tamarind (*Tamarindus indica*):** Known for their drought resistance and services as insurance trees for the marginal and poor farmers in dry tracts. Bears heavily in alternate years like upto 80 years. The seeds are rich in protein and a good source of tannin. Varieties: PKM –1, Pratishthan and Anantapur local)
5. **Jamun (Syzygium cumini):** Varieties: Erkad and Rajmany. All the five fruit species can be grown successfully in moderately deep soils (100 cm soil depth) with an annual rainfall of 600-750 mm. The other fruit species like

guava (Allahabad safed, Lucknow 49), Pomegranate (Ganesh, Muskat, Jyothi and Dolka), Mango (Amrapali, Dassheri, Kesari, Neelam and Benishan), Phalsa (Long & dwarf) and cashewnut can be planted in soils with 150-200 cm depth in regions receiving annual rainfall above 750-1100 mm.

Planting time: Procure seedling / grafts for respective fruit species either from Government nurseries or approved reputed plant nurseries. Planting of fruit seedlings / grafts has to be in early rainy season (July) when soil profile gets charged fully. Seedlings should be healthy, uniform in size and of good quality. After correct root placement good compaction of soil around the seedling has to be done to eliminate air pockets.

Establishment techniques: The planting site for fruit trees should be good in soil depth and well drained. Clear the site from wild bushes and make weed free with appropriate tillage. Align key lines of 0.1 m2 cross section across the slope. The distance between key lines varies with slope and type of species to be planted. Normally fruit trees like custard apple, ber are to be planted at 5-6 m from plant to plant within and also in between rows. Mango, Amla and Jamun are to be planted at 8-10 m distance. Put sticks along the key line at a point where sapling is to be planted. On hilly or undulating lands 68 planting fruit trees along the contour is desirable. Beyond 6% slope contour trenches with mild gradient is useful.

Pit size and filling: Pits of 60 cm diameter and 60 cm depth are to be dug in summer season. Burn the dried leaves in the pit to destroy the egg masses of termites and root grubs.

Microsite improvement: Rainfed lands generally cannot support high biomass associated with fruit tree cultivation. But this can be overcome to some extent by microsite improvement. The planting site has to be improved by suitable amendments. This involves mixing of tank silt or black soil, FYM and native soil in 1:1:1 ratio by volume in case of red soils. About 100 g of DAP and 100 g of 10% Lindane powder per pit are to be added for controlling root grub attack. Placing bentonite clay upto 10-15 cm from bottom of the pit is useful to increase water holding capacity of the site. The amendments are filled in the pit and sticks are placed where sapling is to be planted.

In-situ moisture conservation: The survival of fruit seedlings can be improved by shaping the surface in immediate vicinity of fruit tree to collect the runoff into the root zone.

- On leveled flat lands, 'Saucer' shaped depressions of 15 cm depth at center and 1.0 m diameter are to be made.
- On mild slopy lands (2-4%) 'V' shaped depression with 15 cm depth and 1.0 m dia are to be made. The seedlings has to be planted near the apex of 'V'.
- On slopy lands (4-6%) series of ridges are to be constructed with varied shapes (horse shoe or half moon or fish bone on gentle gradient to intercept and store rain water.

Mulching with small stone pebbles or paddy husk on the surface of the basin with 5-10 cm thickness will be useful to reduce moisture losses and better weed control.

Water management: There is need to protect the fruit seedlings during summer for better survival at least in the first year through protection irrigation. The limited water collected from runoff in ponds or in other sources can be effectively utilized through pitcher and portable drip system.

Pitcher system of irrigation: Earthen pots of 8-10 lit capacity having a wick in a hole made above bottom of the pot have to be placed at 30 cm depth and 15 cm away from the seedling in the basin at the time of planting the seedling. After filling with water, cover the pot with lid. Water drops near to the root dropwise through the wick, which will be useful for the growth of the seedling. The refilling of water in the pot can be done at intervals of 7-10 days in summer.

Management practices: Two weedings in a year are essential to improve the survival and growth of fruit seedlings. Staking with sticks in rainy season is desirable to avoid wind damage and also for straight growth of seelings. Tillage on subsequent summer monsoon shower facilitates better moisture conservation and weed control. Application of manure and fertilizer as recommended are to be followed regularly for maintaining fruit trees in healthy condition.

Agroforestry system with fruit trees: Fruit trees have long gestation period of 4-5 years to get income. Till they develop canopy, the interspaces can be conveniently used for cultivation of surface crops like, groundnut, greengram, blackgram, cowpea, clusterbean, horsegram. The yield from intercrops (400 – 500 kgs) in an additional income and can be obtained upto 2-3 years.

The systems are to be chosen as per land capability i.e. arable and non-arable lands.

For arable lands, the best suited is Agri-Horticulture system

- **Agri.Horticulture system:** This is food-cum-fruit system.
- **Fruit trees:** Custard apple, Ber, guava, Mango, Jamun and Amla.
- **Spacing:** 5-6 m (custard apple, Ber, Guava) 8-10 m (Mango, Jamun, Amla)
- **Under crops:** Horsegram, clusterbean, greengram, blackgram, cowpea, groundnut
- **Yield obtained:** 400 – 500 kg/ha/year.
- **Fruit trees:** Yield fuel wood of 1 to 1.1 tons from prunings in case of Ber.
- **Nutrient management:** Fruit trees are to be provided with plant nutrients on annual basis.

1. **Custard apple:** These plants are to be given @ 250 g N + 125 g P_2O_5 + 125 g K_2O per plant per year for good growth and yield.
2. **Ber:** Annual fertilizers are to be applied. Nitrogen dose may be supplied through organic and inorganic forms 50% each as follows:

Year	Fertilizer (g/tree)			
	FYM (kg/tree)	N	P	K
1st Year	10	100	50	50
2nd year	20	200	100	100

3rd year	30	300	150	150
4th year	40	400	200	200
5th year	50	500	250	250

3. **Guava:** A five year old guava plant needs about 50 kg FYM + 210 g N + 160 g P_2O_5 + 300 g K_2O per tree per year. Applied in two splits i.e. once in June-July & January – February.
4. **Mango:** for adult trees i.e. above 5 years

   N 1-2 kg per/year

   P_2O_5 2-4 kg per/year

   K_2O 1/2-1 kg per/year

   For young 1 year old trees, apply 1/8 of the dose.

   2 year " 1/4 "

   3 year " 1/2 "

   4 year " 3/4 "

   5 year " Full dose
5. **Jamun:** For grown up trees: 500 g N + 600 g P_2O_5 + 300 g K_2O 70
6. **Amla:** At young stage apply: 680-900 g N + 160-200g P_2O_5 + 900-1000 g K_2O/plant/year + FYM 15-20 kg. For mature plants, FYM 30-40 kg in two split doses i.e. April-May & Sept. – October
7. **Undercrops:** All the crops mentioned are leguminous crops. Apply 10 kg N + 30 kg P_2O_5/ha.

The recommended varieties may be same as per the spacing recommended. Intercultivation and weeding operations are to be carried out and plant protection as required i.e. crop management as in its sole crop cultivation.

Plant protection

- **Custard apple:** No serious pests except mealybug. This does not harm the quality of the fruit.
- **Amla:** Seriously damaged by bark borer and Mealybug.
- **Jamun:** Bark borer: The excreta of the larva may be removed with a sharp instrument and cleaned with brush followed by spraying Nuvan @ 1 ml/1 litre or Endosulphan @ 2 ml / 1 litre water.

This is very serious on Amla. To control the pest: (1) summer ploughing, (2) prune the branches touching the ground, (3) apply 2% folidol dust in the basin of the tree, (4) roll alkathene sheet/400 guage on the trunks of the tree upto 1 foot in the months of Nov-December, (5) spray carbaryl 50 W.P. @ 4 g/litre of water or chlorpyriphos @ 2.5 ml/litre of water on tree trunks, branches, twigs thoroughly.

Horti-Pastoral system: Two tier model fruit-cum-fodder based agroforestry system.

Fruit-Trees:

1. Tamarind, jamun, amla, wood apple (8-10 m)
2. Custard apple, ber (5-6 m)

Pastures: Stylosanthes hamata, horsegram, Stylosanthes scabra, Cenchrus ciliaris, Cenchrus setigerus.

Management of fruit trees

- **Tamarind:** The trees are known for their drought resistance and serve as insurance for the marginal and poor farmers in dry tracts. These can be planted in non-arable lands. Trees bear heavily in alternate years and live up to 80 years. The seeds are rich in protein and good source of tannin. Sweet and sour types are available. Yield between 250-1000 kg of ripe pods in a year.
- **Wood apple:** This is one of the slow growing trees found mostly in forests. Propagated by seedlings. Trees attain 40 feet height in ten years. This is a drought tolerant tree. Wood apple has excellent keeping quality.
- **Pastures:** *Stylosanthes hamata*: A legume pasture.
- **Seeding & Fertilization of *Stylo hamata*:** Good soil working leading to thorough seed bed preparation would ensure excellent results. However good results are obtained by light soil working or merely scratching the soil with any light agricultural implement. A seed rate of 4-6 kg/ha mixed with wet soil 10 times of its volume would ensure good establishment when broadcast with the onset of monsoon. Though it establishes and grows well in poor marginal soils, it responds well to the application of single superphosphate (30 kg P_2O_5/ha or 187.5 kg of single super phosphate per ha) normally recommended and should be broadcast before sowing). It is non-specific in rhizobium requirement and nodulates freely with wide range of cowpea rhizobium strains.

Stylosanthus scabra is a shy germinator and scarification of seeds is must. Scarification could be done by rubbing the seeds with coarse river sand.

Both the stylo species viz. *S. hamata* and *S. scabra* grow successfully in association with *Cenchrus ciliaris*, *C. setigerus* etc.

Seeding and fertilization of *Cenchrus ciliaris*

In a well ploughed soil, a seed rate of 5-6 q/ha is broadcast just before the start of monsoon rain. *C. ciliaris* can also be successfully established through transplantation of root slips at 50 cm x 50 cm on a rainy day during monsoon season. It responds well to fertilization. Highest yield is obtained with 40 kg N + 20 kg P_2O_5 /ha applied once before the establishment.

Fruit trees recorded better survival and growth in association with stylo-cenchrus.

- **Yields:** Stylo 6.0 t /ha – Cenchrus 6.6 t/ha. Thus these fruit based systems are useful in stabilizing income of the dryland farmer. Through this system food-fruit-pasture are obtained.

14 TOOLS FOR DETERMINING PRODUCTION AND EFFICIENCIES IN CROPPING AND FARMING SYSTEM

1. **Productivity (Productivity per unit area):** To estimate the productivity of a component and compares with the crop component expressed in terms of equivalent crop yield. Further the production estimation itself varies among the interlinked animal component in IFS i.e. Rice based farming system Productivity in term of grain yield can be recorded and expressed as kg of Grain Equivalent Yield (GEY),

$$GEY = \frac{[\text{Productivity of component/intercrop (kg)} \times \text{Cost of component/intercrop (Rs/kg)}]}{\text{Cost of main crop (Rs/kg)}}$$

2. **Economic analysis:** Parameters like cost of cultivation/production, gross and net return and per day return can be worked out and expressed as Rs/ha.
3. **Employment generation:** Labour required for various activities in crop production given as man days/ha/year (A man working for 8 hours in a day is considered as one man day; A woman working for the same period is treated as 2/3 man day and computed to man days).
4. **Productivity of Livestock Components:** Milk (per day or lactation), dung, urine etc

   Fisheries: Fish weight recorded at harvest and expressed as kg/unit area.

   Poultry: Egg production per day from birds and expressed as total number per month/year from the unit.

   Pigeon: Growth rate at monthly interval and weight at the time of disposal recorded and expressed as kg/unit.
5. **Mushroom:** Yield per day and total yield per year from the unit.
6. **Water requirement:** Water requirement for varying component linkages in the IFS expressed in ha^{-cm}.
7. **Residue addition:** The quantity of residue available from each component (kg). Potential residue addition in terms of N, P and K.
8. **Energy efficiency:** Energy input and output was worked out for individual components based on the input and output energies and energy efficiency.
9. **Nutritive value:** Nutritive value in terms of carbohydrates, proteins and fat (kg).

ASSESSMENT OF YIELD ADVANTAGE IN CROPPING SYSTEMS

Land Equivalent Ratio (LER): Given by Willey (1979)

LER can be defined as the relative land area under sole crop that would be required to produce the equivalent yield under a mixed or an intercropping system at the same level of management.

LER can be mathematically represented as follows:

$$\text{LER} = \frac{\text{Wa} \times \text{WB} + \text{Wb} \times \text{WA}}{\text{WA} \times \text{WB}}$$

Where,

Wa and Wb are the yield per unit area of the component crop A and B in a mixture and WA and WB are their yields when grown alone.

LER = More than 1 indicates yield advantage

Equal to 1 indicates no grain or no loss and

Less than 1 indicates yield loss.

It can be used both for replacement and additives series of intercropping.

Relative yield total (RYT): Given by De wit or Land van Den Bergh (1965)

The mixture yield of a component crop expressed as a portion of its yields as a sole crop from the same replacement series is the relative yields of the crop and sum of relative yields of component crop is called Relativc yield total (RYT).

$$\text{RYT} = \frac{Y_{ab} + Y_{ba}}{Y_{aa} + Y_{bb}}$$

Where,

Y_{aa} = Yields of component a as sole crop

Y_{bb} = Yields of component b as a sole crop

Y_{ab} = Yelds of component a as intercrop in b

Y_{ba} = Yield of component b as intercrop in a

Relative Crowding Coefficient (K or RCC): It is proposed by De Wit (1960). It is used in replacement series of intercropping. It indicates whether a species or crop when grown in mixed population has produced more or less yield than expected in pure stand. In 50:50 mixture Relative crowding coefficient can be defined as –

$$K_{ab} = \frac{\text{Mixture yield of a}}{\text{Pure yield of a} - \text{Mixture yield of a}}$$

$$K_{ab} = \frac{Y_{ab} \times Z_{ba}}{(Y_{aa} - Y_{ab}) \times Z_{ab}}$$

Where,

Y_{aa} = Pure stand yield of species a

Y_{bb} = Pure stand yield of species b

Y_{ab} = Mixture yield of species a (in combination with b)

Y_{ba} = Mixture yield of species b (in combination with a)

Z_{ba} = Sown proportion of species a (in mixture with b)

Z_{ba} = Sown proportion of species b (in mixture with a)

Effective Land Equivalent Ratio (ELER): Mead and Willey (1980) showed that any required ratio could be achieved by growing the intercrop of the land area and one of sole crops the remainder.

$$P = \frac{(La + E)}{LER + E)}$$

LER is a measure of the net advantages from the combined intercrop determined by adding the crop and sole crop relative yields.

$$ELER = \frac{Lb}{((1 - La) + (LER - 1)\times P)}$$

Staple Land Equivalent Ratio (SLER): In situation, where the primary objective is to produce yields of one component (staple crop) usually a cereal and some yields of the legume, the concept of SLER is composed.

$$SLER = \frac{MDA}{SA} + \frac{\{MB\}}{SB}$$

Where, MDA is derived yield of A in mixture yield and P, the proportion of land devoted for intercropping. The two crop yields in the intercrop mixture are MA and MB and yield of crops grown as sole crop are SA and SB.

Land Equivalent Coefficient (LEC):It was proposed by Adetilaye and Ezedinma (1986).

$$LEC = \frac{MA \times MB}{SA \times SB}$$

It can therefore, be regarded as a measure of association or interaction when crop with the strength of relationship for a two crop mixture the minimum expected product by yield advantage is obtained and LEC greater than 0.25 (25%). This was obtained from the 50:50 yields where interspecific competition equaled intraspecific competition.

$$LEC = LA \times LB = 0.5 \times 0.5 = 0.25$$

The LEC is developed to assess the interaction and potential of crop mixture. It is derived from the understanding that in intercrop yields, as compare optimum sole crop yield is due competition assuming that the inter competition is absent.

Area Harvest Equivalency Ratio (AHER): This was proposed by Bal Subramanian and Sekayange (1990).It indicates resources efficiency. The concept of AHER combines the areatime factors in a practical sense quantifying intercrop yield advantage particularly in multi season.

Crop Performance Ratio (CPR): It is defined as the productivity of an intercrop per unit area of ground compared was expected from sole crops sown in the same proportions (Ali *et al*, 1990). Then productivity in the intercrop can be expressed as a partial CPR.

Aggressivity: Proposed by Mc Gilchroist (1965), it is the mixture of how much the relative yield increase in component a is greater than that for component b

$$A_{ab} = \frac{Y_{ab}}{Y_{aa} \times Z_{ab}} - \frac{Y_{ba}}{Y_{bb} + Z_{ba}}$$

$$A_{ab} = \frac{\text{Mixture yield of a}}{\text{Expected yield of a}} - \frac{\text{Mixture yield of b}}{\text{Expected yield of b}}$$

Z_{ba} = Proportion of intercropped area initially allocated to crop "a"

Z_{ab} = Proportion of intercropped area initially allocated to crop "b"

A_{ab} = 0 : component crop equally competitive

A_{ab} = –ve : means dominated

Aggressivity and crowding coefficient can be used only in replacement series where as LER in both replacement and additive series

Competition Index: It is the measure to find out the yield of various crops when grown together as well as separately (Donald, 1963). It indicates the yield per plant of different crops in mixture and their respective pure stand on unit area basis.

If the yield of any crop grown together is less than its respective yield in pure stand then it is harmful association but on increased yield means positive benefit

$$Cl = \frac{(Y_{aa} - Y_{ab}) \times (Y_{bb} - Y_{ba})}{Y_{aa} \times Y_{ab}}$$

Competition Ratio (Willey and Rao, 1980): It is a measure of intercrop competition to indicate that the number of times by which one component crop is more competitive than the other.

$$CRa = \frac{Y_{ab}}{Y_{aa} \times Z_{ab}} \div \frac{Y_{ba}}{Y_{bb} \times Z_{ba}}$$

$$= \left(\frac{Y_{ab}}{Y_{aa}} \div \frac{Y_{ba}}{Y_{bb}}\right) \times \frac{Z_{ba}}{Z_{ab}}$$

$$= \frac{LER_a}{LER_b} \times \frac{Z_{ba}}{Z_{ab}}$$

Competition Coefficient (CC): It is used to find out the relative crowding from which maximum yield can be obtained without any adverse effect on any of the species.

$$CC = \frac{\text{RCC of given species}}{\text{Total RCC of all crops in mixture}}$$

Area Time Equivalent Ratio (ATER): ATER is the ratio of number of hectare-days required in monoculture to the number of hectare- days used in intercropping to produce identical quantities of each of the component crop. Used for comparing the productivity of crops in mixture or monoculture.

$$\text{Ater} = \frac{(RY_a \times t_a) + (RY_b + t_b)}{T}$$

RY_a = Relative yield of species "a" = yield of intercrop / yield of sole crop = Y_{ab}/Y_{aa}

$RY_b = Y_{ba}/Y_{bb}$

t_a = duration of species "a" days

t_b = duration of species "b" days

T = duration of intercropping system

Crop Equivalent Yield (CEY): The yields of different crops are converted into equivalent yield of any one crop based on price of the produce which gives an idea which crop is more economical.

$$\text{CEY} = \frac{\text{YIELD (q/ ha) of crop "b"} \times \text{Price of crop "b" (Rs/ q)}}{\text{Price of a crop into which CEY is to beconverted}}$$

IMPORTANT MULTIPLE CROPPING INDICES TO EVALUATE THE CROPPING SYSTEMS

Land Use Efficiency or Assessment of Land Use: The main objective is to use available resources effectively.Multiple cropping which include both inter and sequential cropping has the main objective of intensification of cropping with the available resources in a given environment.Several indices have been proposed to compare the efficiencies of different multiple cropping system in turns of land use, and these have been reviewed by Menegay et al. 1978.

Multiple Cropping Index or Multiple Cropping Intensity (MCI):

It is the ratio of total area cropped in a year to the land area available for cultivation and expressed in percentage.

$$\text{MCI} = \frac{ai \times 100}{A}$$

Where i = 1, 2, 3, n, n = total number of crops, ai = area occupied by crop and A= total land area available for cultivation. Or MCI is the sum of area planted to different crops and harvested in a single year divided by total cultivable area and expressed as percentage. Or MCI means the sum of areas under various crops raised in a single years divided by net area available for that cropping pattern multiplied by 100. It is similar to cropping intensity.

$$\text{MCI} = \frac{\text{Total number of crops + with their respective area} \times 100}{\text{Net cultivable area}}$$

Cultivated Land Utilization Index (CLUI): It is calculated by summing the products of land area to each crop, multiplied by the actual duration of that crop divided by the total cultivated land times 365 days (Chuang, 1973).

$$\text{CLUI} = \frac{aidi \times 100}{A \times 365}$$

Where, i = 1, 2, 3, n, n = total number of crops. ai = area occupied by the ith crop, di = days that the ith crop occupied ai and A = total cultivated land area available for 365 days.

CLUI can be expressed as a fraction or percentage. This gives an idea about how the land area has been put into use. If the index is 1 (100%), it shows that the land has been left fallow and more than 1, tells the specification of intercropping and relay cropping. Limitation of CLUI is its inability to consider the land temporarily available to the farmer for cultivation.

Crop Intensity Index (CII): Crop intensity index assesses farmer actual land use in area and time relationship for each crop or group of crops compared to the total available land area and time, including land that is temporarily available for cultivation. It is calculated by summing up the product of area and duration of each crop divided by the product of farmers, total available cultivated land area and time periods plus the sum of the temporarily available land area with the time of these land areas actually put into use (Menegay *et al.*, 1978). The basic concept of CLUI and CII are similar. However, the latter offers more flexibility when combined with appropriate sampling procedure for determining and evaluating vegetable production and cropping pattern data.

$$\mathrm{CII} = \frac{\mathrm{Ai.ti}}{\mathrm{AoT} + \mathrm{M(AiTj)}}$$

Where, I = 1, 2, 3……NC, NC = total number of crops grown by a farmer during the time period. T, ai = area occupied by ith crop (months that the crop I occupied an area ai) T = time period under study (usually one year), A_0 = Total cultivated land area available with the farmer for use during the entire time period, T, M= total number of fields temporarily available to the farmer for cropping during time period. T, j=1, 2, 3…….. M, Aj=land area of jth field and Tj= time period Aj is available. When, CII = 1 means that area or land resources have been fully utilized and less than 1 indicates under of resources. CII indicates the number of times a field is grown with crops in a year. It is calculated by dividing gross cropped area with net area available in the farm, region or country multiplied by 100.

$$\mathrm{CII} = \frac{\text{Gross cropped area} \times 100}{\text{Net area}}$$

When long duration crop grown crop remain longer time in field this is the drawback of CII. So time is not considered, thus , when long duration crops like sugarcane and cotton are in grown, the cropping intensity will be low, though crop remain longer period of time in the field.

Simultaneous Cropping Index (SCI): It is computed by multiplying the HDi with 10,000 and dividing the product by MCI (Strout, 1975).

$$\mathrm{SCI} = \frac{\mathrm{HDI} \times 10{,}000}{\mathrm{MCI}}$$

Relative Cropping Intensity Index (RCII): It is the modification of CII and determines the amount of area and time allotted to 1 crop or groups of crop related to area- time actually used in the production of all crops. RCII numerator equal SCII denominator and RCII denominator equal CII numerator.

$$\text{RCII} = \frac{\text{ak.tk}}{\text{ai.ti}}$$

These indices can be used for classifying (i.e a farmer with relative vegetable intensity index 50% would be considered a vegetable grower.) measuring shifts of various crops among farm of different sizes and determining whether the consistent types of cropping pattern occur within various farm size strata. These indices also held to know how intensity cultivated land, area has been utilized. But none of these indices takes productivity into account and cannot be used for comparing different cropping systems and evaluating their efficiency in utilization of the resources other than the land.

Diversity Index (DI)

It was suggested by Strout (1975). It measures the multiplicity of crops or farm products which are planned in a single year by computing reciprocal sum of squares of the share of gross revenue received from each individual farm enterprises in a single year.

$$\text{DI} = \frac{\text{N}}{\text{Yi}}$$

Where,

N = Total number of enterprises (crops or farm products), and

Yi = Gross revenue of the ith enterprises produced within a year.

Harvest Diversity Index (HDI)

It is computed using the same equation as the DI expects that the value of each farm enterprises is replaced by the value of each harvest.

$$\text{HDI} = \frac{1}{\text{Yi}}$$

where,

Yi = Gross value of ith crop planted and harvested within a year.

Crop Performance Ratio and its Types

It is defined as the productivity of an intercrop per unit area of ground compared was expected from sole crops sown in the same proportions (Azam Ali *et al.* 1990). For each same then productivity in the intercrop can be expressed as a partial CPR.

Economic Viability

The indicators like CEY, LER, RYT, etc. give biological suitability of cropping system to an area. At the same time, cropping system should be economically viable and profitable following economic indicates can be used to evaluate profitability of cropping system.

(i) **Gross Returns:** The total monetary returns of the economic produce such as grain, tuber, bulb, fruit etc. And byproducts viz. straw, fodder, fuel, etc.

obtained from the crops included in the system are calculated based on the local market prices. The total return is expressed in terms of unit area, usually one hectare.

The main drawback in this calculation is that the market price of the produce is higher than that actually obtained by the farmer. Generally, gross return calculated is somewhat inflated compared to the actual receipt obtained by the farmer.

(ii) Net Returns or Net Profit: This is worked out by subtracting the total cost of cultivation from the returns. This value gives the actual profit obtained by the farmer. In this type of calculation only the variable cost are considered. Fixed costs such as rent for the land, land revenue, interest on the capital etc. are not included. For a realistic estimate, however, fixed costs should also be included.

(iii) Return for Rupee Invested: This is also called benefit-cost ratio or input-output ratio.

$$\text{Return per rupee invested} = \frac{\text{Gross return}}{\text{Cost of cultivation}}$$

This index provides an estimate of the benefit derived and expenditure incurred by the farmer in adopting a particular cropping system. Anything above the value of 2.0 (meaning that the farmer can get Rs. 2 as return for every rupee invested) can be considered worthwhile.

(iv) Per Day Return: This is called as income per day and can be obtained by dividing the net return by number of cropping period (days).

$$\text{Per day return} = \frac{\text{Net returns}}{\text{Cropping period (days)}}$$

This gives the efficiency of the cropping system in terms of monetary value. If the system is stretched over one year, the denominator can be replaced by 365 days and per day for the whole year can be calculated.

No single index is capable of giving good comparison of different cropping system and so a number of indices are used together to assess the economic viability of the system.

REFERENCES

Agbonlabor, M. U., Aromolaran, A. B. and Aiboni, V. I. 2003. Sustainable soil management practices in small farms of Southern Nigeria: A poultry-food crop integrated farming approach. *Journal of Sustainable Agriculture*, **22**: 51-62.

Agelet A., Bonet M.A. and Valles J. 2000. Homegardens and their role as a main source of medicinal plants in mountain regions of Catalonia (Iberian Peninsula). *Econ Bot* **54**: 295-309.

Anonymous 2008. Fishery statistical yearbook of Bangladesh 2007-08. Fisheries survey system, Department of Fisheries, Bangladesh.

Balasubramanian M.A. 1986. Yongchak as shade tree. *Planters' Chronicle* **80**(4): 134.

Balusamy M, Shanmugham PM, Baskaran R. Mixed farming an ideal farming. *Intensive Agric.* **41** (11–12):20–25.

Balusamy, M., Shanugham, P. M. and Bhaskaran, R. 2003. Mixed farming- An ideal farming. *Intensive Agri.*, **41**(11-12): 20-25.

Banik P and Sharma R.C. 2009 Yield and resource utilization efficiency in baby corn-legume-intercropping system in the Eastern Plateau of India. *J Sustain Agric* **33**:379-95.

Bargaza A, Noyceb G L, Fulthorpe R, Carlsson G, Furze J R, Jensen E S, Dhiba D and Isaac M E 2017 Species interactions enhance root allocation, microbial diversity and P acquisition in intercropped wheat and soybean under P deficiency. *Appl Soil Eco* **120**:179-88.

Basavaraj, N. B. 2008. A study on impact of income generating activities on sustainable rural livelihoods of KAWAD project beneficiaries. M. Sc. Thesis, University of Agricultural Sciences, Dharwad.

Behera, U. K. and Mahapatra, I. C. 1999. Income and employment generation for small and marginal farmers through integrated farming systems.*Indian Journal of Agronomy*, **44**(3): 431-439.

Biswas B C 2010 Farming System Approach to Improve IUE, Employment and Income in Eastern India. *Fertiliser Marketing News* **41**: 6-12.

Bosma R H, Udo H M J, Verreth J A J, Visser L E and Nam C.Q. 2005. Agriculture Diversification in the Mekong Delta: Farmers' Motives and Contributions to Livelihoods. *Asian Journal of Agriculture and Development* **2**: 49-66.

Brar A S and Walia U S 2010 Rice residue position and load in conjunction with weed and wheat (*Triticum aestivum* L.). *Indian J Weed Sci* **42**:163-7.

Brar A S, Gill M S, and Walia S S 2007 Productivity, profitability and stability of rice based cropping systems at cultivators' fields under assured input conditions. *J Res* **44** (2):87-89.

Brown C, Waldron S, Yuman L and Longworth J 2009. Forage-livestock policies designed to improve livelihoods in Western China: A critical review. *China Agric Econ Rev* **1**: 367-381.

Ceccolini, L. 2002. The homegardens of Soqotra island, Yemen: an example of agroforestry approach to multiple land-use in an isolated location. *Agroforest Syst* **56**: 107-115.

Chahal P S, Brar H S and Walia U S 2005 Soil seed bank dynamics of Phalaris minor in relation to different cropping systems. *J Res Punjab Aric Univ* **42** (1): 13-18.

Chand, R., Prasanna, P. A. L. and A. Singh. 2011. Farm size and productivity: Understanding the strengths of smallholders and improving their livelihoods. *Economic and Political Weekly*, **46** (26 & 27): 10-14.

Chandler P. 1994. Adaptive ecology of traditionally derived agroforestry in China. *Hum Ecol* **22**: 415-442.

Channabasavanna A S and Biradar D P 2007 Relative performance of different rice-fish-poultry integrated farming system models with respect to system productivity and economics. *Karnataka J. Agric Sci* **20** (4): 706-709.

Channabasavanna A.S, Biradar D P, Prabhudev K N and Mahabhaleswar Hegde 2009 Development of profitable integrated farming system model for small and medium farmers of Tungabhadra project area of Karnataka. *Karnataka J. Agric. Sci.*, **22**:25-27.

Channabasavanna A S, Itnal C J and Patil S G 2002 Productivity, economic analysis and changes in physicochemical properties of soil as influenced by integrated rice (*Oyrza sativa*) based farming systems. *Indian Journal of Agronomy* **47**: 1-5.

Chhokar R S, Sharma R K, Pundir A K and Singh R K 2007 Evaluation of herbicides for control of *Rumex dentatus*, *Convolvulus arvensis* and *Malva parviflora. Indian J Weed Sci* **39:** 214-8.

Chhokar R S, Singh S, Sharma R K and Singh M 2009 Influence of straw management on *Phalaris minor* Retz. control. *Indian J Weed Sci* **41:**150-6.

Chnnabasavanna AS, Biradar DP 2007. Relative performance of different rice-fish-poultry integrated farming system models with respect to system productivity and economics. Karnataka *J Agric Sci*. **20** (4) : 706-709.

Choudhary S K, Singh R N, Upadhyay P K, Singh R K, Yadav R I and Choudhary H R 2014 Growth and available nutrient in winter maize (*Zea mays* L.) with vegetable intercrops in Eastern Uttar Pradesh. *The Bioscan* **9**:151-4.

Choudhary V K and Kumar P S 2014 Influence of mulching on productivity, root growth and weed dynamics of maize (Zea mays)-based cropping systems. *Ind J Agron* **59**(3): 364-70.

Colecchia S A, Rinaldi M, and De Vita P 2015 Effects of tillage systems in durum wheat under rainfed Mediterranean conditions. *Cereal Res Commun* **43:** 4.

Crossman S.M.A., Palada M.C and Davis A.M. 1999. Performance of West Indian hot pepper cultivars in the Virgin Islands. Proceedings Caribbean Food Crops Society **35**: 169-176.

De Silva T. 1997. Industrial utilization of medicinal plants in developing countries. p. 38-48. In G. Bodeker, K.K.S. Bhat, J. Burley and P. Vantomme (eds.) Medicinal Plants for Forest Conservation and Healthcare. Non-Wood Forest Products No. 11, FAO, Rome.

Devasenapathy, P., Mytswamy, V., Christopher Louduraj, A. and Rabindran, R. 1995. Integrated farming systems for sustained productivity.*Madras Agricultural Journal*, **82**: 306-307.

Dobermann A and Fairhurst T H 2002 Rice straw management. Better crops. Intern 16 (Sp. Supp. May), 7-9.

Dolli, S. S. 2006. Sustainability of natural resources management in watershed development project. Ph. D. Thesis, University of Agricultural Sciences, Dharwad.

Dwivedi, R. P., Tewari, R. K., Kareemulla, K., Chaturvedi, O. P. and Rai, P. 2007. Agrihorticultural system for household livelihood - A case study.*Indian Research Journal of Extension Education*, **7** (1): 22-26.

Edwards, P. 1997. Sustainable food production through aquaculture.Aquaculture Asia. Vol. 2.School of Environment, Resources and Development, Asian Institute of Technology (AIT), Pathumthani, Thailand.

Esther, Shekinah D , Jayanthi C and Sankaran N 2005 Physical Indicators of sustainability–A Farming Systems Approach for the Small Farmer in the Rainfed Vertisols of the Western Zone of Tamil Nadu. *J. Sustainable Agriculture* **25** (3): 43-65.

FAO. 1990. Guidelines for Soil Profile Description. Third edition (revised). Soil Resources, Management and Conservation Service, Land and Water Development Division, FAO, Rome.

Farooq M, Flower K C, Jabran K, Wahid A and Siddique K H M 2011 Crop yield and weed management in rainfed conservation agriculture. *Soil Till. Res.***117**:172-183.

Farrell M J. 1957. The Measurement of Productive Efficiency. *Journal of Royal Statistical Society* 120:253-281.

Fleming E, Villano R, Farrell T and Fleming P 2007 Efficiency and productivity analysis. Final Rep. benchmarking sub-project, Australian sheep industry CRC, Armidale.

Follett J. 1997. Ginseng production in NZ forests: experiences from Tiketere. *New Zealand Tree Grower* **18**(3): 19-21.

Fresco L O and Westphal E 1988 A hierarchical classification of farm systems. *Expl Agric* **24**: 399-419.

Frey R. 1995. Ipomoea carnea ssp. fistulosa (Martius ex Chiosy) Austin: Taxonomy, biology and ecology reviewed and inquired. *Trop Ecol* **36**: 21-48.

Gadi R, Kulshrestha U C, Sarkar A K, Garg S C and Parashar D C 2003 Emissions of SO_2 and NOx from bio-fuels in India. *Tellus B Chem Phys Meteorol* **55:**787-95.

Ganeche C, Swaminathan, C and Vairavan K 2000, Intensive farming system for semi-arid region : Today's indispensable venture. LEISA India, **2** (3): 21-22.

Gangwar B 1993 Farming systems research for accelerating agricultural development in less developed countries – A review. *Agric. Reviews* **14** (3): 149-159.

Gathala M K, Ladha J K, Kumar V, Saharawat Y S, Kumar V, Sharma P K, Sharma S and Pathak H 2011a. Tillage and Crop Establishment Affects Sustainability of South Asian Rice-Wheat System. *Agronomy Journal* **103**: 961-672.

Gathala M K, Ladha J K, Saharawat Y S, Kumar V, Kumar V, Sharma P K 2011b Effect of Tillage and Crop Establishment Methods on Physical Properties of a Medium-Textured Soil under a Seven-Year Rice – Wheat Rotation. *Soil Science Society of America Journal* **75** 1851-1862.

Gautam, U. S., Paliwal, D. K. and Seema, N. 2007. Improvement in livelihood security for small and marginal farmers through front line demonstrations on oilseed and pulse crops in central India.*Indian Research Journal Extension Education*, **7** (1): 1-5.

Getahun Legesse, Marianna Siegmund-Schultze, Girma Abebe and Anne Valle Zarate 2010 Economic performance of small ruminants in mixed-farming systems of Southern Ethiopia. *Trop Anim Health Prod* **42**: 1531-1539.

Ghimire R, Lamichhane S, Acharya B S, Bista P and Sainju U M 2017 Tillage, crop residue, and nutrient management effects on soil organic carbon in rice-based cropping systems: A review. *J Integr Agric* **16:** 1-5.

Gill, M. S. 2004. Methodologies for farming system approach- A case study. In: Proceedings of National Symposium on Alternative Farming Systems held at PDCSR, Modipuram, 16-18 September, 2004, pp. 95-108.

Gokhale S B, Gokhale R B, Phadke N L and Desale R J 2002. Status of village goat management practices in Maharashtra. *Indian J Anim Sci* **72**: 810-814.

Gopinath, K. A., Dixit, S., Chary, R., Srinivasarao, G., Osman, M., Raju, B. M. K., Ramana, D. B. V., Venkatesh, G., Grover, M., Maheswari, M. and Venkateswarlu, B. 2013. Improving the rainfed farming systems of small and marginal farmers in Anantapur and Adilabad districts of Andhra Pradesh.Central Research Institute for Dryland Agriculture, Hyderabad, Andhra Pradesh.46 p.

Goswami, R. and Malay, P. 2012.Using sustainable livelihoods framework for assessing the impact of extension programmes: An empirical study in the context of joint forest management. *Indian Research Journal Extension Education*, **12** (3): 28-36.

Goulding K W T, Jarvis S C J and Whitmore A P 2008. Optimising nutrient management for farm systems. Philosophical Transactions of the Royal Society of London, Series B, **363**, 667-680.

Gupta R. 1986. Integration of medicinal plants cultivation in forest and forest plantations of North-Western Himalaya. p.59-67. In Agroforestry Systems: A New Challenge. Indian Society of Tree Scientists, Solan, India.

Hager N. and Otterstedt J. 1996. Coope San Juan, a farmers' cooperative. Sustainable use of the natural forest-past and future: a minor field study. Working Paper No. 302. International Rural Development Centre, Swedish University of Agricultural Sciences, Uppsala, Sweden. 59 pp.

Herzog F. 1994. Multiple shade trees in coffee and cocoa plantations in Coted'Ivoire. *Agroforest Syst* **27**: 259-267.

Hill D.B. and Buck L.E. 2000. Forest farming practices. p. 283-320. In H.E. Garrett, W.J. Rietveld and R.F. Fisher (eds.) North American Agroforestry: An Integrated Science and Practice. American Society of Agronomy Inc., Madison, WI, USA.

Hu F, Gan Y, Chai Q, Feng F, Zhao C, Yu A, Mu Y, Zhang Y 2016 Boosting system productivity through the improved coordination of interspecific competition in maize/pea strip intercropping. *Field Crops Res* **198**:50-60.

Hua F, Fenga F, Zhaoa C, Chaia Q, Yua A, Yina W and Gan Y 2017 Integration of wheat-maize intercropping with conservation practices reduces CO_2 emissions and enhances water use in dry areas. *Soil Till Res* **169**:44-53.

Huang S, Liao P, Zeng Y, Shang Q, Zeng Y and Shi Q 2017 Interactive effects of liming and straw retention on yield and nitrogen uptake in a double rice-cropping system of subtropical China. *Proceedings of the 18th Australian Society of Agronomy Conference*, 24- 28 September 2017, Ballarat, Australia.

ICRAF 1992. A Selection of Useful Trees and Shrubs in Kenya: Notes on their Identification, Propagation and Management for Use by Farming and Pastoral Communities. International Centre for Research in Agroforestry, Nairobi, Kenya, 226 pp.

Itnal C J, Hundekar S T, Warad S M and Itnal M C 1999 Farming system –a rational approach for sustainable agriculture, pp. 53-73. In. Lecture notes of summer short course on Farming System for sustainable production, Univ. Agric. Sci., Dharwad, 24th May to 2nd June, 1999.

Jagadeeshwara K, Nagaraju Y, Bhagyavathi and Nagaraju K 2011 Livelihood improvement of vulnerable farmers through Integrated Farming Systems of Southern Karnataka. pp: 145-146.

Jain N, Bhatia A and Pathak H 2014 Emission of air pollutants from crop residue burning in India. *Aerosol Air Qual Res* **14**: 422-30.

Jaswal, S.C., Mishra, V.K. & Verma, K.S. 1993, Intercropping ginger and turmeric with poplar (Populus deltoides 'G-3' Marsh.) *Agroforest Syst*. **22**: 111

Jat M L, Malik R K, Saharawat Y S, Gupta R, Bhag M and Raj Paroda 2012. Proceedings of Regional Dialogue on Conservation Agricultural in South Asia, New Delhi, India, APAARI, CIMMYT, ICAR, p 32.

Jat M L, Singh R G, Saharawat Y S, Gathala M K, Kumar V, Sidhu H S and Gupta R 2009. Innovations through conservation agriculture: progress and prospects of participatory approach in the Indo-Gangetic plains. In Pub Lead Papers, 4th World Congress on Conservation Agriculture (pp. 60-64). 4-7 February, 2009, New Delhi India.

Jat M L, Singh S, Rai H K, Chhokar R S, Sharma S K and Gupta R K 2005. Furrow Irrigated Raised Bed Planting Technique for Diversification of Rice-Wheat System of Indo-Gangetic Plains. *Journal of Japan Association for International Cooperation for Agriculture and Forestry* **28** (1): 25-42.

Jayanthi C 2002. Sustainable Farming System and lowland farming of Tamil Nadu. IFS Adhoc scheme. Completion Report.

Jayanthi C, Baluswamy M, Chinnusamy C and Mythily S 2003 Integrated nutrient supply system of linked components in lowland integrated farming system. *Indian Journal of Agronomy* **48**: 241- 246.

Jayanthi C, Mythili S and Chinnusamy C 2002. Integrated farming systems - A viable approach for sustainable productivity, profitability and resource recycling under lowland farms. *J Ecobiol* **14** (2):143-148.

Jayanthi C, Mythili S. and Chinnasamy C 2002 Integrated farming systems – A viable approach for sustainable productivity, profitability and resource recycling under low land farms. *J. Ecobiology* **14** (2): 143-148.

Jayanthi C, Rangasamy A and Chinnusamy C 2002. Integrated nutrient management in rice based cropping systems linked with lowland integrated farming system. *Fertilizer News.* **42** (3): 25-30.

Jayanthi C, Rangasamy A, Chinnusammy C, Purushothaman S and Planiappan S P 1994 Integrated farming systems for smallholdings. *Indian Journal of Agronomy* **39**:1-7.

Jayanthi, C., Rangasamy, A., and Chinnusamy, C. 2000. Water budgeting for components in lowland integrated farming systems. *Agricultural Journal*, **87**: 411- 414.

Jayanthi, C., Vennila, C., Nalini, K. and Chandrasekaran, B. 2009. Sustainable integrated management of crop with allied enterprises- Ensuring livelihood security of small and marginal farmers. Special Feature: Sustainable Agriculture.

Jha G K, Burman R R, Dubey S K and Singh G 2011 Yield gap analysis of major oilseeds in India. *J Comm Mobiliz Sustain Develop* **6** (2):209-16.

Jirli, B., Bhati, D. S. and De, D. 2008. Diversifying cropping system with rapeseed mustard – An approach to sustain livelihood. In proceedings of international seminar on strategies for improving livelihood security of rural poor, Goa, India, pp: 201-202.

Kar, A., Singh, M. and Ray, A. K. 2003. Linkages between crop diversification and poverty in indian states. Agricultural Situation in India. Directorate of Economics and Statistics. Ministry of Agriculture, Government of India, New Delhi.

Korikanthimath V.S. and Hegde R. 1994. Cardamom and arecanut mixed-cropping systems. *Indian Cocoa, Arecanut and Spices J* **18** (4): 109-112.

Korn M 1996. The dyke- pond concept: Sustainable agriculture and nutrient recycling in China. *Ambio* **25** (1): 6-13.

Kumar A, Jat M L, Kumar A, Tomar J, Kumar S and Kushwaha S R 2016 Rice residue management in wheat under different tillage practices and nitrogen doses. *Ann Agric* Res **37:** 49-55.

Kumar A., Bisht P.S. and Kumar V. 2002. Traditional medicinal plants of Uttaranchal Himalayas. *Asian Agri-History* **6**: 167-170.

Kumar B.M. and Nair P.K.R. 2004. The enigma of tropical homegardens. In: P.K.R. Nair, M.R. Rao, and L.E. Buck (eds). NEW VISTAS IN AGROFORESTRY: A Compendium for the 1st World Congress of Agroforestry 2004. Kluwer Academic Publishers, Dordrecht, The Netherlands.

Kumar K. and Gupta C. 1991. Intercropping of medicinal plants with poplar and their phenology. *Indian Forester*: 535-544.

Kumar S, Sharma K L, Kareemulla K, Chary G R, Ramarao C A, Rao C S and Venkateswarlu B 2011. Techno-economic feasibility of conservation agriculture in rainfed regions of India. *Current Science*. **10**:1171-81.

Kumar, S. and Jain, D. K. 2005. Are linkages between crops and livestock important for the sustainability of the farming system? *Asian Economic Review*, **47** (1):90-101.

Kurothe R S, Kumar G, Singh R, Singh H B, Tiwari S P, Vishwakarma A K, Sena D R and Pande V C 2014 Effect of tillage and cropping systems on runoff, soil loss and crop yields under semiarid rainfed agriculture in India. *Soil and Tillage Research*. **140**:126-34.

Kutu FR 2012 Effect of conservation agriculture management practices on maize productivity and selected soil quality indices under South Africa dryland conditions. *African Journal of Agricultural Research* **7** (26): 3839-46.

Lal R and Miller F P 1990. Sustainable farming system for tropics. In: Sustainable agriculture: Issues and prospective. Singh R P (Ed.). Indian Society of Agronomy, IARI, New Delhi. Vol.I pp.68-89.

Lamont S.R., Eshbaug W.H. and Greenberg A.M. 1999. Species composition, diversity and use of homegardens among three Amazonian villages. *Econ Bot* **53**: 312-326.

Lightfoot C 1990. Integration of aquaculture and agriculture: a route to sustainable farming systems. NAGA: *The ICLARM Quarterly* **13** (1): 9-12.

Lightfoot, C., and Minnick, D. R. 1991. Farmer-first qualitative methods: Farmers diagrams for improving methods of experimental design in integrated farming systems. *Journal for Farming Systems Research and Extension*, **2**: 11-34.

Lindsay Bell, W and Andrew Moore, D 2012. Integrated crop–livestock systems in Australian agriculture: Trends, drivers and implications. *Agric Syst* **111**: 1-12.

Maheswarappa H.P., Hegde M.R. and Nanjappa H.V. 1998. Kacholam (Kaempferia galanga L.) - a potential medicinal-cum-aromatic crop for coconut gardens. Indian *Coconut J* **29** (5): 4-5.

Maheswari S.K., Dhantonde B.N., Yadav S. and Gangrade S.K. 1985. Intercropping of Rauvolfia serpentina for higher monetary returns. *Indian J Agri Sci* **58**: 108-111.

Malik R K, Gupta R K, Singh C M, Yadav A, Brar S S, Thakur T C, Singh S S, Singh A K, Singh R and Sinha RK 2005. Accelerating the adoption of resource conservation technologies in rice wheat system of the Indo-Gangetic Plains. InProceedings of Project Workshop, Directorate of Extension Education, Chaudhary Charan Singh Haryana Agricultural University (CCSHAU) June 1-2, 2005. Hisar, India

Mall U S, Manjhi R P and Thakur R 2014 Intensification and diversification of rice (*Oryza sativa*) based cropping systems for productivity, profitability and water expense efficiency in Jharkhand. *Int J Agric Sci* **10** (1):124-9.

Mangala, B. 2008. Impact of agricultural interventions on socio-economic status of Bharatiya Agroindustries Foundation (BAIF) beneficiary farmers. M. Sc. Thesis, University of Agricultural Sciences, Dharwad, Karnataka.

Melero S, Vanderlinden K, Ruiz J C and Madejon E 2008 Long-term effect on soil biochemical status of a Vertisol under conservation tillage system in semi-arid Mediterranean conditions. *European journal of soil biology* **44** (4):437-42.

Midega C A O, Pittchar J, Salifu D, Pickett J A and Khan Z R 2013 Effects of mulching, N-fertilization and intercropping with *Desmodium uncinatum* on *Striga hermonthica* infestation in maize. *Crop Prot* **44**:44-9.

Mishra A 2014 Effect of winter maize-based intercropping systems on maize yield, associated weeds and economic efficiency. *Comunicata Scientiae* **5**:110-7.

Miyuki Iiyama, Joseph Maitima and Patrick Kariuki 2007 Crop-livestock diversification patterns in relation to income and manure use: A case study from a Rift Valley Community. *Kenya Afr J Agril Res* **2** (3): 58-66.

Mkhabela, T 2005. Technical efficiency in a vegetable based mixed-cropping sector in Tugela Ferry, Msinga district, Kwazulu-Natal. *Agricultural economics research, policy and practice in Southern Africa, Agrekon,* **44** (2): 187-204.

Mullins G 2009 Phosphorus, Agriculture and The Environment. Virginia Polytechnic Institute and State University, Petersburg.

NAAS (National Academy of Agricultural Sciences) 2009. State of Indian Agriculture, New Delhi.

Nadeau I., Oliver A., Semard R.R., Coulobe J. and Yelle S. 1999. Growing American ginseng in maple forests as an alternative land use system in Quebec, Canada. *Agroforest Syst* **44**: 345-353.

Nageswaran M, Selvaganapathy E, Subbiah V R and Nair S 2009 Demonstration and Replication of Integrated Farming Systems at Chidambaram. Report of M.S. Swaminathan Research Foundation (MSSRF), Chennai, pp. 16 -53.

Nageswaran, M., Selvaganapathy, E., Vijay, R. and Sudha, N. 2009. Demonstration and replication of integrated farming systems at Chidambaram, M. S. Swaminathan Research Foundation, Chennai.

Nair G.S., Sudhadevi P.K. and Kurian A. 1989. Introduction of medicinal and aromatic plants as intercrops in coconut plantations. p. 163-165. In S.P. Raychauduri (ed.), Recent Advances in Medicinal, Aromatic and Spice Crops. Today and Tomorrow's Printers and Publishers, New Delhi, India, Vol. 1.

Nair P.K.R. 1993. An Introduction to Agroforestry. Kluwer Academic Publishers, Dordrecht, The Netherlands, 499 pp.

Narayana H and Setty SVS 1977. Studies on the incorporation of mulberry (Morus indica) leaves in layers mash on health, production and egg quality. *Indian Journal of Animal Science* **47**: 212-215.

Narkhede W N, Deshmukh M S and Gangwar B 2015 Alternate cropping system for central plateau zone of Maharashtra. *Ind J Agron* **60**(2):192-6.

NCAER (National Council of Applied Economic Research) 2009. Publications on agriculture, New Delhi.

Ngambeki D S , Deuson, R R and Preckel P V 1992 Integrating livestock into farming systems in northern Cameroon. Agricultural Systems **38**: 319-338.

Nhan D K, Phong L T, Verdegem M J C, Duong L T , Bosma R H and Little D C 2007 Integrated freshwater aquaculture, crop and animal production in the Mekong Delta, Vietnam: determinants and the role of the pond. *Agricultural system* **94**:445-458.

Notenbaert A, Herrero M, Kruska R, You L, Wood S, Thornton P and Omolo A 2009. Classifying livestock production systems for targeting agricultural research and development in a rapidly changing world. ILRI Discussion Paper 19. *Int Livestock Res Inst, Nairobi, Kenya*. p. 41

O'Kting'ati A. Maghembe J.A., Fernandes E.C.M. and Weaver G.H. 1984. Plant species in the Kilimanjaro agroforestry system. *Agroforest Syst* **2**: 177-186.

Olayide S O and Heady EmO 1982. Introduction to Agricultural Production Economics. Department of Agricultural Economics University of Ibadan. University Press Ibadan. pp. 9-247.

Padoc C. and de Jong W. 1991. The house gardens of Santa Rosa: diversity and variability in an Amazonian agricultural system. *Econ Bot* **45**: 166-175.

Palada M.C. and Williams M.E. (eds.) 2000. Utilizing Medicinal Plants to Add Value to Caribbean Agriculture. Proceedings of the Second International Workshop on Herbal Medicine in the Caribbean. University of the Virgin Islands, St. Croix, U.S. Virgin Islands. 217 pp.

Pandey R K, Sharma S K and Hegde D M 1992. Approach to farming system research in ICAR concept and methodology, PDCSR Bulletin No. 2, Meerut, Uttar Pradesh.

Pandey S.B.S., Jadeja D.B., Manohar Narendra Singh and Tandel M.B. , 2016 Economic comparison of intercropping of ginger and turmeric under sapota-jatropha based agro-forestry systems in south gujarat , *International Journal of Science, Environment and Technology* **5** No 5, 2016, 3635 – 3642

Parthasarathy Rao P and Hall A J 2003. Importance of crop residues in crop-livestock systems in India and farmers perceptions of fodder quality in coarse cereals. *Field Crop Res* **84**: 189-198.

Pathak H, Saharawat Y S, Gathala M and Ladha J K 2011 Impact of resource-conserving technologies on productivity and greenhouse gas emission in rice-wheat system. *Greenhouse Gases: Science and Technology* **1:** 261-277

Pittelkow C M, Liang X, Linquist B A, Van Groenigen K J, Lee J, Lundy M E, van Gestel N, Six J, Venterea R T and van Kessel C 2015 Productivity limits and potentials of the principles of conservation agriculture. *Nature* **517**(7534):365.

Plunkett D L 1993 Modern crop production technology in Africa: The conditions for sustainability. In: CASIN/SAA/Global 2000. Africa's Agricultural Development in the 1990s: Can it be Sustained? CASIN, Geneva, Sasakawa, Africa Foundation, Tokyo and Global 2000, Georgia.

Poorani A, Jayanthi C and Vennila C 2011 Farmer participatory research on Integrated Farming Systems. In: National seminar on "Innovations in farming systems research and extension for inclusive development" 24 -25 Nov, Madras Veterinary College, Chennai, pp. 153.

Pradhan A, Idol T and Roul P K 2016 Conservation agriculture practices in rainfed uplands of India improve maize-based system productivity and profitability. *Frontiers in plant science***7**:1008.

Prajapati N.D., Purohit S.S., Sharma A.K. and Kumar T. 2003. A Handbook of Medicinal Plants. Agribios (India), 553 pp.

Prasad D,Yadav M S and Singh C S 2013 Diversification of rice (*Oryza sativa*)-based cropping systems for higher productivity, profitability and resource-use efficiency under irrigated ecosystem of Jharkhand. *Indian J Agron* **58** (2):77-83.

Prein M 2002 Integration of aquacultural into crops-animal systems in Asia. *Agricultural system* **71**:127-146.

Price, T. 2000. Cultivation our futures.Final paper.OECD Publications No. 2.Organization for Economic Co-operation and Development.

Radha, Y., Eshwaraprasad, Y. and Vijayabhinandana, B. 2000. Study on income and employment generation on agricultural based livestock farming system. Paper presented at VIII Annual Conference of AERA at TNUASU, Chennai, 28-29 December, 2000.

Radhamani S, Balasubramanian A, Ramamoorthy K and Geethalakshmi V 2003 Sustainable integrated farming systems for dry lands: A review. *Agricultural Reviews* **24**: 204-210.

Radhammani, S., Balasubramanian, A., Ramamoorthy, K., and Geethalakshmi, V. 2003. Sustainable integrated farming systems for dry lands: A review. *Agricultural Reviews*, **24**: 204210.

Ram H, Dadhwal V, Vashist K K and Kaur H 2013 Grain yield and water use efficiency of wheat (*Triticum aestivum L.*) in relation to irrigation levels and rice straw mulching in North West India. *Agric Water Manag* **128:** 92-101.

Ramasamy C, Natarajan S, Jayanthi C and Kumar D S 2008 Intensive Integrated Farming System to boost income of farmers. Paper presented in the 32nd IAUN VC's Annual convention held at RAU, Ranchi

Ramrao W Y, Tiwari S P and Singh P 2006 Crop-livestock integrated farming system for the Marginal farmers in rain fed regions of Chhattisgarh in Central India. *Livestock Research for Rural Development* **18** (7)

Ramrao W Y, Tiwari S P and Singh P 2005 Crop-livestock integrated farming system for augmenting socio-economic status of smallholder tribal of Chhattisgarh in central India. *Livestock Research for Rural development* **17** (90).

Rangasamy A 1995. Integrated farming systems research in Tamil Nadu. Country paper on development of sustainable farming systems. Organised at National Productivity Organization, Srilanka. June 5 -10.

Rangasamy A, Premsekhar M and Venkitaswamy R 1994. Integrated farming system for gardenland conditions of Coimbatore district - An overview. Adv. *Agric Res India* **1**: 70-77.

Rangasamy A, Venkitasamy R, Jayanthi C, Purshothaman S and Palaniappan SP 1995. Rice based farming system: A viable approach. *Indian Farming* **44** (11): 27-29

Rangaswamy A, Venkatswamy R, Premshekhar M, Jayanthi C and Palaniappan S P 1996 Integrated farming systems for rice based ecosystem. *Madras Agricultural Journal* **82**: 290-293.

Rao K.S., Maikhuri R.K. and Saxena K.E. 1999. Participatory approach to rehabilitation degraded forest lands: a case study in a high altitude village of Indian Himalayas. *Int Tree Crops J* **10**: 1-17.

Rautaray, S. K., Dash, P. C. and Sinhababu, D. P. 2005. Increasing farm income through rice fish based integrated farming system in rain fed lowlands of Assam. *Indian journal of Agricultural Sciences*, **75** (2): 79-82.

Ravi, K. 2004. Integrated farming: Boon for small and marginal farmers. *Kurukshetra*, **47** (9): 2223.

Ray M, Roy D C and Zaman A 2016 Evaluation of rice (*Oryza sativa*)-based cropping systems for increasing productivity, resource-use efficiency and energy productivity in coastal West Bengal. *Indian J Agron* **61** (2):131-7.

Ray, D.P. 2009. Livelihood Security in rice based farming systems. In Invited Papers and Abstracts of the National Seminar on Managing Rural Livelihood in India: Challenges and Opportunities. Orissa University of Agriculture and Technology, Bhubaneswar, Odisha: 1-5 pp (2009).

Reddy, V. R. 2001. Watershed development and livelihood security: An assessment of linkage and impact, Project Report Centre for Economic and Social Studies, Hyderabad.

Ruddle K and Chung K F 1988. Integrated agriculture- aquaculture in South China: The dyke- pond system of the Zhujiang Delta. Cambridge University Press, Cambridge.

Saha R, Singh J and Singh K 2010 Effect of crop diversification and intensification in rice-wheat cropping system on system productivity, profitability and energy use efficiency. *Int J Agri Sci* **52**:149-52.

Saharawat Y S, Ladha J K, Pathak H, Gathala M, Chaudhary N, and Jat M L 2012 Simulation of resource-conserving technologies on productivity, income and greenhouse gas emission in rice-wheat system. *Journal of Soil Science and Environmental Management* **3** (1): 9-22.

Samal, P., Barah, B. C. and Padney, S. 2006. An analysis of rural livelihood system in rainfedricebased farming systems of coastal Orissa, *Agricultural Economics Research Review*, **19**: 281292.

Santhakumari K., Vishnu P. K., Regi, J. T., George V. T. and Maria, L. G. 2010. Improving food and nutritional security of small and marginal coconut growers through diversification of crops and enterprises. *Journal of Agriculture and Rural Development in the Tropics and Subtropics*, **111**(2): 101-109.

Schippmann U., Leaman D.J. and Cunnningham A.B. 2002. Impact of cultivation and gathering of medicinal plants on biodiversity: global trends and issues. In Biodiversity and the Ecosystem Approach in Agriculture, Forestry and Fisheries. Ninth Regular Session of the Commission on Genetic Resources for Food and Agriculture. FAO, Rome, Italy, pp. 1-21.

Schultz R. C., Isenhart T.M., Simpkins W.W. and Colletti J.P. 2004. Riparian forest buffers in agroecosystems: Lessons learned from the Bear Creek Watershed Project in Central Iowa, USA. In: P.K.R. Nair, M.R. Rao, and L.E. Buck (eds). NEW VISTAS IN AGROFORESTRY: A Compendium for the 1st World Congress of Agroforestry 2004. Kluwer Academic Publishers, Dordrecht, The Netherlands.

Shah M S, Verma N and Vishwkarma S K 2015 Diversification of rice (*Oryza sativa*)-based cropping systems for higher production efficiency, water productivity and economic viability in Madhya Pradesh. *Indian J Agron* **60** (1):25-30.

Shah S N, Shroff J C, Patel R H and Usadadiya V P 2011 Influence of intercropping and weed management practices on weed and yields of maize. *Int J Sci Nat* **2**:47-50.

Sharma A R, Singh R and Dhyani S K 2005 Conservation tillage and mulching for optimizing productivity in maize-wheat cropping system in the outer western Himalaya region – a review. *Indian Journal Soil Conservation* **33** (1): 35-41

Sharma B K, Das M K 1988. Integrated fish livestock corp farming system. In Joseph M (ed.) The First Indian Fisheries Forum, Proc., Asian Fisheries Society, Indian Branch, Mangalore pp 27-30.

Sharma K L, Rao C S, Chandrika D S, Nandini N, Munnalal K, Indoria A K and Kumar T S 2016 Assessment of GMean biological soil quality indices under conservation agriculture practices in rainfed Alfisol soils. *Current science* **111** (8):1383.

Sharma L R, Bhati J P and Singh R 1991 Emerging farming systems in Himachal Pradesh: Key issues in sustainability. Indian Journal of Agricultural Economics **46** (3): 422-427.

Sharma N K, Singh R J, Mandala D, Kumar A, Alama N M and Keesstra S 2017 Increasing farmer's income and reducing soil erosion using intercropping in rainfed maize-wheat rotation of Himalaya, India. *Agric Ecosyst Environ* **247**:43-53.

Sharma N K, Singh R J, Mandala D, Kumar A, Alama N M and Keesstra S 2017 Increasing farmer's income and reducing soil erosion using intercropping in rainfed maize-wheat rotation of Himalaya, India. *Agric Ecosyst Environ* **247**:43-53.

Shen L. 1998. Management considerations and economic benefits of intercropping tea with Ginko biloba. *J Zhejiang Forest Sci Technol* **18** (2): 69-71.

Sidhu H S, Singh M, Humphreys E, Singh Y, Singh B, Dhillon S S, Blackwell J, Bector V M and Singh S 2007 The happy seeder enables direct drilling of wheat into rice straw. *Australian Journal of Experimental Agriculture* **47**: 844-854.

Sidhu H S, Singh M, Singh Y, Blackwell J, Lohan S K, Humphreys E, Jat M L, Singh V and Singh S 2015 Development and evaluation of the Turbo Happy Seeder for sowing wheat into heavy rice residues in NW India. *Field Crops Res* **184:**201-12.

Singh Balwinder, Eberbach P L, Humphreys E and Kukal S S 2011a The effect of rice straw mulch on evapotranspiration, transpiration and soil evaporation of irrigated wheat in Punjab, India. *Agric Water Manag* **98:**1847-55.

Singh Balwinder, Humphreys E, Eberbach, P L, Katupitiya A, Yadvinder-Singh, Kukal S S 2011b. Growth, yield and water productivity of zero till wheat as affected by rice straw mulch and irrigation schedule. Field Crops Res **121**:209-25

Singh K B, Jalota S K and Gupta RK 2015 Soil water balance and response of spring maize (Zea mays) to mulching and differential irrigation in Punjab**.** *Indian J Agron* **60**:279-84.

Singh K P, Singh S N, Kumar H, Kadian V S and Saxena K K 1993 Economic analysis of different farming systems followed on small and marginal land holdings in Haryana. *Haryana Journal of Agronomy* **9**: 122-125.

Singh K, Bohra J S, Singh Y and Singh J P 2006 Development of farming system models for the north-eastern plain zone of Uttar Pradesh. *Indian farm* **56**: 5-11

Singh M, Sidhu H S, Singh Y and Blackwell J 2011 Effect of rice straw management on crop yields and soil health in rice-wheat system. *Conser Agric* **18:**1-12.

Singh N K, Rajbir Singh and Singh R P 1996. Income and employment potential of mixed farming in irrigated area. *J Res Birsa Agric Univ* **8** (2): 143-148.

Singh R, Riar T S and Gill J S 2017 Integrated farming systems and socio-economic characteristics of Punjab Agricultural University awardee farmers. *Asian Journal of Agricultural Extension, Economics & Sociology* **16**(3): 1-5.

Singh Rajender, Singh Narinder, Phogat SB, Sharma UK, Singh R, Singh N. 1999 Income and employment potential of different farming system. *Haryana Agr Univ J Res*;**29** (3-4): 143–145.

Singh U., Wadhwani A.M. and Johri B.M. 1996. Dictionary of Economic Plants in India. Indian Council of Agricultural Research, New Delhi, India. 288 pp.

Singh V K, Dwivedia B S, Shuklaa A K, Chauhanb Y S and Yadav R L 2005 Diversification of rice with pigeonpea in a rice-wheat cropping system on a Typic Ustochrept: effect on soil fertility, yield and nutrient use efficiency. *Field Crops Res* **92**:85-105.

Singh Y and Sidhu H S 2014 Management of cereal crop residues for sustainable rice-wheat production system in the Indo-Gangetic Plains of India. *Proc Indian Natl Sci Acad.* Vol 80, pp 95-114, Calcutta, West Bengal.

Singh Yadvinder, Singh Bijay, Ladha J K, Khind C S, Khera T S and Bueno S S 2004 Effect of residue decomposition on productivity and soil fertility in rice-wheat rotation. *Soil Sci Soc American J* **68:**854-64.

Singh Yadvinder, Singh M, Sidhu H S, Humphreys E, Thind H S, Jat M L, Blackwell J and Singh V 2015 Nitrogen management for zero till wheat with surface retention of rice residues in north-west India. *Field Crops Res* **184:**183-91.

Singh, J. P., Gangwar, B., Pandey, D. K. and Kochewad, S. A. (2011).Integrated farming system model for small farm holders of western plain zone of Uttar Pradesh. PDFSR Bulletin No. 05, pp. 58. Project Directorate for Farming Systems Research, Modipuram, Meerut, India.

Singh, K. P. 2002. Integrated farming systems for smallholders- A key issue for research-education extension linkages in semi-arid tropical conditions. Paper submitted to 17th symposium of the international farming association, November, 17-20, Oriando, Florida, USA.

Sinha V R P 1979 New trends in fish farm management. In: Pillay TVR and Dill Wm A (eds). Advances in aquaculture, Fishing News Books Ltd., England pp123- 126.

Sinha V R P 1985 Crop-fish-production systems. Orientation cum training programme on improving the income and employment potential of rice farming systems. Organized by International Rice Research Institute (IRRI) and University of Philippines, Manila 21–30 January.

Sivamurugan.2001. Sustainable farming systems under irrigated upland situation. Ph. D. Thesis, TNAU, Coimbatore (TN), India.

Smitha B. (2005). Livelihood security of rural community: A critical analysis. Ph. D. Thesis, IARI, New Delhi.

Soni M, Upadhyay V B, Vishwakarma S K and Singh P 2012 Productivity and sustainability of rice (*Oryza sativa*) based cropping systems in Kymore plateau and Satpura hills zone of Madhya Pradesh. *Indian J Agron* **57** (1):8-13.

Swella G B, Ward P R, Siddique K H M and Flower K C 2005 Combinations of tall standing and horizontal residue affect soil water dynamics in rainfed conservation agriculture systems. Soil & Tillage Research **147**: 30–38.

Teel W.S and Buck L.E. 2002. Between wildcrafting and monocultures: agroforestry options. p. 199-222. In E.T. Jones, R.J. McLain. and J. Weigand (eds.) Non-Timber Forest Products in the United States. University Press of Kansas, Lawrence, KS, USA.

Tellez-Rio A, Vallejo A, García-Marco S, Martin-Lammerding D, Tenorio J L, Rees R M, and Guardia G 2017 Conservation Agriculture practices reduce the global warming potential of rainfed low N input semi-arid agriculture. *European Journal of Agronomy* **84**: 95-104.

Thamrongwarangkul A 2001 For out Thailand. Annual report on sustainable community development for good livelihoods and environmental project. Khon Kaen University.

Tipraqsa P, Craswell, E T, Noble A D and Schmidt V D 2007 Resource integration for multiple benefits: multifunctionality of integrated farming systems in Northeast Thailand. Agricultural Systems **94**: 694-703.

Tiwari SP, Ravi R, Nandeha KL, Vardia HK, Sharma RB, and Rajgopal S. 1999 Augmentation of economic status of Bastar tribals through integrated (crop, livestock, poultry, duck, fish) farming system. *Indian J Animal Sci.*;**69** (6): 448–52.

Tripathi H, Tomar S S, Pandey R, Solanki V S, Singh R, Meena K L, Tomar M and Adhikari D S 2010 Economic feasibility of Integrated Farming System models with respect to productivity and economics. In: 22nd national seminar on "Role of Extension in Integrated Farming Systems for sustainable rural livelihood, 9th -10th Dec, Maharashtra, pp. 42-43

Ugwumba, C. O. A., Okoh, R. N., Ike, P. C., Nnabuife, E. L. C. and Orji, E. C. 2010. Integrated farming system and its effect on farm cash income in Awka south agricultural zone of Anambra state, Nigeria. *American-Eurasian Journal Agricultural & Environmental Science*, **8** (1): 1-6.

Veluswamy, R. and Manoharan, M. 1998. Perceived impact of NGOs in rural development. *Journal of Extension Education*, **9** (3): 2084-2088.

Vinod, Y. (2006). A study on dynamics of farming systems in central Uttar Pradesh. Ph. D. Thesis, Chandra Shekhar Azad University of Agriculture and Technology, Kanpur.

Vinodkumar S N, Desai B K, Channabasavanna A S, Rao S, Patil M G and Patil S S 2017 Resource recycling and their management under integrated farming system for North-East Karnataka. *Internat. J. Agric. Sci.* **13**: 321-326

Vyas S. and Nein S. 1999. Effect of shade on the growth of Cassia ungustifolia. *Indian Forester* **125:** 407-410.

Walia S S, Gill M S, Bhushan B, Phutela R P and Aulakh C S 2011 Alternate cropping systems to rice (*Oryza sativa*)-wheat (*Triticum aestivum*) for Punjab. *Indian J Agron* **56** (1):20-7.

Walia S S, Gill M S, Dhaliwal V and Phutela R P 2009 Evaluation of different cropping systems for increased yields, water productivity and profitability in Punjab. *J Res Punjab agric Univ* **46** (3&4):131-6.

Wanyama M, Mose L O, Odendo M, Okuro J O, Owuor G and Mohammed L 2010. Determinants of income diversification strategies amongst rural households in maize based farming systems of Kenya. *Afr J Food Sci* **4** (12): 754-76.

Wezel A. and Bender S. 2003. Plant species diversity of homegardens of Cuba and its significance for household food supply. *Agroforest Syst* **57**: 39-49.

Yadav S K, Babu S, Singh Y, Yadav G S, Singh K, Singh R and Singh H 2013 Effect of organic nitrogen sources and biofertilizers on production potential and energy budgeting of rice (*Oryza sativa*)-based cropping systems. *Indian J Agron* **58** (4):459-64.

Yoshino K. and Ando K. 1999. Utilization of plant resources in homestead (bari-bhiti) in floodplain in Bangladesh. Japanese *J Trop Agric* **43**: 306-318.

Zhou S.Q. 1993. Cultivation of Amomum villosum in tropical forests. *Forest Ecol Manag* **60**: 157-162.

Zou X and Sanford R.L. 1990. Agroforestry systems in China: a survey and classification. *Agroforest Syst* **11**: 85-94.

TERMINOLOGY

- **Absorption losses:** Loss of water from a canal or a reservoir by capillary action and percolation and in case of canal during the process of delivery.
- **Acid equivalent (a.e.):** The differential yield of parent acid from an active ingredient.
- **Acid soil:** A soil which is deficient in available bases, particularly Ca and which give an acid reaction when tested by a standard method.
- **Acre foot water:** The amount of water that would cover an area of land to a depth of one foot assuming no seepage evaporation and run off.
- **Acre inch day:** Term used principally in irrigated section of United State for measuring quantity of flow of water. It is equal to a flow which will cover one acre to a depth of one inch in a 24 hours period or 0.042 cubic feet per second.
- **Acre inch:** It is a measure of quantity of flow of water and is equal to the flow which will cover one acre to a depth of one inch.
- **Acre:** (43560 sq. ft) an area of land about 220 feet long and 198 feet wide.
- **Active ingredient (a.i.):** The chemical in a product that is responsible for the effects.
- **Adaptation:** The modification in structure and function that increases the probability that a crop will survive at reproduce in a particular environment is termed as adaptation.
- **Adobe soil:** These soils are formed by the broken material of rocks transported by both wind and water.
- **Agriculture:** It is an art, science and business of raising crops and rearing of animals through exploring the natural resources with the coordination of socio economic infrastructure to meet the basic necessities of life .i.e. .food, feed, fiber and shelter.
- **Agrisilviculture:** Agrisilviculture is a production technique which combines the growing of agricultural crops with simultaneously raised and protected forest crops. Here it is intended to grow tree crops along with the agricultural crops.
- **Agrisilvipastoral systems:** ASPS is a collective name for land-use **systems**, implying the combination or deliberate association of a woody component (trees or shrubs) with cattle in the same site.

- **Agroforestry:** Agroforestry is a land use management system in which trees or shrubs are grown around or among crops or pastureland. It combines shrubs and trees in agricultural and forestry technologies to create more diverse, productive, profitable, healthy, ecologically sound, and sustainable land-use systems.
- **Agronomic efficiency (AE):** It refers to the additional produce obtained in kg per kg of an applied nutrient.
- **Agronomy:** The science of crop production and soil management. The name is derived from the Greek words *agros*(field) and *nomos* (to manage). It incorporates the basic sciences (Biology, Chemistry, Physics, Geology, Microbiology, etc.) into an applied science which is the foundation for most agriculture.
- **Allelopathy:** Phenomenon involving the release of certain chemicals from plant parts into the environment which may when present in sufficient amounts, inhibit or suppress the germination or growth of the plants in the neighbourhood.
- **Alluvial soil:** These are the soils which are formed by the deposition of broken material of rocks transported and deposited by water of streams and rivers.
- **Arable farming:** The term arable farming refers to system in which only crops that require cultivation of the soil are grown.
- **Arboriculture:** Intensive cultivation of individual trees possibly for fruits gums and resins.
- **Arid/Aridity:** A term applied to land that is dry. Also means deficient in rainfall.
- **Autoecology:** It deals with the ecology of an individual species and its population including the effect of other organisms and environmental conditions on every stage of its life cycle.
- **Autotrophs:** Organism able to manufacture their own food from inorganic materials. Using energy from outside sources. Most green plants are completely autotrophs.
- **Available water:** The amount of water between field capacity and permanent wilting point is considered as available water for plant use.
- **Band placement:** The placement of fertilizer in the soil in continuous bands or narrow ribbons in between the plant roes or around each plant. The fertilizer applied in bands is covered by the soil but is not mixed with it.
- **Barani soil:** When the source of irrigation to crop is only the rain water.
- **Basal treatment:** An application to the stems of the plants at and just above the ground line.
- **Base period:** Period of time in days from the first watering of crop before sowing and the last watering before harvesting.

- **Basic seed:** Is the progeny of pre – basic seed produced so as to maintain genetic purity and identity.
- **Basin:** Flat area of land surrounded by low ridges or bunds
- **Bed:** A narrow flat topped ridge on which crops are grown with a furrow on each side for the drainage of water.
- **Blind hoeing:** Hoeing before the crop emerges, usually done in sugarcane.
- **Biofertilizers:** Microbial inoculants probably known as biofertilizers refer to living or latent cells of efficient strains of microorganisms capable of fixing atmospheric nitrogen.
- **Biological yield:** It is the total dry matter produced by a plant as a result of photosynthesis and nutrient uptake minus that lost by respiration.
- **Block planting:** Here the preference is for closer planting of trees 1m × 1m to 2m × 2m in case of eucalyptus, and 2m x 2m to 3m x 3m in case of poplars in the entire field/ block.
- **Bolting:** Formation of elongated stem or seed stalk, it is usually takes place during the second season of the growth in biennial plants.
- **Botanical variety:** When a group of plant occurring in nature is different from the general species originally described and the botanical binomial name is not enough to identify it is called as botanical variety.
- **Boundary/ line planting:** Under this system, tree species are planted in single or double rows, at an inter tree spacing of 1-4 m, on the farm and field boundaries, approach paths, water channels etc.
- **Broadcasting:** Manual spreading of seed in the field and mixing of the spread seed by ploughing or planking the field.
- **Brush control:** Control of woody plants.
- **Bulk density of mineral soils:** It is defined as mass of a unit volume of dry soil. This volume includes both solids and pores.
- **Capillary water:** It is held in the pores of capillary size. It includes most of the water taken up by growing plants.
- **Catchment area:** The area which drains the rain water falling on it, via streams and rivers, eventually to the sea or into a lake.
- **Cereal crops**: A cereal is defined as crop grown for its edible seed. These crops are also known as grain crops e.g. wheat, rice, maize etc.
- **Certified seed:** It is the progeny of basic seed and is produced by registered growers of seed producing agencies.
- **Clayey soil:** A soil is known as clayey which contain at least 30% clay particles and in most cases not less than 40%, usually it contain 45% clay , 30% silt and 25% sand.
- **Climate:** Aggregate of atmospheric condition over a long period of time.
- **Clone:** A cultivar propagated by vegetative method is called a clone.

- **Colluvial soil:** Are those which are form from the material transported by the force of gravity.
- **Command area:** Area which can be economically irrigated by an irrigation system.
- **Commercial farming system:** In this type of farming system, crops are raised on a commercial scale for marketing.
- **Companion crops:** The two crops grown together are called companion crops. e.g. berseem and barley.
- **Component crop:** It is used to refer to either of the individual crops making up the intercropping situation.
- **Composite variety:** A mixture of genotype from general sources maintained by normal pollination.
- **Composting:** Composting is a biological process in which micro-organisms of both types namely, aerobic and anaerobic, decompose the organic matter and lower the carbon nitrogen ratio of the refuse. The final product of the composting is the well rotten manure known as compost.
- **Condiment crops:** Crops which are grown and consumed as condiments e.g. coriander, mint.
- **Conditioner:** A material added to a fertilizer to prevent caking and to keep it free flowing.
- **Consumptive use (Cu):** It is use to designate the losses due to ET and the water that is used by the plant for its metabolic activities. Cu is taken equivalent to ET, because water used in metabolic processes does not exceed 1 per cent.
- **Contact herbicide:** Which kill only those parts of plant with which they come into contact and used mostly to control annual weeds when seedling, they have little residual effect.
- **Cover crops:** The crops, which are planted to cover the ground and to reduce the soil erosion and nutrients losses by leaching. e.g grasses and rye.
- **Critical hydration level of seed:** The minimal water content at which the seed begins to germinate and the hydration level at which seed water uptake becomes biologically irreversible.
- **Critical period of competition:** During the crop period there is a certain time when crop plants are most sensitive to competition by weeds, this time is known as the critical period of competition.
- **Critical Threshold Level (CTL):** A weed, insect pest density capable of causing significant damage to crop is termed as critical threshold level.
- **Crook stage:** The stage after bean seedlings have broken through the soil but before the stem has become erect.

- **Crop ecology:** It is the study of the interrelationships between crop plants and the environment. Factors controlling the habitat of crop during its entire life cycle such as climate, physiography, soil and biota are considered.
- **Crop Growth Rate (CGR):** CGR represents total dry matter productivity of the community per unit land area over a certain time span.

  CGR = gm^{-2} land area day^{-1}
- **Crop rotation:** It is the process of growing different crops in succession on piece of land in a specific period of time, with an object to get maximum profit from least investment without impairing the soil fertility.
- **Crop system:** An arrangement of crop populations that tansform solar energy, nutrients, water and other inputs into useful biomass. The crop system is a subsystem of cropping system. For example, in maize crop system, maize is the dominant crop which is grown in association with other crops.
- **Crop water use efficiency:** It is the ratio of crop yield (Y) to the amount of water depleted by the crops in the process of evapotranspiration (ET).
- **Crop water requirement:** The amount of water required to raise a crop to maturity within a given period of time.
- **Crop:** A crop is a community of plants grown under field condition for its economic value.
- **Cropping index (Cropping intensity):** The number of crops grown per annum on a given area of land multiplied by 100.
- **Cropping pattern:** The yearly sequence and spatial arrangement of crops or of crops and fallow on a given area.
- **Cropping scheme:** The cropping scheme is a plan according to which crops are grown on individual plots of a farm during a given period of time with the object of obtaining maximum return from each crop without impairing the soil fertility.
- **Cropping system:** Cropping system is a land use unit comprising soils, crop, weed, pathogen and insect subsystems that transform solar energy, water, nutrients, labour and other inputs into food, feed, fuel and fibre. The cropping system is a subsystem of the farming system.
- **Crown:** The point where stem and root joints in a seed plant.
- **Cultivator:** Which only cut and stir the soil.
- **Delta of water:** The depth of irrigation water required for the full crop period.
- **Determinate plants:** Those plant which initiate their reproductive stage after completing vegetative growth, e.g. wheat and barley.
- **Development:** Development refers to the whole series of change which an organism goes through during its life cycle, but it may equally be applied to individual organs, to tissues or even to cells. Plant development involves both growth and differentiation.

- **Differentiation:** During development there appear not only quantitative differences in number and arrangement of cells within different organs, but also qualitative differences between cells, tissues and organs to which the term differentiation is applied.
- **Diversified farming:** This is an expanded type of farming system in which varieties of crops are produced and many types of animals are reared.
- **Dobari crops:** A crop grown on residual moisture after the harvest of rice.
- **Dormancy:** Seed dormancy is the state of inhibited germination of seeds with viable embryos in condition conducive to plant growth.
- **Double, triple and quadruple cropping:** Growing two, three or four crops respectively, on the same land in a year in sequence.
- **Drainage:** It is the removal of excess surface or ground water from the root zone of a crop by means of surface or sub – surface drains.
- **Drought resistance/ Drought escape:** The ability of crop to grow satisfactorily in areas subjected to water deficit has been termed as drought resistance.
- **Dry farming:** In which crops and livestock are raised on land which does not receive sufficient rainfall for water intensive crops and no irrigation facilities are available to crops which fall into this category.
- **Duty of water:** The relationship between irrigation water flow and its commanded area where crops mature fully with that amount of water within a base period.
- **Earthing up:** The operation of pulling up soil from the center of crop rows to the bottom of the plants, this helps in uprooting weeds and supporting plants.
- **Ecology:** It is defined as the study of plants and animals in reciprocal relationship with their environment or external world.
- **Economic yield:** That part of the crop for which crop is grown e.g. stem in sugarcane, root in the case of sugar beet.
- **Ecosystem:** Any collection of organisms that interact has the potential to interact along with the physical environment in which they live, form an ecological system or ecosystem.
- **Effective rainfall:** Precipitation which falls during the growing period of the crop and is available to meet the evapotranspiration requirements of crops.
- **Efficiency of applied water:** Efficiency in any system is a measure of output obtainable from a given input under a particular production environment.
- **Efficient cropping zone:** A zone where both yield and spread indices are maximum.
- **Emergence:** It is the process of emerging out of the plumule from the soil after seed germination.
- **Eolian soil:** The soil which is formed by the material transported by winds from one place to another.

- **Epigeal germination:** It is derived from two words epi "above" and geas "earth". In this type of germination the cotyledons come out above the soil surface and generally turn green and act as first foliage leaves. e.g. bean,cotton.
- **Epinasty:** The twisting or curling of leaves and stems caused by uneven growth of cells, especially in leaves, in which the upper surface grows faster than the lower surface and thus cause the leaf edges to bend down.
- **Equivalent acidity:** A term used to express the calcium carbonate equivalent of acidic residue left by a fertilizer material.
- **Equivalent basicity:** A term used to express the calcium carbonate equivalent of basic residue left by a fertilizer material.
- **Evapotranspiration (ET):** Evaporation and transpiration are combined in one term evapotranspiration. It is difficult to separate these two losses in cropped fields.
- **Exhaustive crops:** Crops which feed heavily on the soil and deplete soil nutrients e.g. sorghum, tobacco.
- **Extensive farming or extensive agriculture:** It is an agricultural production system that uses small inputs of labor, fertilizers, and capital, relative to the land area being farmed.
- **Facultative weed:** A weed found growing both wild and in association with crops.
- **Farm forestry:** Forest farming is the cultivation of high-value specialty crops under a forest canopy that is intentionally modified or maintained to provide shade levels and habitat that favour growth and enhance production levels. Forest farming encompasses a range of cultivated systems from introducing plants into the understory of a timber stand to modifying forest stands to enhance the marketability and sustainable production of existing plants
- **Farm Yard Manure (FYM):** It refers to the decomposed mixture of dung and urine of farm animals along with the litter and left over material from fodder fed to cattle. Well rotten FYM contains about 0.5 % N, 0.3 % P_2O_5 and 0.5% K_2O.
- **Farming system:** Farming system is a decision making and land use unit comprising the farm household, cropping and livestock systems that produces crop and animal product for consumption and sale. The farming system is a subsystem of a higher level land use system, such as village or water shed, that, in turn, forms a component of the agricultural sector of the regional system.
- **Fertilizer grade:** An expression that indicates the percentage of plant nutrients in a fertilizer. For example, a 10-5-5 grade of fertilizer indicates 10% N, 5% P_2O_5 and 5%K_2O.
- **Fertilizer ratio:** The fertilizer ratio designates the relative proportion of the three major plant nutrients, keeping the percentage of N as one in the ratio.

- **Fertilizer Use Efficiency (FUE):** It is the measure of the extent to which the fertilizers and management practices have worked together with the soil climate complex in producing high yield.
- **Fiber crops:** The crops, which are grown for their fiber and are used in making textiles, ropes. e.g. jute, sun hemp, cotton.
- **Field water use efficiency:** This is the ratio of crop yield (Y) to the total amount of water used in field (WR).
- **Field capacity:** When a fully saturated soil is allowed to drain after a day or so, rapid downward movement in response to hydraulic gradient will become negligible. The soil is then said to be at its field capacity.
- **Field carrying capacity:** It is the field capacity. It is also called normal moisture capacity.
- **Foliar fertilization:** Fertilization of plants or feeding nutrients to plants by applying chemical fertilizers to the foliage. This is also known as foliar feeding or spray fertilization.
- **Forage crops:** Those crops, which are grazed by animals and harvested for green chop, hay, silage are classified as forage crops e.g. maize and sorghum.
- **Fruit farming:** In which orchards are planted and the objectives are to maximize fruit production, enhance quality and increase income.
- **Functional allelopathy:** It is the case when the chemicals are toxic after being transformed by microorganism.
- **Garden crops:** Vegetable crops, which are grown for their edible leaves, shoots, flowers, fruit and seed. e.g cabbage and okra.
- **Germination:** Is the emergence and development from the seed embryo of those essential structures which, for the kind of seed provided, indicate the ability to produce a normal plant under favourable conditions.
- **Granular fertilizer:** A fertilizer composed of particles of roughly the same composition and about $1/10^{th}$ of an inch in diameter.
- **Grassland farming:** These systems are mainly concerned with growing grasses for consumption by livestock kept for milk or meat production.
- **Gravitational potential:** It is produced by gravitational forces operating on soil water.
- **Gravitational water:** Water in excess of field capacity is termed as gravitational water.
- **Green manure crops:** Some crops are grown and ploughed in the soil in green form in order to improve soil fertility e.g. berseem, guara, dhaincha etc
- **Green manuring:** It is a practice of ploughing or turning into the soil undecomposed green plant material for improving the physical condition of the soil or for adding nitrogen when the green manure is a legume.
- **Gross irrigation requirement:** The total amount of water, inclusive of losses, applied through irrigation is termed as gross irrigation requirement.

- **Growth analysis:** Growth analysis is a technique for quantifying components of plant growth from a series of mathematical equations.
- **Growth:** The term growth is applied to quantitative changes occurring during development. It may be defined as an irreversible change in the size of cell, organ or whole organism.
- **Hard pan:** A hard semi-impervious layer usually developed due to continuous ploughing to a depth of about 15 cm, with cultivators, or with continuous deposition of salts due to soil or surface irrigation water.
- **Harvest Index (HI):** It is the ratio of economic yield to biological yield. It is expressed in percentage H.I. = economic yield x 100 / Biological yield.
- **Hay:** The herbage of grasses or fine-stemmed plants cut and cured for forage.
- **Herbicides:** Chemicals used for killing or inhibiting the growth of higher plants.
- **Herd system:** The arrangement of (domesticated) animals populations and their feed sources.
- **Hill placement:** This refers to applying fertilizer either in bands or near the plants. This method is usually preferred when relatively small quantities are to be applied.
- **Homeostasis:** It refers to the tendency of the ecosystem to resist change and remain in a state of equilibrium. The science of cybernetics or of controls is legally dependent on the feedback system and has its application in understanding homeostasis.
- **Husk:** The coarse outer envelope of a fruit, such as the glumes of an ear of maize.
- **Hybrid vigour:** Qualities in a hybrid not present in either parent.
- **Hydrophyte:** Plant which grows in water, or which loves water.
- **Hygroscopic water:** Hygroscopic water is that tightly bound by the soil solids at suction value greater that 31 atm.
- **Ideotype:** An ideal plant type developed through breeding.
- **Indeterminate plants:** In these plants, the vegetative and reproductive stages continue simultaneously e.g. okra, tomato. Soybean is the only crop, which has determinate and indeterminate as well as semi – determinate growing types.
- **Inoculant:** The bacteria containing material used to introduce N – fixing Rhizobium bacteria into soil.
- **Integrated weed management:** The concept of IWM involves the planned use of all possible direct and indirect measures rather than relying on a single method to combat weeds.
- **Interception:** When the drops of rainfall or precipitation are intercepted by plant leaves it is called interception.

- **Intercrop yield:** It is a yield of the component crop when grown in intercropping and expressed over the total intercropped area (i.e. area occupied by both the crops).
- **Intercropping:** Intercropping is the cultivation of two or more crops simultaneously on the same field, with or without row arrangement (row intercropping or mixed cropping).
- **Inter-cultivation:** Inter cultivation is a process involving operations in the rows of standing crops. Weeding, tiling and cultivating are the examples for the same. For intercropping, tree lines are kept apart at 4-5 m to allow the movement of farm implements. The distance between the trees within a tree line is kept at 1.5-1.8 m.
- **Irrigated soil:** The soil, which receives irrigation water from well and tube wells, are known as irrigated or chahi soil.
- **Irrigation efficiency:** It is a term used to indicate how efficiently the available water supply is being used for crop production.
- **Irrigation scheduling:** It refers to the number of irrigations for a crop and their timing.
- **Irrigation water requirement (IR):** The irrigation water requirement is the total amount of water applied to the land surface in supplement to the water supply through rainfall and soil profile, to meet the water needs of crops for optimum growth.
- **Irrigation:** Irrigation is the artificial application of water to soil or crop plants to assist crop production.
- **Kera:** Manual sowing of seeds in lines in furrows.
- ***Kharif crops*:** Those crops, which are planted in the summer month from the March to July and harvested in autumn and winter, are called *Kharif crops* e.g. rice and cotton.
- **Lacustrine soil:** When the material transported by streams and river, if deposited in lakes the soils are called Lacustrine soil.
- **Layerage:** Methods of plant propagation by asexual means in which a portion of stem, shoot or branch has covered with soil or some other medium in which root can develop , after which the rooted portion get detached from the parent plant.
- **Leaching:** Removal of plant nutrients in solution by the passage of water through soil. Among the major nutrients, N is lost in large quantities by leaching.
- **Leaf area duration (LAD):** Leaf area integrated over time is called leaf area duration. It is equal to leaf area x time.
- **Land equivalent ratio:** This is the relative land area under sole crops that is required to produce the yields achieved in intercropping.

Yi = is the yield of ith component from a unit area of intercrop expressed as a fraction of the yield.

Yij = is the yield of that component grown as sole crop over the same area and n is the number of crops involved.

- **Light soil:** A term used for sandy or porous texture soil. Wherever, practicable, green manuring should be done to improve the physical condition of the light soil.
- **Line:** A cultivar propagated by seed.
- **Livestock and poultry farming:** This category includes farming system in which various kinds of livestock are reared for meat, milk, wool and eggs.
- **Livestock system:** A land use unit comprising pastures and herbs and auxiliary feed sources transforming plant biomass into animal products.
- **Loamy soil:** The soil is more or less than midway clay and sandy soil and ideal loam soil may defined as a mixture of sand , silt and clay particles which shows the properties of sandy, silt and clayey in equal proportion.
- **Localized placement of fertilizers:** This refers to the application of fertilizers into the soil close to the seed or plant. This method is specifically adopted with phosphatic fertilizers.
- **Lodging:** The bending or breaking over of a plant before harvesting.
- **Main crop/base crop:** This is the crop which is planted at its optimum sole crop population in an intercrop situation and the second crop is planted in between rows of the main or base crop, with a view to obtain some extra intercrop yield without sacrificing the main crop yield.
- **Matric potential:** It is produced by capillary and surface forces.
- **Maximum potential soil moisture deficit:** Is the greatest value of potential soil moisture deficit attained during the growth of a crop.
- **Maximum water holding capacity:** When the water is continuously applied to the soil. At a stage when its all (macro and micro) pores are filled with water. The soil is said to be saturated w.r.t water and is at its maximum retentive capacity.
- **Mixed intercropping:** Growing two or more crops simultaneously with no distinct row arrangement. Also referred to as mixed cropping.
- **Monocropping:** A cropping system in which the populations of crops, other than the main crop, grown in relay, sequential or intercropping with the main crop are greatly reduced.

  Or

  Monoculture is the repetitive growing of the same crop on the same land.
- **Muck soil:** If the quantity of organic matter exceeds 20 % but less than 50% are called muck soil.

- **Mulch:** Any material or practice which is used to check the loss of water by evaporation is called mulch.
- **Multiple cropping:** The intensification of cropping in time and space dimensions, growing two or more crops on the same field in a year.
- **Narcotic or drug crops:** This category includes those crops, which have some narcotic and drug value e.g. poppy, tobacco.
- **Natural erosion:** The erosion of the soil under natural condition.
- **Natural soil:** Strictly speaking a soil having pH of 7, in practice a soil having PH 6.6 and 7.3.
- **Necrosis:** Death of organs of a plant, either as blight or death of tissue in localized areas, usually inside fruit and stems or die back or death of stems or branches.
- **Net plot:** Area from which yield and other characteristics are measure. It is also known as the net area of the plot.
- **Nutrient Recovery:** It refers to the actual amount of nutrient taken up from the fertilizers.
- **No tillage crop (zero tillage crop):** Crop grown without any tillage to prepare seed bed or row.
- **Node**: A slightly enlarge portion of stem where leaves and bud arises and where branches originate.
- **Nucleus seed:** Seeds obtained from selected individual plants of a particular variety which needs to be purified and multiplied in such a way as to maintain its genetic purity.
- **Nut cycle:** The regeneration/cycling of nutrients.
- **Nutrients budget:** A quantitative data of the major nut flowing to retained within the discharge from the system.
- **Nutrients:** The food for microbial and plant life mainly composed of nitrogen and phosphorous but also of potassium , Mg, Fe, Ca, Co, Cu, Zn and others elements.
- **Oil seed crops:** These are the crops, which are grown for the purpose of extracting oil from their seed e.g. mustard and groundnut.
- **Olericulture:** Branch of horticulture which deals with cultivation of vegetables.
- **Onfarm water management:** The planned use of irrigation water at the farm level or more efficient utilization in agriculture.
- **Organic farming:** System of farming which avoid the use of artificial fertilizers, pesticides or herbicides, and concentrate on methods of crop rotation and the use of home growth feed , organic fertilizer.
- **Osmotic potential:** It is also called solute potential. It is produced by various solutes in soil water.

- **Particle density of mineral soils:** Mass of unit volume of soil solid is called particle density. It is expressed as g/cm^3. It varies between 2.6 and 2.75.
- **Pasteurization:** Process of killing organisms in a product, commonly milk by heating to a controlled temperature.
- **Patch cropping:** The component crops are planted in patches.
- **Peak period consumptive use:** The average daily water use rate during few days of the highest consumptive use of the season is called the peak period consumptive use.
- **Peat soil:** If the quantity of organic matter in soil is more than 50%.
- **Pedology:** The study of soil.
- **Percolation:** Downward movement of water with in the soil profile.
- **Planker:** Implement used to crush clods on land where a roller can be used consisting of number of fixed over lapping plates, showed with iron bars along the working edges which is pulled over the land.
- **Plant development:** Plant development is the progress of plant from germination to maturity through a series of stages.
- **Plant growth:** It is the increase in the dry weight of a plant over time mainly as a consequence of photosynthesis.
- **Plough pan:** A dense, compacted layer about 5 to 7 cm thick formed beneath the surface soil by repeated ploughing in the same path.
- **Plough:** Its function is to cut, stirr, invert, and pulverize the soil.
- **Plough-sole placement:** With the plough-sole method, the fertilizer is placed in a continuous band on the bottom of a furrow in the process of ploughing. Each band is covered as the next furrow is turned.
- **Pore space:** It is a space between soil particles occupied by air and water; it is largely controlled by the texture of soil.
- **Potential Evapotranspiration (PET):** Potential evapotranspiration is the ET that occurs when the ground is completely covered by actively growing vegetation and there is no limitation of soil moisture.
- **Potential soil moisture deficit:** It is the difference between a crop potential evapotranspiration and the amount of rainfall received by a crop plus the quantity of water delivered to it in irrigation.
- **Prebasic seed:** It is the progeny of nucleus seed, and is handled so as to maintain specific genetic purity and identity as completely as possible.
- **Pressure potential:** It is produced by actual hydrostatic pressure.
- **Puddling:** Puddling is the reduction in apparent specific volume (void ratio) of a soil caused by mechanical work done on soil.
- **Pulses or grain legumes:** The crops belonging to Leguminoseae family are grown for their edible seed e.g chick pea, lobia.

- ***Rabi* crops:** These crops are planted in winter from October to December and harvested in summer from March to May e.g., wheat,
- **Ratoon cropping:** The cultivation of crop regrowth after harvest, although not necessarily for grain.
- **Readily available water:** The portion of the available water that is most easily extracted by a plant is called readily available water.
- **Relay crops:** A relay crops is one which is planted as a second crop after the first crop has reached its reproductive stage of growth but before it is ready for harvest. e.g. planting of sugar cane in sugar beet.
- **Relative Growth Rate (RGR):** It is defined as increase in weight per unit of original weight of time interval.
- **Relay intercropping:** Growing two or more crops simultaneously during the part of life cycle of each. A second crop is planted after first crop has reached its reproductive stage of growth much before it is ready for harvest. Often referred to as relay cropping.
- **Relief:** When rising ground or mountains running at right angle to the prevailing wind.
- **Root and tuber crops:** These are vegetable crops grown for their under ground parts like roots, bulbs, rhizomes, corms and stem tubers e.g. carrot and onion.
- **Row intercropping:** Growing two or more crops simultaneously where one or more crops are planted in a row. Often referred to as intercropping.
- **Rumber:** The process in which planker or leveler is used to conserve moisture at watter condition before the preparation of land after rauni.
- **Run off:** When water flows out the field by breaking the bunds of the field or flows to the sloppy areas from the high level is called run off..
- **Rural compost:** It is prepared by using weeds, crop stubble, cotton stalks, groundnut husk, sugarcane trash, urine soaked earth and manure from cattle shed, waste fodder etc. It contains about 0.5 % N, 0.15 % P_2O_5 and 0.5% K_2O.
- **Sandy soil:** Soil which contains 2.00 to 0.2 mm diameter soil particles , it contain 85% sand , and >15% , silt and clay. These soils are poor in plant material.
- **Saturation capacity:**This term refers to the amount of water present in the soil when it is completely saturated with water.
- **Scarification:** Any physical or chemical treatment that makes the seed coat permeable is known as scarification.
- **Seasonal consumptive use:** The total amount of water used in evaporation and transpiration by a crop during the entire growing season is called seasonal consumptive use. It is expressed in depth of water in cm.
- **Seed certification:** Is the process to secure, maintain and make available high quality seed and propagating materials of superior crop varieties, so grown

and distributed as to ensure desirable standards of genetic identity, physical purity and quality attributes.

- **Seed dressing:** The chemical treatment of seeds particularly cereals, with fungicides and some time insecticides to protect them against soil and seed -borne disease and pest.
- **Seed:** Seed may be defined as a fertilized ovule consisting of intact embryo, stored food and seed coat which is viable and has got a capacity to germinate. This is also known as true seed because in some of the crops, vegetative parts are used for planting and are also called as seeds.
- **Seedling:** A young plant that has grown from a seed.
- **Seeling:** Mechanical weeding process carried out by ploughing with bullocks or tractors.
- **Seepage:** Losses through the beds and banks of canals and water channels take place mainly by unlined canals.
- **Sequential cropping:** Growing two or more crops in sequence in the same field in same farming year. The succeeding crop is planted after the preceding crop has been harvested.
- **Sett:** A piece of seed cane with two – four buds(eyes).
- **Short day plants:** Plants which changes from vegetative to reproductive stage and produces flowers and fruits, when the days become shorter.
- **Silage crops:** Those crops, which are harvested when still green and succulent and are fed directly to animals without curing, are called silage crops. e.g. berseem, shaftal, maize.
- **Silage:** Forage preserved in a succulent condition by partial fermentation in a tight container.
- **Silvipastoral systems:** The combined use of forestland or woodland for both wood production and animal production by grazing of the coexisting indigenous forage, or vegetation that is managed like indigenous forage.
- **Social forestry:** Social forestry is the practice of forestry for society, by the society, through the society, especially, for fuel, fodder, fruits and small timber requirements of the nearby surroundings.
- **Soil fertility:** It is defined as the inherent capacity of soil to supply amount of adequate nutrients to plants in balanced form.
- **Soil productivity:** It is defined as the capacity of the soil to yield crops per unit area or it can be defined as the total crop production from the field. A soil can be fertile but may not be productive.
- **Soil:** A natural body developed from variable mixtures of broken and weathered minerals and decaying organic matter which covers the earth crust in thin layers and supplies proper amount of nutrients and air water and mechanical support to plants.

- **Soiling crops:** Crops harvested when green and succulent condition is fed to animals shortly after cutting, it is neither dried nor stored for future feeding.
- **Sole cropping:** One crop/variety grown alone in pure stands at normal density.
- **Stratification:** Is the practice of exposing imbibed seeds to cool temperature conditions for a few days prior to germination in order to break their dormancy.
- **Strip intercropping:** Growing two or more crops simultaneously in strips wide enough to permit independent cultivation but narrow enough for the crops to interact agronomically.
- **Subsistence farming:** In which basic necessities like food, clothing, and shelter are produced for the family to live on.
- **Sugar crops:** These are the crops, which are grown for sugar purposes. e.g. sugar cane and sugar beet.
- **Synecology:** It deals with the ecology of plant communities. It involves the study of structure, nature, organization and development of plant communities.
- **System:** By system we mean a unified whole made of regularly interacting and interdependent components.
- **Temporary wilting percentage:** Soil water content at which plants wilt during the hot windy part of the day but regain turgidity during the cooler part of the day is called TWP.
- **Tillage:** It is usually defined as the mechanical manipulation of the soil aimed at improving its physical condition.
- **Tiller:** To put forth shoots other than the mother shoot from roots axis.
- **Tilth:** It is the final condition of soil when all the tillage operations are carried.
- **Topping:** Topping in tobacco is the removal of the terminal bud with or without some of the small top leaves just before or after the appearance of the flower head.
- **Town compost:** This utilizes night soil, street and dustbin refuse. Town compost is rich in nutrients (1.4 % N, 1.0 % P_2O_5 and 1.4% K_2O) as compared to FYM and rural compost.
- **Transpiration ratio:** It is the ratio of the weight or volume of water transpired by the plant during its growth period to the weight of dry matter produced by the plant.
- **Transpiration:** The loss of water from exposed parts of plants mainly through leaves is called transpiration.
- **Transplanting:** It is a process of planting seedlings, grown in separate nursery, in the puddle fields.
- **Trench layering**: An asexual reproductive method of plant propagation involving laying down the whole stem, the new shoots are thus forced to push their way through a layer of soil which prevents the bark from coloring and favors root formation.

- **Truck gardening/farming:** It refers to the system in which the bulk of the outputs are produced and marketed. Growing of crops like potato, onion and cabbage on large scale for distinct market.
- **True Allelopathy:** When the allelochemical is toxic in their original form is called true Allelopathy.
- **Unit Leaf Rate (ULR):** It may be defined as rate of increase in dry weight per unit leaf area, assuming that both dry weight and leaf area are increasing exponentially.
- **Variety:** In general the term variety has been used to refer to a group of similar plant with in a particular species that is distinguished by one or more then one character and given the name.
- **Vernalization and chilling:** Many biennials and temperate annuals, as well as certain fruit trees, require exposure to cold temperature before they can flowers. This is known as Vernalization requirement for annuals and biennials and chilling requirement for fruit trees.
- **Viability of seed:** It represents the capacity of seed to germinate or capacity of seed to remain capable of germination for some specific period of time.
- **Vines:** Are plants, which have tender stems, and requires some support for upward growth.
- **Water holding capacity:** Ability of soil to retain water.
- **Water potential:** Refers to the chemical potential of water.
- **Water requirement of crops:** Water requirement may be defined as the quantity of water, regardless of its source required by a crop of diversified pattern of crops in a given period of time for its normal growth, under field conditions at a place.

  WR = ET or Cu + application losses + special needs

  Source of water: IR = irrigation water

  ER = effective rainfall

  S= soil profile contribution
- **Watter:** Describes the condition when soil moisture level is suitable for cultivation.
- **Weathering:** The process by which soil disintegrates and decomposes, eventually producing soil particles by exposure to the physical and chemical effects of atmospheric agents.
- **Weed control:** Weed control aims at checking the growth or reducing the number of weed to a minimum so that they do not affect crop yields.
- **Weeds:** Weeds are that species of plant which grow unwanted or are not useful, often prolific and persistent, interfere with agricultural operations, increase labour cost and reduce the crop yields.
- **Wilting coefficient:** As the soil dries out, plant will begin to show the effects of reduced soil moisture. During the day time they will tend to wilt. At first this

day time wilting will be associated with renewed night time turgor. Ultimately, the rate of supply of water to the plants will be so slow that the plants will remain wilted day and night. The soil moisture content at this stage is called wilting coefficient or critical moisture.

- **Wilting point:** The point at which the water content of a soil reaches such a level that it is firmly held by soil and unavailable to plant roots, so that the plants wilt permanently and die.
- **Yield index of crop** = (average yield of state or zone/ average yield of country) × 100
- ***Zaid Kharif* crops:** These are the crops, which are planted in August to September and harvested in December to January e.g. toria.
- ***Zaid Rabi* crops:** These are crops, which are planted in February and harvested in May-June e.g. tobacco.